Biotreatment of Industrial and Hazardous Waste

Other Environmental Engineering and Management Books

CHOPEY • *Environmental Engineering for the Chemical Process Industries*

CORBITT • *Standard Handbook of Environmental Engineering*

FREEMAN • *Hazardous Waste Minimization*

FREEMAN • *Standard Handbook of Hazardous Waste Treatment and Disposal*

JAIN • *Environmental Impact Assessment*

MCKENNA & CUNEO, TECHNOLOGY SCIENCES GROUP, INC. • *Pesticide Regulations Handbook*

MAJUMDAR • *Regulatory Requirements for Hazardous Materials*

NANNEY • *Environmental Risks in Real Estate Transactions*

WALDO • *Chemical Hazard Communication Guidebook*

Biotreatment of Industrial and Hazardous Waste

Morris A. Levin   Editor
Maryland Biotechnology Institute
University of Maryland
College Park, Maryland

Michael A. Gealt   Editor
Department of Bioscience and Biotechnology
Drexel University
Philadelphia, Pennsylvania

McGraw-Hill, Inc.
New York   San Francisco   Washington, D.C.   Auckland   Bogotá
Caracas   Lisbon   London   Madrid   Mexico City   Milan
Montreal   New Delhi   San Juan   Singapore
Sydney   Tokyo   Toronto

Library of Congress Cataloging-in-Publication Data

Biotreatment of industrial and hazardous waste / Morris A. Levin, editor,
  Michael A. Gealt, editor.
    p.   cm.
  Includes bibliographical references and index.
  ISBN 0-07-037554-2 (alk. paper)
    1. Hazardous wastes—Biodegradation. 2. Factory and trade waste—
  Purification. I. Levin, Morris A. II. Gealt, Michael A.
  TD1061.B55 1993
  628.4—dc20   93-3130
                                                                 CIP

1 2 3 4 5 6 7 8 9 0   DOC/DOC   9 9 8 7 6 5 4 3

ISBN   0-07-037554-2

*The sponsoring editor for this book was Gail F. Nalven, the editing
supervisor was Paul R. Sobel, and the production supervisor was
Suzanne W. Babeuf. It was set in Century Schoolbook by McGraw-Hill's
Professional Book Group composition unit.*

Printed and bound by R. R. Donnelley & Sons Company.

Contents

Contributors

Piero M. Armenante *Department of Chemical Engineering, Chemistry, and Environmental Science, New Jersey Institute of Technology, Newark, New Jersey*

Ronald M. Atlas *Department of Biology, University of Louisville, Louisville, Kentucky*

Rita Colwell *Maryland Biotechnology Institute, University of Maryland, Baltimore, Maryland*

D. L. Crawford *Center for Hazardous Waste Remediation Research, Department of Bacteriology and Biochemistry, University of Idaho, Moscow, Idaho*

R. L. Crawford *Center for Hazardous Waste Remediation Research, Department of Bacteriology and Biochemistry, University of Idaho, Moscow, Idaho*

B. D. Ensley *Envirogen, Lawrenceville, New Jersey*

S. B. Funk *Center for Hazardous Waste Remediation Research, Department of Bacteriology and Biochemistry, University of Idaho, Moscow, Idaho*

Michael A. Gealt *Department of Bioscience and Biotechnology, Drexel University, Philadelphia, Pennsylvania*

David E. Giamporcaro *Section Chief, Biotechnology Program, Office of Pollution Prevention and Toxics, U.S. Environmental Protection Agency, Washington, D.C.*

R. H. Kaake *Center for Hazardous Waste Remediation Research, Department of Bacteriology and Biochemistry, University of Idaho, Moscow, Idaho*

A. Keith Kaufman *Bioremedial Services Division, RESNA Industries, Los Angeles, California*

Cheryl C. Krueger *Bioremedial Services Division, RESNA Industries, Los Angeles, California*

Morris A. Levin *Maryland Biotechnology Institute, University of Maryland, College Park, Maryland*

Carol D. Litchfield *Chester Environmental, Monroeville, Pennsylvania*

P. H. Pritchard *U.S. Environmental Protection Agency, Environmental Research Laboratory, Sabine Island, Gulf Breeze, Florida*

Bruce E. Rittmann *John Evans Professor of Environmental Engineering, Northwestern University, Evanston, Illinois*

D. J. Roberts *Center for Hazardous Waste Remediation Research, Department of Bacteriology and Biochemistry, University of Idaho, Moscow, Idaho*

Pablo B. Sáez *Associate Professor of Environmental Engineering, Pontificia Universidad Catolica de Chile, Santiago, Chile*

Gregory D. Sayles *U.S. Environmental Protection Agency, Risk Reduction Engineering Laboratory, Cincinnati, Ohio*

Malcolm Shields *Center for Environmental Diagnostics, University of West Florida, Pensacola, Florida*

Makram T. Suidan *Department of Civil and Environmental Engineering, University of Cincinnati, Cincinnati, Ohio*

G. J. Zylstra *Rutgers University, New Brunswick, New Jersey*

Preface

One of the characteristics of modern society is the rapidly increasing environmental release of chemicals. There are two main bases for these releases, chemicals created for environmental use and waste material.

The increased production of synthetic materials which are difficult to degrade and/or toxic to the environment is one of the hallmarks of modern times. To enhance functionality the compounds have been specifically designed for both long life in the field and nonreactivity with common environmental chemicals. These very characteristics deter degradation. Compound structures which deter chemical attack, hence yielding longevity and maintenance of effective structure—sought-after characteristics—are resistant to the enzymes of most microbes. In some cases, e.g., polychlorinated biphenyls (PCBs), these characteristics have been responsible, in part, for their continued presence in areas where their production and use have been discontinued.

Wasteful manufacturing processes have also led to accidental release of products and toxic (sometimes highly toxic) precursors. Even when toxicity is documented and production of a particular compound, or class of compounds, is terminated, the very nature of the material assures that disposal and destruction problems will be present for many more years.

Legitimate and permanent solutions to disposal problems are actually quite limited. Burial at hazardous waste landfills has been a reasonably cost-effective process for many years. However, problems have arisen from landfills where leakage to groundwater has occurred. In newer landfills this problem has been alleviated, at least temporarily, by the use of a complex system of geosynthetic liners as well as drainage and containment systems. A more serious landfill problem has been the rapid decrease in available space—especially in the heavily industrialized and populated eastern region of the United States. Removal of toxic materials and their transport to more distant sites adds significantly to the total disposal costs.

While incineration solves some of the landfill problems, especially that of storage space, it raises problems of its own. Incineration of liquid and solid waste converts at least some of the material into an air pollution hazard. The heat may convert compounds into gaseous forms of extremely toxic compounds, e.g., dioxin, unless extensive (and expensive) scrubbing occurs. If incineration is not performed on-site (and federal, state, and local ordinances may preclude this), then the cost of removal and transport need to be added to the basic cost of burning.

This brings us to the harnessing of natural degradation processes, generally referred to as biodegradation. In this method microorganisms, usually bacteria, but also fungi and algae, are responsible for breaking the toxic compounds into nontoxic by-products. In some cases degradation may result in the production of CO_2 and H_2O. The advantages are obvious: microbes are cheap, their reactions are predictable, and many times the reactions can be carried out on-site, if not in situ. This saves the costs of transport and, with in situ degradation, even the cost of removal from the ground or groundwater. If everything proceeds optimally, costs can be one-fourth to one-third that of landfill or incineration.

A major potential advantage of developing a better understanding of the natural degradation process is the ability to select a microbe that can degrade a toxic substance more efficiently than organisms from the native population. In addition, some workers are designing genetically engineered bacteria which are specific for certain compounds or avoid the induction problems.

However, problems do exist with biodegradation. There may not be appropriate indigenous microbes to degrade the toxic compound. The level of toxic compound may be so high as to preclude in situ degradation because of extensive lethality. Degradation may be induced by a compound which is, itself, toxic. There may be insufficient nutrients, cofactors, or O_2 present for in situ degradation. And most serious for proponents of strains designed with modern biotechnology, there are serious regulatory problems which must be addressed.

This book attempts to approach the topic of biodegradation of hazardous wastes in a holistic fashion. The issues of science, engineering and regulation are all addressed. As much as possible, both theoretical and practical considerations have been dealt with. Selection of bacteria for the specific purpose of degrading compounds is discussed at the bench-scale (Ensley and Zylstra; Gealt et al.) to the field level (Roberts et al). Engineering theory as applied to growth on toxic substances is discussed by Rittman and reactor technology by Sayles. The legal issues are covered by Giamporcaro. There are also several examples of field studies indicating the current usage of biodegradation, both within reactors and in situ. The use of biodegradation is compared with other mechanisms of disposal, in terms of time limitations, degradation limitations and, perhaps most important, cost. We believe the scope of this volume is certainly unique.

Morris A. Levin
Michael A. Gealt

1

Overview of Biotreatment Practices and Promises

Morris A. Levin

Maryland Biotechnology Institute
University of Maryland
College Park, Maryland

Michael A. Gealt

Department of Bioscience and Biotechnology
Drexel University
Philadelphia, Pennsylvania

The Basic Problem: Release of Hazardous Materials

Waste materials released directly from industry and agriculture are responsible for considerable contamination of soil and water in the United States. There are approximately 14,000 industrial sites in this country producing about 265 tons of hazardous waste annually. Types, quantities, and toxicity of hazardous waste found in waste dumps vary greatly. A typical waste dump may contain material ranging from mixtures of oyster shells and copper, paint and paint-associated material, as well as boxes of potassium cyanide, arsenic, candy, and cookies. The U.S. Environmental Protection Agency (U.S. EPA) has instituted a program to treat these sites Superfund Innovation Technology Evaluation Program (SITE), and efforts are underway in many other countries.

Although most of the commercial biotreatment efforts have been conducted in the United States,[61] many locations in Europe have also been successfully treated.[2,11,58,63] Major firms such as Dow Chemical are also involved in research to improve the biodegradation process. These efforts are without the use of genetically engineered organisms, although the potential abilities of these modified strains suggest that greater efforts in the future will achieve even greater degrees of remediation.[13,65] Recently Bhamidimarri et al.[10] reported that over 98 percent of a landfill leachate containing a daily loading of 1.6 kg of phenoxies and 0.5 kg of phenols was degraded by a semicontinuous method using a mixed soil microbial population.

A group of smaller companies has recently formed an association and produced a compendium describing successful instances of biotreatment on a commercial scale.[2] William Reilly, while head of the U.S. EPA, has emphasized the importance of bioremediation for the treatment of hazardous wastes.[29]

In many cases the toxic materials which need remediation are present because they were used to solve other technical problems. For example, use of polychlorinated biphenyls (PCBs) in transformers was at one time standard practice because it eliminated the fire hazard of other lubricants. Release sometimes occurred in the course of compound production, or the construction or disposal of the transformer. This does not in any way make the remediation any easier.

Attracting the most media attention have been the spills and other accidental releases which have occurred. The headline-grabbing *Exxon Valdez* accidental spill will forever exemplify the potential difficulties which arise from transportation of hazardous materials. What is often forgotten is the more constant and consistent release of oil, albeit in much smaller quantities, which occurs every time ballast is released from an oil-carrying ship. A similar situation exists with unscheduled releases from trucks or rail cars. Thus, when a tank car carrying cresol is ruptured, it is not necessary to excavate tons of soil to prevent contamination of groundwater. An appropriate strain or mixture of bacteria, coupled with appropriate nutrients, can be applied to degrade the offending material.[2] While dramatic, such major releases are also sporadic, and therefore may pose a smaller problem for remediation than the persistent unintentional releases that accompany pumping into or out of a tanker.

Mechanisms of Remediation

Various methods, both biological and nonbiological, are available for the remediation of contaminated sites. Each offers advantages and disadvantages, and, at times, a combination of approaches is the most rapid and cost-effective means of treatment. Thorough examination of the site is necessary to select the proper combination of technologies. Evaluation of the site must include a careful physical and biological workup, including hydrogeology, soil composition, characterization of the microbial population, and climate consideration, in addition to determining the range of pollutants present.

Nonbiological treatment

Several nonbiological treatments have been effective in removing exposure to hazardous wastes. These physical/chemical methods include use of long-term storage, landfills, incineration, and air stripping. While many other techniques have demonstrated degradation potential at the laboratory scale, the crucial ability to scale up to remediate hundreds of contaminated acres at reasonable cost has made these methods very useful.

Long-term storage and landfilling have similar advantages and disadvantages. Since there is no manipulation of the toxic material, cost is generally minimal as long as a reasonably close locale can be identified for the location of a storage locker or landfill site. The major cost involved is the removal and transport of the hazardous waste. This cost can be extreme if the distance is very great. Further complication is now frequently encountered because of local public resistance to the development of long-term storage facilities. The possibility of insufficient maintenance followed by possible leaching into soil or groundwater has increased the public concern. With increased concern has come legislation and regulation on container and landfill design, and thus increasing costs.

Incineration and air stripping have the advantage of decreasing the amount of toxic material existing. Incineration consumes energy and may also lead to production of additional toxic materials, e.g., dioxin, which requires costly scrubbing before release into the atmosphere. Air stripping also results in environmental release of small amounts of toxics, and, depending on the method used, may also involve the transfer of toxic materials onto activated carbon, which will then require treatment. In many cases combinations of biological and physical/chemical treatment are preferable because of economic and time considerations. The data to make these decisions should be gathered during the initial site inspection.

Biological treatment

Other technologies take advantage of the ability of microorganisms engineered to detoxify compounds. Mizrahi[41] has reviewed the various treatment methods and the modifications in biogas digesters, anaerobic digestion technology, and managerial aspects that result in more efficient sewage plant operation. Early studies led to the use of gravel as a percolating filter. The toxic wastes were retained by the gravel, in many cases enabling indigenous organisms to degrade them. Although simple in practice, these percolating filters were often very slow to detoxify. This work was followed by the development of more efficient anaerobic digesters and then aerobic digesters. Aerobic digestion is simply the addition of air to the digestion mixture, thus increasing growth and oxidative abilities of the microbes. One of the earliest practices of genetic selection was the discovery that the use of an inoculant from previous digestions greatly enhanced the degradation process. Modifications in the process for degradation shortened the time period needed, sometimes to only 5 to 10 days.

Many different procedures have been developed to permit the goal of enhanced contact between microbes and the target pollutants to maximize the efficiency of the procedure and minimize any hazards. Table 1.1 (see also Roberts et al., Chap. 10) describes some of the most common procedures and the safety issues associated with each. Some of the procedures (e.g., land farming) present difficulties similar to those encountered with nonbiological treatment: i.e., the potential for leaching from the treatment site. In the case of biological treatment, untreated chemicals, intermediate breakdown products, and the microbes themselves may escape from the site. However, it must be pointed out that the time period during which adverse effects from the wastes or the breakdown products could occur is much shorter. The microbial degradation procedure will be complete in a finite period of time: the storage approach has no upper limit. The potential adverse effects of added microbes have been reviewed before allowing the initial release. The microbes are ones which are normally present (indigenous).

Compounds which can be biologically degraded.   Arguments abound as to whether microbes are limited in their digestive abilities, or whether they can reasonably degrade any compound that humans can devise.[62]

TABLE 1.1   Comparison of Treatment Methods

Type of treatment	Cost per cubic yard ($)	Time required (months)	Additional factors/ expense	Safety issues
Incineration	250–800	6–9	Energy	Air pollution
Fixation	90–125	6–9	Transport; long-term monitoring	Leaching
Landfill	150–250	6–9	Long-term monitoring	Leaching
Biotreatment	40–100	18–60	Time commitment of land	Intermediary metabolites and polymerization

The truth undoubtedly lies somewhere between these two opinions. Without doubt, however, microbes are able to degrade many compounds under many different conditions. Many synthetic compounds are also able to be modified or transformed by a bacterium, fungus, or some diverse microbial population working as a consortium.[5] These processes range from the rotting of foodstuffs to the cleaning of spilled oil from coastal beaches. Many cases are beneficial and essential. After all, much of the organic and inorganic cycling that is necessary for ecosystem maintenance is a result of microbial activity.[23]

In addition to being able to modify or degrade a compound, it is necessary that the throughput be reasonably high. It has been estimated that 6000 tons of sulfur cycle between organic and inorganic compounds annually. The large amount of xenobiotic (synthetic) compounds released must also be readily amenable to microbial digestion through complex modifications of natural cyclic pathways.[43] Each year the United States produces well over 50 million metric tons of federally regulated hazardous wastes.[17] In addition to new production, there are many (estimates range as high as 50 percent) petroleum product storage tanks which are leaking. Chemical contaminants include toluene, benzene, di-, tri-, and tetrachloroethylene; and parathion. Concentrations can be high. For example, trichloroethylene (TCE) has been found as concentrated as 27 ppm in groundwater.[54]

Short time frames do not permit the evolution of microbial systems capable of rapidly and facilely coping with the onslaught of all xenobiotic chemicals.[30,56] Many xenobiotic chemicals are resistant to microbial attack and/or are toxic to the microbes. Nevertheless, microbes that can degrade many xenobiotic compounds with different degrees of ease and at different rates have been isolated from locations contaminated with various xenobiotic compounds. Some can degrade more than one compound and do so at different rates. Abramowicz[1] demonstrated such results in soils contaminated with PCBs. He proposed combining genetic material from 26 different isolates to produce a single more useful bacterium. Molecules which have been demonstrated to be biodegraded include ethyl chloride;[52] PCBs;[1,42,12] gasoline and other petroleum products;[7,28] dinitro and trinitro compounds, including nitrogenous herbicides[3,60] and trinitrotoluene;[22] polyaromatic hydrocarbons, including pentachlorophenol,[38,39,43] tetrachloroethylene (PCE), TCE, dichloroethylene (DCE), and vinyl chloride (VC);[37,46,66,67,71] tetrachloroethene;[19] creosote; [21,45] and fluoranthene.[32,44]

Organisms involved. The organisms which are capable of carrying out the degradation process are varied. In some cases they have been identified and characterized, whereas in others even growth in isolation has proved to be extremely difficult. One of the best-studied organisms has been *Pseudomonas* G4.[47,48] This bacterium is effective in degradation of TCE. The organism is able to be induced under more general environments with a great number of cosubstrates.[59] Because it can degrade several compounds and be induced by several other compounds, G4 may have great utility in remediation of wastes from the many sites where a mixture of toxics is present. Genetic modifications of this bacterium should enhance its abilities.

Genetic modifications have been investigated, as discussed in Chap. 3 by Ensley and Zylstra. Modifications to increase the range of capabilities of the microbes as well as the rate of degradation and stability under environmental stresses have been attempted. While excellent results have been obtained in laboratory situations, scale-up studies are still in the developmental stage. As discussed by Giamporcaro (Chap. 6), the definition of modified organisms from a regulatory perspective is still developing, ranging from organisms created via classical techniques (e.g., conjugation, transformation, mutagenesis) to those resulting from molecular biology approaches involving the use of vectors or electroporation, ballistics, and site-directed mutagenesis. Giamporcaro also discusses specific problems associated with the use of engineered microbes in detail from a regulatory perspective.

Other examples of the usefulness of *Pseudomonas* species include the many strains containing the TOL plasmid, which enables the degradation of toluene. The location of the genes involved in toluene degradation on a plasmid has greatly enhanced the ability to dissect the regulatory mechanisms which enable enhanced remediation to occur. The TOL plasmid is naturally occurring, which suggests that evolution of the ability to degrade xenobiotic compounds may have resulted, in part, from interaction with molecules produced by indigenous microbes.

Among the many other strains in which degradation capability has been demonstrated are a pseudomonad (strain LB400) that has been isolated which has activity against PCBs.[12] An isolate of *Clostridium* has proved capable of transformation of trichloroethane, trichloromethane, and tetrachloromethane.[25] Dinitrophenol herbicides have been shown to be degraded by an *Azotobacter* sp.[68]

While it has been possible to find single organisms which are able to demonstrate degradation of simple compounds or groups of compounds, it usually requires a consortium of bacteria to carry out degradation of a mixed waste stream. Under many circumstances, consortia are more effective even for single wastes.[20,26,27,33] Further developments in natural and constructed consortia may result in the greatest enhancement of degradation practices. As described by Roberts et al.

(Chap. 10), selection practices of natural populations can easily and reproducibly develop an inoculum which is capable of degradation in the field of the pesticide dinoseb.

In many cases indigenous organisms require augmentation with nutrients for relatively rapid and complete degradation of introduced hazardous wastes.[4,24] Organisms under natural conditions are generally starved for phosphorus, nitrogen, and sulfur. By adding these compounds, growth of the natural population is stimulated and, perhaps more importantly, their metabolism is enhanced so that transport across cell membranes and metabolic attack can proceed. Since most degradations of toxic material occurs by cometabolism, it may be necessary at times to also add a carbon source.

Bioremediation Techniques

Once the organisms have been chosen, or the decision has been made to use indigenous organisms, the next task is to determine the best technique for degradation. Methods vary in both biological and engineering requirements, and safety considerations (Table 1.2). The method chosen will depend on the site, the time frame, and the amount of money available.

Land farming is perhaps the simplest and most common method to treat wastes. It is done directly on site and simply involves addition of nutrients and moisture as required to achieve microbial growth and metabolism of pollutants. Large amounts of soil can be treated, requiring only minimal transportation, usually done with a backhoe; aeration is frequently accomplished with the same backhoe. Only minimal environmental control of physical parameters (pH, temperature, etc.) is possible and products of microbial reactions are dispersed freely. It is possible to speed up land farming by recycling some of the soil used, thus enhancing the microbial population able to degrade specific wastes. Anaerobic land farming is possible by oversaturating the soil with nutrients, thus decreasing the amount of oxygen available to bacteria and fungi (see Chap. 10, by Roberts et al.). Linings have been used to contain the process, but these are known to fail, and monitoring is difficult.

Removal of contaminated soil or water, e.g., groundwater, into a reactor where it is acted upon by microbes is frequently used methodology. Construction of a reactor in which the hazardous wastes and microbes are mixed with nutrients is more efficient in terms of control, but more costly in terms of overall construction and operation.[14,15] Temperature, time of contact, nutrient levels, and concentration of waste can all be optimized. Optimal reactor construction and theory

TABLE 1.2 Types of Biotreatment Processes

Type	Principle	Comments	Safety issues
Land farming	Soil mixed with nutrients and tilled in situ.	Requires lining to contain microbes and material.	Lining and cap have leakage and aging problems; monitoring and treating can be difficult.
Soil slurry (tank or lagoon)	Soil and water agitated together in reactor.	No temperature control.	Little control over degradation process; effluent can be monitored and treated.
Subsurface reclamation	Water, nutrients, and oxygen (electron acceptor) pumped through soil.	Enhanced growth of entire indigenous population. Primary applications: oil and gasoline spills.	Organic contamination of groundwater as a result of mobilization of compounds; no control over dispersal of microbes or degradation products.
Soil treatment system	Wash procedure to solubilize adsorbed contaminants.	Pretreatment necessary to maximize efficacy.	Effluent goes to SBR; washed soil can be monitored before replacing at site.
Sequencing batch reactor (SBR)	Microbial digestion in liquid suspension.	Allows control of reaction conditions.	Release of microbes to environment; can monitor for microbes and pollutants.
Aqueous treatment system	Immobilized microbes or enzymes in flow-through system.	Requires soluble organic material.	No microbial release; effluent can be monitored and treated.
Fixed-film bioreactor	Microbes/enzymes on plastic media in column to maximize surface area and nutrient exchange.	Can treat low concentrations of organic material.	No microbial release; recycling of pollutants permits enhanced degradation and monitoring.

are discussed by Armanente (Chap. 4) and by Sayles and Suidan (Chap. 11). However, dispersal of microbes and degradation products is reduced and faster degradation is possible. The use of immobilized microbes or fixed-film bioreactors results in greater efficiency, and the effluent can be monitored to minimize release of microbes to the environment. Recycling of the material to be treated is possible, thus increasing the overall efficiency of the treatment process. One of the major advantages is that reactors can often be made small enough to be transported on a truck; this makes it possible to use the same reactor at several sites, thus decreasing costs. Posttreatment is frequently necessary before release of wastes to municipal systems. Genetically modified microbes are logical organisms to use in reactors because of the enhanced ease of containment, but also because the polluted material can be passed through a reactor which contains the pollutant-microbe mixture and monitored for nutrient, pH, and oxygen levels. Release to the environment can be permitted after the pollutant levels have reached acceptable levels and the effluent can be treated to minimize release of the microbial population.

Modifications in reactor technology can enhance degradation. For example, the use of sequencing batch reactors (SBRs) to treat leachate is described by Pierce[52] and by Irvine.[31] Irvine's efforts focused on a leachate from a contaminated industrial site. Initially the leachate was placed in storage tanks in contact with "non-sterile raw waste feed" from a wastewater plant for up to 19 days prior to being filtered through granular activated-carbon (GAC) columns. Selected specific organisms were added to the reactors. In one case,[65] a unique strain of *Pseudomonas putida* that both contained a specific degradative ability not found in the original strain and was uniquely adapted to SBRs was isolated, cultivated, and added to the existing SBR microbial mix. The SBRs were operated as closed systems. All volatile organic material was trapped on GAC and recycled annually, and reduction of compounds monitored. The SBR process had the greatest effect on total organic carbon (TOC) and phenol, achieving greater than 99 percent reduction from starting levels of 10,575 and 1,553 mg/L, respectively. The SBR-treated leachate still required GAC treatment to meet discharge standards. Because of the carbon needed to capture TOC and phenol, the cost of treatment was reduced by approximately $30 per cubic meter of water following SBR treatment.

Composting is a treatment process common to all farming communities, serving as a method for converting wastes into reusable soil enrichers. Basically it is a microbial conversion carried out under conditions which limit the types of bacteria and fungi, and thereby the available activities.[6] Optimal composting requires stringent regulation of moisture (50–60 percent) and temperature. The temperature profile of a compost heap generally allows for a succession of microbes able to attack different units of the hazardous waste. Aeration is also critical, and much attention is frequently paid to the method of air introduction. Small compost heaps are frequently aerated by mechanically turning over the rotting material. Better regulation of aeration is accomplished by either forcing air through the compost pile or drawing it through the pile by vacuum. Oiled shoreline waste has been successfully treated by aerated composting.[34]

One of the most exciting areas of bioremediation is in situ treatment of hazardous wastes. By treating not only at the site, but in the ground or water, there are substantial savings in terms of digging, pumping, transportation, etc. On the other hand, there is also a loss of control over environmental factors and microbes which are able to interact with the waste material. In all cases, it is necessary to add nutrients to the soil or water in order to get sufficient degradation. Oxygen is frequently introduced either directly or by addition of H_2O_2 with the nutrients. The need for a cosubstrate implies that most of the degradation is by way of cometabolism or cooxidation. Sometimes it is necessary to add a structural analog (e.g., aniline for 3,4-dichloroaniline or biphenyl for PCBs) to induce enzymes necessary for degradation. This would be necessary even in cases where nutrients and oxygen are sufficient. The toxicity of the inducer substance is one of the difficulties that may arise when this technique is proposed. Many examples of techniques for in situ biodegradation are found in the chapters by Atlas (Chap. 2) and by Litchfield (Chap. 7).

In situ degradation usually requires cometabolism, in which the compound being degraded serves only poorly or not at all as an energy or carbon source.[6] The work of Pfaender and Alexander[51] and Sakazawa[57] illustrates the fact that where cometabolism is involved, often the species designation of the organism(s) involved is not known, although in most cases the genus is specified. When consortia of microbes are involved, the end products of metabolism are identified and the microorganisms are often not specified.[49] The degradation process often involves a consortium of microbes including strains in the genera *Nocardia, Pseudomonas, Acinetobacter,* and *Flavobacterium,* without identification to the species level. One company reported that as many as 32 different microbes were involved in degrading a specific gasoline spill.[11] In general, the more complex the mixture, the more complex the consortium of microbes.[11,50]

It may be necessary to add more microbes to the in situ situation. These are generally naturally occurring. Local or prestressed microbes tend to work best (see Chap. 10). Research to better understand the relationships between the genetic capability of the entire microbial population at a given site and the phenotypic expression of biodegradation is ongoing. The goal is to develop a microbial population which is better able to remediate wastes at a specific site, and, perhaps, to commercialize the consortium for similarly contaminated sites. The research will also identify activities which, when augmented, could further enhance degradation capabilities (see Ref. 65 and Chap. 8).

Evaluation for Biodegradation

Practical considerations in developing a remediation or restoration plan include:

Time constraints

The amount of time which can be devoted to remediation is frequently a consideration in the choice of treatment processes. In general, in situ bioremediation can be a fairly rapid process, especially when compared to processes which require more extensive construction. However, there are significant time differences for the various biodegradation methodologies. The use of selected (or recycled) microbes can speed up processes from months down to a few weeks. It is possible, however, that different treatments will be speedier for certain xenobiotics.

Cost constraints

Costs to communities for remediation are expected to reach $10 billion by the year 2000 and $30 billion over the following decade.[53] This assumes cleanups accomplished by a mixture of physical, chemical, and biological methods. Currently biological methods are used in 15 to 20 percent of the cases. Table 1.1 compares costs of various treatment methods and is based on published cost figures.[11,40,58,55,61,64,70] This comparison is problematic because it is necessary to consider more than just the direct estimation of actual expenses. Biotreatment requires the least energy and is the only method that can achieve mineralization of the waste material to innocuous products. However, biotreatment takes longer and does not necessarily result in cleanup to the level required by federal or local regulations. This results in the need for additional treatment, and therefore added costs. However, biotreatment also results in significant waste volume, which significantly reduces the cost of additional treatment.

Regulatory constraints

The *Federal Register* describes the U.S. government policy on regulation of biotechnology products. The policy was to be based on existing laws which covered safety in the workplace, the environment, agriculture, food processing, and production of pharmaceuticals, and public health. Relative to bioremediation there are four laws that must be considered in order to assure compliance with existing regulations. These are the Toxic Substances Control Act (TSCA), the Comprehensive Environmental Response, Compensation, and Liability Act (CERCLA), the Resource Conservation and Recovery Act (RCRA), and the Federal Plant Pest Act (FPPA).[8,36] All of these statutes fall under the jurisdiction of the U.S. EPA. The EPA has announced its intention to include coverage of microbes under TSCA and has stated that the definition of "chemical" in TSCA includes microbes. RCRA requires the EPA to define hazardous wastes and to set standards for treatment, while CERCLA regulates the management of the wastes. These statutes defining types of wastes and describing waste treatment procedures are discussed in detail by Giamporcaro (Chap. 6).

Site constraints

The physical composition of the site containing the hazardous material will have a significant effect on the ability of microbes to carry out remediation. It is important to determine the hydrogeology, location of underground pipes, and other factors which will affect the ability to get added nutrients and oxygen to the degrading bacteria. Keystone Environmental Resources spent two years studying the soil beneath and immediately under a contaminated area.[16] During this time the physical aspects, such as site hydrology, soil type, subsurface conditions, and climate characteristics were defined, while laboratory studies to determine the characteristics of the microbial flora and impact of the pollutants on the flora were carried out. Details of the site requirements for in situ bioremediation are given in the chapter by Litchfield (Chap. 7). Perhaps the most important considerations are the exact nature of the contaminants, the extent of contamination, and whether the contamination is mostly in the vadose soil layer or has reached the groundwater.

The location of the contaminated site will affect prospects for biodegradation because of temperature, pH, and water availability conditions. Temperature, even seasonal variation, will affect the ability of microbes to degrade wastes. However, it is clear that bacteria and fungi exist which are able to carry out remediation at all locations, so that even in Alaska sufficient bacteria were present for in situ bioremediation to occur (see Chap. 12). Temperature affects not only enzyme activity rates, but also bioavailability, membrane transport, etc. The acidity or alkalinity of the environment will also affect remediation, affecting, for example, organism growth and molecular species of xenobiotic present. Perhaps most important is water availability. Soil water content is necessary for biological activity.

In many cases indigenous microbes can degrade contaminating material if appropriate nutrients, electron acceptor, and moisture are supplied. The nutrients involved are nitrogen, phosphorus, and minerals. Carbon is rarely the limiting factor in contaminated subsurfaces. Usually oxygen is added as an electron acceptor, although occasionally nitrate is added as an alternate, as was done in the Upper Rhine Valley of Germany.[69] Hydrogen peroxide (H_2O_2) may be added instead of oxygen. There may be cases (see Chap. 10) where anaerobic degradation is desired. This can be achieved by the addition of nutrients without the addition of oxygen.

If the treatment is to be conducted in situ, monitoring procedures, including sampling times, locations, and duration, must be established before the project begins. Selection of the compounds to monitor will depend on the regulatory requirements as well as an understanding of the specific compounds involved and their breakdown products. Standards for release, monitoring procedures, and statistical analysis have been set by U.S. EPA (Table 1.3). In the case of Keystone Environmental Resources, the chloride content was monitored as an indication of mineralization and direct pollutant measurements were made at three upstream and three downstream wells. After 12 weeks of treatment, approximately 90 percent of the contaminant had been removed. In other field applications a 98–99 percent reduction in levels of carbon tetrachloride, chlorobenzene, and ethyl benzene has been observed.

Typically, to degrade ~1,000 gal of hydrocarbon, 10,000 lb of oxygen and 875 lb of ammonia nitrogen would be required. This would result

TABLE 1.3   Measurements Required to Monitor Treatment Effectiveness

Parameter	Acceptable value
Total organic carbon	2.3–2.4 mg/L[†]
Total organic halide	297–353 mg/L
pH	5–10[‡]
Concentration of:	
Benzoate	<2.5–34 mg/L
Phenol	4.9–7.5 mg/L[§]
Chlorobenzoate	
ortho	<3.5 mg/L
meta	<5.0 mg/L
para	40–47 mg/L
Het Acid*	279–246 mg/L

*1,4,5,6,7-hexachloro-5-norbonene-2-3-dicarboxylic anhydride.
†U.S. EPA priority pollutant: acceptable level = 300 mg/L.
‡U.S. EPA priority pollutant: acceptable level = 5–10.
§U.S. EPA priority pollutant: acceptable level = 1 mg/L.

in the production of ~7,000 lb of bacteria. The success of nutrient augmentation was demonstrated in the efforts to clean up the Alaskan coastline after the *Exxon Valdez* spill[18] and is discussed by Pritchard in Chap. 12.

Problems associated with the fluctuations in environmental conditions include partial degradation resulting in the presence of intermediate metabolites. For example, the pathway for the degradation of PCE, a known animal carcinogen, can result in the accumulation of VC, which is also a human carcinogen.[9,35] McCall et al. (1981) reported that during the degradation of PCE, concentrations of 2,4,5-trichlorophenol and 2,4,5-trichloroanisoie were found in the soil. Vira and Fogel propose a combination of anaerobic and aerobic treatments with the addition of nutrients (including oxygen and methane) to establish control over the degradation process in environmental settings.

Technical Considerations in Bioremediation

Technical problems related to biological interaction with hazardous wastes are often a function of the nature of the final product. A common example is the increase in VC which frequently accompanies the degradation of TCE. Since the VC is much more toxic than the TCE, it would have been better to do nothing than to have "remediated" the problem into a more serious one.

Microbes have also been used to induce polymerization of hazardous wastes. Polymers will gradually degrade, with the result that the release of the hazardous substance may merely be delayed, not prevented. This may be a case of in situ storage, rather than true remediation. Another type of common problem is the production of more toxic intermediates if the material is not mineralized completely. When biotransformation occurs, one must determine the toxicity of new products (or intermediates that may appear en route to complete mineralization). These transformed compounds are most likely to occur when dealing with PCB congeners greater than 1248, with polyaromatic hydrocarbons with rings containing more than five carbons, and in mixed wastes involving radioisotopes.

There are several factors that need to be considered when developing a proposal for the use of biodegradation. Of immediate concern is the presence or absence of a population of bacteria at the site which are able to degrade the hazardous waste. While there are some waste materials that defy (at least for practical purposes) degradation by microbes, in general, organisms are obtainable which can attack most wastes. The organisms may need to be enhanced in terms of cell density or metabolic activity. Time and effort (expense) may be required to develop strains adapted for the particular site; the site may need modification to maximize the degradative potential of the microbes; the costs associated with these processes may be greater than those available for the project. This is especially true if the modification needed would fall into the realm of genetic engineering, which is capable of yielding great increases in degradative ability, but generally requires a very long lead time involving extensive research. Another consideration is the requirement for a cosubstrate or other inducing substance (see Chap. 2 by Atlas and Chap. 3 by Ensley and Zylstra for extended discussion) to enhance metabolism of the specific waste. If this substance is too costly or toxic, it may not be feasible to use, especially if in situ treatment or land farming are the proposed methodologies.

Scale-up considerations

The steps involved in implementing a biotreatment program begin with feasibility determinations (i.e., can it be done) and end with engineering and economic determinations (i.e., how can it be accomplished on the scale needed at this site for a reasonable cost). The Keystone project cited above is one example of scale-up activities. Wick and Pierce[69] have described an integrated approach to the development and implementation of bioremediation which involves the careful examination of the site, estimation of the amount and type of material to be degraded, large-scale demonstration and cost analysis of the process, consideration of the regulatory issues (i.e., will the use of microbes—engineered or natural—be permitted; what are the regulatory limits imposed by the appropriate authorities for the starting and end products?), estimation of time required to complete the remediation process, and evaluation of whether any other procedures may be combined with biotreatment to speed up the cleanup or decrease the cost. These authors point out that a detailed examination of all of the above factors is essential for successful completion of a bioremediation project.

In general, the use of biological treatments enables the user to proceed with a cost-effective treatment process which can bring the level of toxic materials down below that required by federal and state agencies. While the process occasionally takes longer than nonbiotic processes such as incineration, it usually results in complete mineralization with less possibility for exposure of human populations to potentially toxic and carcinogenic wastes.

References

1. Abramowicz, D. A. 1989. Biodegradation of PCB contaminated soil using recombinant bacteria. Proc. A&WMA/EPA International Symp. Haz. Waste Treatment. Cincinnati, OH.
2. Applied Biotreatment Association. 1989. Compendium of Biotreatment Applications. ABTA, Washington, D.C.

3. Adrian, N. R., and J. M. Suflita. 1990. Reductive dehalogenation of a nitrogen hete-rocyclic herbicide in anoxic aquifer slurries. *Appl. Environ. Microbiol.* 56:292–294.
4. Aggarwal, P. K., J. L. Means, and R. E. Hinchee. 1991. Formulation of nutrient so-lutions for in situ biodegradation, pp. 51–66. In R. E. Hinchee and R. F. Olfenbuttel (eds.), In Situ Bioreclamation: Applications and Investigations for Hydrocarbon and Contaminated Site Remediation. Butterworth-Heinemann, Boston.
5. Alexander, M., 1991. Research needs in bioremediation. *Environ. Sci Technol.* 25:1972–1973.
6. Atlas, R. M., and R. Bartha. 1973. Stimulated biodegradation of oil slicks using oleophilic fertilizers. *Environ. Sci. Technol.* 7:538–541.
7. Atlas, R. M., and M. C. Atlas. 1991. Biodegradation of oil and bioremediation of oil spills. *Curr. Opinion Biotechnol.* 2:440–443.
8. Bakst, J. S. 1990. Impact of present and future regulations on bioremediation. *J. Ind. Microbiol.* 8:13–22.
9. Barrio Bell, R. H., and A. H. Hoffman. 1991. Gasoline spill in fractured bedrock ad-dressed with in situ bioremediation, pp. 437–443. In R. E. Hinchee and R. F. Olfenbuttel (eds.), In Situ Bioreclamation. Applications and Investigations for Hydrocarbon and Contaminated Site Remediation. Butterworth-Heinemann, Boston.
10. Bhamidimarri, S. M. R., D. Catt, and C. Mercer. 1990. Semi-continuous biotreatment of a landfill leachate containing phenoxy herbicide chemical. CHEMECA 90. Processing Pacific Resources. 18th Australasian Chemical Engineering Conference, Auckland, New Zealand, Vol. II, pp. 1039–1044.
11. Bluestone, M. 1986. Microbes to the rescue. *Chem. Week,* Oct. 29, 1986, pp. 34–40.
12. Bopp, L. H. 1986. Degradation of highly chlorinated PCBs by Pseudomonas strain LB400. *J. Ind. Microbiol.* 1:23–29.
13. Bourquin, A. W. 1990. Bioremediation of hazardous waste. *Biofutur* 93:24–37.
14. Brauer, H. 1987. Development and efficiency of a new generation of bioreactors. Part 1. *Bioproc. Eng.* 2:149–159.
15. Brauer, H. 1988. Development and efficiency of a new generation of bioreactors. Part 2, Description of new bioreactors. *Bioproc. Eng.* 3:11–21.
16. Campbell, J. R., J. K. Fu, and R. O'Toole. 1989. Biodegradation of PCP contaminated soils using in situ subsurface reclamation. 2nd National Conference on Biotreatment; Proceedings, pp. 17–29. Hazardous Materials Control Research Institute, Washington, D.C.
17. Chiras, D. D. 1988. Environmental Science: A Framework for Decision Making, 2nd ed., Benjamin Cummings, Reading, MA.
18. Crawford, M. 1990. Bacteria effective in Alaska cleanup. *Science* 247:1537.
19. DiStepheno, T. D., J. M. Gossett, and S. H. Zinder. 1991. Reduction dechlorination of high concentrations of tetrachloroethene to ethene by anaerobic enrichment culture in the absence of methanogenesis. *Appl. Environ. Microbiol.* 57:2287–2292.
20. Dolfing, J., and J. M. Tiedje, 1987. Growth yield increase linked to reductive dechlo-rination in a defined 3-chlorobenzoate degrading methanogenic coculture. *Arch. Microbiol.* 149:102–105.
21. Evangelista, R. A. Treatment of phenol and cresol contaminated soil. *J. Haz. Mat.* (Amsterdam) v25:343–360.
22. Fernando, T., J. A. Bumpus, and S. D. Aust. 1990. Biodegradation of TNT (2,4,5-trinitrotoluene) by Phanerochaete chrysosporium. *Appl. Environ. Microbiol.* 56:1666–1671.
23. Fliermans C. B., and T. D. Brock. 1972. Ecology of sulfur oxidizing bacteria in hot acid soils. *J. Bacteriol.* 111:343–350.
24. Floodgate, G. D. 1979. Nutrient limitation, pp. 107–119. In A. W. Bourquin and P. H. Pritchard (eds.), Microbial Degradation of Pollutants in Marine Environments. EPA-66019-012. Environmental Research Laboratory, Gulf Breeze, FL.
25. Galli, R., and P. L. McCarty. 1989. Biotransformation of 1,1,1-trichloroethane, trichloromethane, and tetrachloromethane by a Clostridium sp. *Appl. Environ. Microbiol.* 55:837–844.
26. Genthner, B. R. S., W. A. Price, and P. H. Pritchard. 1989a. Anaerobic degradation of chloroaromatic compounds in aquatic sediments under a variety of enrichment conditions. *Appl. Environ. Microbiol.* 55:1466–1471.

27. Genthner, B. R. S., W. A. Price, and P. H. Pritchard 1989b. Characterization of anaerobic dechlorinating consortia derived from aquatic sediments. *Appl. Environ. Microbiol.* 55:1472–1476.
28. Hoeppel, R. E., R. E. Hinchee, and M. F. Arthur. 1991. Bioventing soils contaminated with petroleum hydrocarbons. *J. Ind. Microbiol.* 8:141–146.
29. Hoyle, R. 1991. EPA moves on bioremediation. *Biotechnology* 9:10–34.
30. Hutzinger, O., and W. Vertkamp. 1981. Xenobiotic chemicals with pollution potential, pp. 3–45. In Microbial Degradation of Xenobiotic and Recalcitrant Compounds, T. Leisenger, A. M. Cook, R. Hutter, and J. Neusch (eds.). Academic Press, London.
31. Irvine R. L., S. A. Sojka, and J. F. Colaruotolo. 1982. Treating landfill by pure strain inoculations of SBR's, pp. 96–107. In Impact of Applied Genetics in Pollution Control, C. F. Kulpa, R. L. Irvine, and S. J. Sojka (eds.), University of Notre Dame, Notre Dame, IN.
32. Kelley, I., and C. E. Cerniglia. 1991. The metabolism of fluoranthene by a species of Mycobacterium. *J. Ind. Microbiol.* 7:19–26.
33. Kröckel, L., and D. D. Focht. 1987. Construction of chlorobenzene utilizing recombinants by progenitive manifestation of a rare event. *Appl. Environ. Microbiol.* 53:2470–2475.
34. Labrie, P., and B. Cyr, 1990. Biological remediation of shoreline oily waste from marine spills, pp. 339–387. In Proceedings of the Thirteenth Annual Arctic and Marine Oil Spill Program Technical Seminar. Environment Canada, Ottawa, Canada.
35. Lage et al. 1986. Lee, M. D., and R. L. Raymond, Sr. 1991. Case history of the application of hydrogen peroxide as an oxygen source for in situ bioreclamation, pp. 429–437. In R. E. Hinchee and R. F. Olfenbuttel (eds.), *In Situ Bioreclamation: Applications and Investigations for Hydrocarbon and Contaminated Site Remediation.* Butterworth-Heinemann, Boston.
36. Levin, M. A. 1992. *J. of Clear Technol. and Environ. Sci.* V2.1 31–39. Impact of regulations on oil spill clean up: The Alaska story. In press.
37. Little, C. D., A. V. Palumbo, S. E. Herbes, M. E. Lindstrom, R. L. Tyndall, and P. J. Gilmer. 1988. Trichloroethylene biodegradation by a methane-oxidizing bacterium. *Appl. Environ. Microbiol.* 54:951–956.
38. Mahaffey, W. R., and R. A. Sanford. 1991. Bioremediation of PCP-contaminated soil: Bench to full-scale implementation. *Remediation* (Summer):305–323.
39. McCormick, D. 1985. One bug's meat. *Bio/Technology* 3:429–435.
40. Mikesell, M. D., and S. A. Boyd. 1988. Enhancement of pentachlorophenol degradation in soil through induced anaerobiosis and bioaugmentation with anaerobic sewage sludge. *Environ. Sci. Technol.* 22:1411–1414.
41. Mizrhahi, A. 1989. Biological waste treatment. *Adv. Biotechnol. Proc.* 1:1–310.
42. Mondello, F. J. 1989. Cloning and expression in *Escherichia coli* of *Pseudomonas* strain LB400 genes encoding polychlorinated biphenyl degradation. *J. Bacteriol.* 171:1725–1731.
43. Morgan P., and R. J. Watkinson. 1989. Hydrocarbon degradation in soils and methods for soil biotreatment. *CRC Crit. Rev. Biotechnol.* V8.4:305–333.
44. Mueller, J. G., P. J. Chapman, and P. H. Pritchard. 1989. Action of a fluoranthene-utilizing bacterial community on polycyclic aromatic hydrocarbon components of creosote. *Appl. Environ. Microbiol.* 55:3085–3090.
45. Mueller, J. G., D. P. Middaugh, S. E. Lantz, and P. J. Chapman. 1991. Biodegradation of creosote and pentachlorophenol in contaminated groundwater: Chemical and biological assessment. *Appl. Environ. Microbiol.* 57:1277–1285.
46. Nelson, M. J. K., S. O. Montgomery, W. R. McHaffey, and P. H. Pritchard. 1987. Biodegradation of trichloroethylene and involvement of an aromatic biodegradative pathway. *Appl. Environ. Microbiol.* 53:949–954.
47. Nelson, M. J., J. V. Kinsella, and T. Montoya. 1990. In situ biodegradation of TCE contaminated groundwater. *Environ. Progr.* 9:190–196.
48. Nelson, M. J. K., and A. W. Bourquin. U.S. Patent 4,925,802. Ecova Corp., Redmond, WA.
49. Neilson, A. H., A. S. Allard, C. Lindgrin, and M. Remberger. 1987. Transformations of chloroguaiacols, chloroveratroles, and chlorochatecols by stable consortia of anaerobic bacteria. *Appl. Environ. Microbiol.* 53:2511–2519.

50. Olson, B. H. 1991. Tracking and using genes in the environment. *Environ. Sci. Technol.* 25:(4)604–610.
51. Pfaender, F. K., and M. Alexander. 1972. Extensive degradation of DDT in vitro and DDT metabolism by natural communities. *J. Agric. Food Chem.* 20:842–846.
52. Pierce, G. E. 1982. Diversity of microbial degradation and its implications in genetic engineering, pp. 20–25. In Impact of Applied Genetics in Pollution Control, Kulpa, C. F., R. L. Irvine, and S. J. Sojka (eds.), University of Notre Dame, Notre Dame, IN.
53. Porta, Augusto. 1991. A review of European bioreclamation practice. *Biotreatment News,* 1:(9)6–8.
54. ReVelle, P., and C. ReVelle. 1988. The Environment, Issues and Choices for Society, 3rd ed. Jones and Bartlett, Boston.
55. Risel, H. L., Boston, T. M., and Schmidt, C. J. 1984. Costs of Remedial Response Actions at Uncontrolled Hazardous Waste Sites. Noyes, Park Ridge, NJ.
56. Rochkind-Dubinsky, M. L., J. W. Blackburn, and G. S. Sayler, 1986. Microbial Decomposition of Chlorinated Aromatic Compounds. EPA/600/2-86/090, Washington, DC.
57. Sakazawa, C., M. Shimao, Y. Taniguchi, and N. Kato. 1981. Symbiotic utilization of polyvinyl alcohol by mixed cultures. *Appl. Environ. Microbiol.* 41:261–267.
58. Savage, P. 1987. Bacteria pass a Houston cleanup test. *Chem. Week,* Nov. 11, 1987, pp. 55–56.
59. Shields, M. 1991. Treatment of TCE and degradation products using *Pseudomonas cepacia.* EPA Symposium on Bioremediation of Hazardous Wastes, Apr. 16–18, McClean, VA.
60. Spanggord, R. J., J. C. Spain, S. F. Nishino, and K. E. Mortelmans. 1991. Biodegradation of 2,4-dinitrotoluene by a Pseudomonas sp. *Appl. Environ. Microbiol.* 57:3200–3205.
61. Stapps, J. J. 1988. Developments in in situ biorestoration of contaminated soil and groundwater in the Netherlands, pp. 379–390. In Z. Filip (ed.), Biotechnologische In situ-Sanierung. Fischer Verlag, Stuttgart/New York.
62. Sterritt, R. M. and J. N. Lester 1988. Microbiology for Environmental and Public Health Engineers. E. F. and W. Spon LTD, London and New York, 4–13.
63. Stone, Michael. 1984. Superbugs devour poisonous wastes. *Eur. Chem. News Chemscope,* November, 36–37.
64. Thayer, A. M. 1991. Bioremediation: Innovative technology for cleaning up hazardous waste. *Chem. Eng. News* 69:23–25.
65. Timmis, K. N., F. Rojo, and R. J. Ramos. 1988. Prospects for laboratory engineering of bacteria to degrade pollutants. In Environmental Biotechnology: Reducing Risks from Environmental Chemicals through Biotechnology, G. S. Omenn (ed.), Plenum Press, New York.
66. Uchiyama, H., T. Nakajima, and O. Yagi. 1989. Aerobic degradation of trichloroethylene at high concentration by a methane-utilizing mixed culture. *Appl. Environ. Microbiol.* 55:1019–1024.
67. Vogel, T. M., and P. L. McCarty, 1985. Biotransformation of tetrachloroethylene to trichloroethylene, dichloroethylene, vinyl chloride and carbon dioxide under methanogenic conditions. *Appl. Environ. Microbiol.* 49:1080–1083.
68. Wallnoefer, P. R., W. Ziegler, G. Engelhardt, and H. Rothmeier. 1978. Transformation of dinitrophenol herbicides by Azotobacter sp. *Chemosphere* 7:967–972.
69. Werner, P. 1985. A new way for the decontamination of polluted aquifers by biodegradation. *Water Supply* 3:41–47.
70. Wick, C. B., and Pierce, G. E. 1990. An integrated approach to development and implementation of biodegradation systems for treatment of hazardous organic wastes. *Dev. Ind. Microbiol.* 31:81–93. McCall, P., S. Urona, and S. Kelly, 1981. Fate of uniformly carbon-14 ring labelled 2,4,5 trichlorphenoxyacetic acid and 2,4 dichlorophenoxyacetic acid, *J. of Agricul. and Food Chem.* 29:100–107.
71. Wilson, J. T., and B. H. Wilson, 1985. Biotransformation of trichloroethylene in soil. *Appl. Environ. Microbiol.* 49:242–243.

Bioaugmentation to Enhance Microbial Bioremediation

Ronald M. Atlas

Department of Biology
University of Louisville
Louisville, Kentucky

The two major approaches to bioremediation of hazardous/oily wastes and environmental pollutants are microbial inoculation (seeding) and the bioaugmentation of naturally occurring microbial activities. In most cases *bioaugmentation* consists of environmental modification to eliminate some limiting factor that is restricting the rates of microbial growth and metabolism of a polluting substance. For this approach to work the pollutant must not be *recalcitrant*—that is, microorganisms must have the genetic and physiological capability to degrade the substance. The most common factors controlled to stimulate biodegradative activities by bioaugmentation are nutrient concentrations—usually nitrogen and phosphorus concentrations, molecular oxygen concentration, redox potential, and moisture levels. Additionally, cosubstrates can be provided as growth-supporting substances.

The potential use of bioaugmentation for bioremediation of hazardous and oily wastes and pollutants is an outgrowth of basic studies on microbial metabolism and ecology. Studies beginning in the 1950s focused on elucidating pathways of microbial metabolism of hydrocarbons with the aim of developing single-cell protein bioreactors using petroleum-based substrates. During this period the basic understanding of metabolism of hydrocarbons was developed. These studies expanded to examine the metabolism of chlorinated hydrocarbons, many of which are used as pesticides or otherwise wind up as persistent and/or toxic environmental contaminants. These basic physiological studies on microbial metabolism were supplemented beginning in the late 1960s with an examination of the ecological factors controlling the distributions and activities of microorganisms in the environment. Some of these studies aimed at elucidating the factors controlling the rates of microbial degradation of pollutants in the environment. During this period it was recognized that microorganisms had limitations and sometimes fail to degrade wastes and pollutants at rates fast enough to preclude adverse environmental impacts. The fallibility of microorganisms was revealed, particularly in terms of their ability to degrade xenobiotics and to exhibit rapid degradative activities under all conditions.

While bioaugmentation could do little to overcome the limitations of microbial evolution to develop enzymatic degradation pathways, it could overcome limitations imposed by environmental restrictions on microbial degradation. Conditions could be optimized in bioreactors, as they had been for decades in industrial fermentors and sewage and other waste treatment facilities, to achieve microbial degradation of wastes and pollutants. In situ treatments could also be designed to overcome factors limiting microbial degradative activities in nature.

Cosubstrates

The use of microorganisms in bioremediation generally depends upon the abilities of particular microbial strains to degrade pollutants to less toxic compounds that do not cause environmental harm or have adverse human health effects. For this purpose the microorganisms usually have to grow on the waste or polluting substance. Microorganisms gain energy and/or materials for the production of new cells as a result of the catabolic metabolism of growth-supporting substrates. In some cases, however, microorganisms are unable to metabolize a substance as the sole source of carbon and energy but can transform that substance if provided an alternate growth substrate called a *cosubstrate*. This phenomenon is known as *cometabolism, cooxidation,* or *cotransformation.*

In a formal sense, *cometabolism* describes the phenomenon that occurs when a compound is transformed by a microorganism, yet the organism is unable to grow on the compound and does not derive energy, carbon, or any other nutrient from the transformation. Usually the products of cometabolism are partially oxidized and/or dehalogenated metabolites. If the necessary growth-supporting cosubstrates are supplied, microorganisms can simultaneously gratuitously transform waste and polluting substances that cannot alone support microbial growth. The addition of cosubstrates thus can be used to support the cometabolic degradation of some wastes and environmental pollutants.

The use of cometabolizing microorganisms shows promise for promoting the biodegradation of some chemical wastes, assuming that appropriate cosubstrates can be found and delivered in a cost-effective manner to the sites where degradation of the waste or pollutant needs to occur. If a xenobiotic organochlorine compound fails to support microbial growth, a microorganism may be enriched for on a nonchlorinated or less chlorinated structural analog. If the organism has sufficiently broad-spectrum enzymes, it may cometabolize the xenobiotic chemical even though it cannot grow on it. It stands to reason that soil contaminated with a recalcitrant organochlorine compound may be detoxified if a biodegradable structural analog is provided. This was in fact demonstrated in the case of polychlorinated biphenyls (PCBs) and 3,4-dichloroaniline. In the former case, biphenyl, and in the latter, aniline, was used as the stimulating substrate (You and Bartha, 1982; Brunner et al., 1985). The disappearance of the xenobiotics was accelerated with this approach by an order of magnitude or more. Degradation of DDT (n-[dichloro-diphenyl-trichloroethane]) was first observed using cultures enriched on the cosubstrate biphenyl.

Since methanotrophic bacteria through their production of methane monooxygenase (MMO) are able to degrade trichloroethylene (TCE), dichloroethylene (DCE), and vinylchloride by cometabolism (Fogel et al., 1986), several investigators have considered using methane-oxidizing bacteria for the bioremediation of sites contaminated with these halogenated compounds. The idea is to enrich for methanotrophs by supplying them with methane in natural gas or another substrate such as acetate to support their growth requirements. Extensive aerobic degradation of TCE and other halogenated C_2 compounds by a methane-utilizing microbial consortium has been demonstrated (Fogel et al., 1986; Moore et al., 1986; Little et al., 1988).

The low specificity of methane monooxygenase allows the conversion of TCE to TCE epoxide, which subsequently spontaneously hydrolyzes to products utilizable by microorganisms. *Methylococcus capsulatus* was reported to convert chloro- and bromomethane to form aldehyde dichloromethane to CO and trichloromethane to CO_2 while growing on methane (Dalton and Stirling, 1982). Cometabolic degradation of TCE by methane-utilizing bacteria has been documented in a number of studies (Little et al., 1988; Palumbo et al., 1991; Uchiyama et al., 1989; Vogel and McCarty, 1985; Wilson and Wilson, 1985).

McCarty et al. (1991) found that they could stimulate indigenous methanogenic populations. They developed a model in situ bioremediation treatment which would require 5200 kg of methane and 19,200 kg of oxygen in order to convert 1375 kg of chlorinated hydrocarbons from an aquifer of 480,000 m³ containing a contaminant load of 1617 kg of halogenated compounds. Palumbo et al. (1991) found, however, that the presence of perchlormethene (PCE) inhibited the methanogens and suggested that anaerobic PCE removal would be necessary prior to stimulating methanogens to remove TCE. Methanotrophic bacteria show some promise for bioremediation of halocarbon-polluted aquifers.

In addition to the methane-oxidizing bacteria, the anaerobic methanogens are candidates for cometabolic transformations of halogenated wastes and environmental pollutants. Some anaerobic microorganisms carry out reductive dehalogenation, producing substances that can be degraded further by other microorganisms. Tetrachloroethylene degradation has been demonstrated for a methanogenic bacterial consortium growing on acetate in an anaerobic reactor (Vogel and McCarty, 1985; Galli and McCarty, 1989). Chlorinated pesticides including DDT are subject to reductive dehalogenation, that is, removal of the chlorine substituents by anaerobic microorganisms (Genthner et al., 1989a, 1989b). A nitrogen-heterocyclic herbicide, for example, has been shown to be attacked by reductive dehalogenation (Adrian and Suflita, 1990). Chlorobenzoates and chlorobenzenes are also degraded by this mechanism (Dolfing and Tiedje, 1987; Fathepure et al., 1988; Stevens and Tiedje, 1988; Stevens et al., 1988; Linkfield et al., 1989; Mohn and Tiedje, 1990a, 1990b).

Pfaender and Alexander (1972) succeeded in demonstrating the in vitro mineralization of DDT ring carbon by sequentially using cell-free extracts of a hydrogen-oxidizing bacterium, whole cells of *Arthrobacter,* addition of cosubstrates, and a regimen of alternating anaerobic and aerobic conditions. Nevertheless, they came to the conclusion that biochemical DDT mineralization in the environment either does not occur or does so at exceedingly slow rates. Other than the unfavorable energetics of the dechlorination steps, no satisfactory explanation could be found for this fact.

Degradation of PCBs typically is by cometabolism and is enhanced by addition of less chlorinated analogs such as dichlorobiphenyl (Adriaens et al., 1989; Brunner et al., 1985; Novick and Alexander, 1985). Semprini et al. (1991) found that sites contaminated with carbon tetrachloride, trichloroethane, and Freon could be bioremediated by stimulating indigenous denitrifying populations through the addition of acetate. Carbon tetrachloride biodegradation was demonstrated in contaminated soils using cometabolism when bioaugmentation was performed using acetate in the presence of sulfate and nitrate in the absence of oxygen.

Oxygen Supplementation

The growth potentials of microorganisms and their specific metabolic activities depend upon the availability of molecular oxygen and the redox potential. Some processes occur only under aerobic conditions, whereas others are strictly anaerobic. The initial steps in the biodegradation of hydrocarbons by most bacteria and fungi, for example, involve the oxidation of the substrate by oxygenases, for which molecular oxygen is required (Atlas, 1984). Hydrocarbons are abundant pollutants, occurring in environments contaminated by wood treatment products (creosote contains high concentrations of polynuclear aromatic hydrocarbons), oil spills (petroleum contains a vast diversity of aliphatic, alicyclic, and aromatic hydrocarbons), and leakages from underground storage tanks (benzene, toluene, and xylenes—BTX—are principal contaminants from hundreds of thousands of leaking tanks).

Anaerobic degradation of aromatic petroleum hydrocarbons by microorganisms has been reported (Ward and Brock, 1978; Grbic-Gallic and Vogel, 1987; Vogel and Grbic-Gallic, 1986; Zeyer et al., 1986, 1990). Hambrick et al. (1980) followed the $^{14}CO_2$ release from radiolabeled n-hexadecane and naphthalene in estuarine sediment slurries incubated for 1 month at present and controlled redox potentials ranging from -250 to $+510$ mV. At the lowest redox potentials, naphthalene biodegradation was undetectable during the 1-month experiment, and hexadecane biodegradation was at least 4 times lower than at the high redox potential. This laboratory study, along with field measurements (Ward and Brock, 1978; Ward et al., 1980; Delaune et al., 1980), leads to the general conclusion that the ecological and environmental significance of anaerobic hydrocarbon biodegradation is very low as compared to aerobic biodegradation.

While oxygen usually is not rate limiting in the upper levels of the water column in marine and freshwater environments (Floodgate, 1984; Cooney, 1984), the availability of oxygen in soils, sediments, and aquifers is often limiting and dependent on the type of soil and whether the soil is waterlogged (Jamison et al., 1975; Huddleston and Cresswell, 1976; von Wedel et al., 1988; Bossert and Bartha, 1984; Lee and Levy, 1991). When pollutants reach the water table and have contaminated aquifers, oxygen availability is the major problem in bioremediation. Oxygen solubility in water is low (at saturation, around 8 mg/L), and the oxygen demand for hydrocarbon degradation is very high. The microbial degradation of petroleum hydrocarbons in some groundwater and soil environments is severely limited by oxygen availability. The oxidation of 1 L of hydrocarbon will exhaust the dissolved oxygen (8 mg/L) in 385,000 to 400,000 L of water.

It is possible to overcome oxygen limitations by supplying oxygen to microorganisms in situ or by placing contaminated materials in an aerobic bioreactor where oxygen is supplied. In surface soil, oxygenation can be achieved by providing adequate drainage. Air-filled pore spaces in the soil facilitate the diffusion of oxygen, while in waterlogged soil oxygen diffusion is extremely slow and cannot keep up with the demands of heterotrophic decomposition processes. Substantial concentrations of decomposable organic wastes and pollutants create a very high oxygen demand in soil, and the rate of diffusion is inadequate to satisfy it even in well-drained and light-textured soils. Unsaturated subsurface (vadose) soil is normally aerobic, but strong hydrocarbon biodegradation activity may exhaust oxygen faster than it can be resupplied by diffusion. In such a situation, periodic raising and lowering of the groundwater table can facilitate air exchange (Beraud et al., 1989).

Cultivation (ploughing, rototilling) has been used to turn the soil and assure its maximal access to atmospheric oxygen (Kincannon, 1972; CONCAWE 1980a; 1980b). In laboratory soil columns, microbial proliferation in response to jet fuel contamination and bioremediation was three to five orders of magnitude close to the surface of the columns, but was less than one order of magnitude in the deeper portions of the columns (Song and Bartha, 1990).

Composting and aerobic bioreactors

Composting is an aerobic microbial process that has long been used for the degradation of organic wastes, such as leaves, that also can be used for the disposal of various hazardous and oily wastes and for the treatment of polluted soils. For optimal composting, several conditions are critical. Adequate moisture (50–60 percent water content) must be present, but excess moisture (70 percent or above) should be avoided, since it interferes with aeration and lowers self-heating because of its large heat capacity. Composting, with the aid of largely inert bulking materials such as wood chips, rice hulls, or peanut shells, can also be used in chemical waste biodegradation.

Oiled shoreline waste from a marine crude oil spill was successfully treated in an aerated compost heap (Labrie and Cyr, 1990). Oily sand, mud, and organic debris were placed on an impermeable sheet at a slight slope to allow leachate collection. A gravel layer below the oily material contained perforated pipes connected to a reversible air compressor. After 180 days of operation, initial oil concentrations of up to 30 percent were reduced below 1 percent. At that stage, the residue was landfilled. During the same treatment, the contaminant's contents of polycyclic aromatic hydrocarbons (PAHs) was reduced from an initial 283 ppm to 8 ppm.

Because of a strong settling tendency, slurry-type bioreactors for soil are difficult to design and to operate, but supernatants from soil-washing processes can be effectively bioremediated in activated-sludge-type aerobic sewage treatment units (CONCAWE, 1980a; Morgan and Watkinson, 1989). It should be possible to process pollutant-contaminated soils in solid-phase-composting-type bioreactors (Pavoni et al., 1975), but the typically very long residence times required for hydrocarbon biodegradation and the unfavorable effects of elevated temperature on the process at present do not favor this approach.

Aerobic bioreactor treatment systems, modified from those typically used for sewage treatment to reduce biological oxygen demand (BOD), can also be used for the biodegradation of pollutants. Many industrial facilities have aerobic waste treatments that reduce the overall BOD of organic compounds in wastewaters and also remove the concentrations of specific pollutants.

Bioreactors used at industrial sites include the film-flow-type trickling filter, in which an absorbed microbial community mineralizes dissolved organic nutrients that pass over the immobilized microorganisms. Aeration is provided passively by the porous nature of the bed. Overload may lead to excess microbial slime, reducing aeration and percolation rates and necessitating a renewal of the trickling-filter bed. Cold winter temperatures strongly reduce the effectiveness of these outdoor treatment facilities. A more advanced aerobic film-flow-type treatment system is the rotating biological contactor or "biodisc" system. Closely spaced discs, usually manufactured from plastic material, are rotated in a trough containing the effluent. Azo dyes, components of munitions such as TNT, and various other wastes can be treated in this manner. Oxidation ditches and lagoons are low-cost treatment systems that are frequently employed for the disposal of impounded pollutants. They tend to be inefficient and require large holding capacities and long retention times. As oxygenation is usually achieved by diffusion and by the photosynthetic activity of algae, they need to be shallow. In some bioreactors, forced aeration is used to supply the necessary oxygen. A lagoon can have forced-air spargers, circulators, and baffles added to ensure better aeration. Ballast water treatment facilities, used to reduce BTX before discharge, are usually designed in this manner. A more advanced type of liquid waste treatment system with forced aeration that often is employed is the activated-sludge process. A waste liquid, containing dissolved organic compounds, is introduced into an aeration tank. Aeration is provided by air injection and/or mechanical stirring. Microbial activity is maintained at very high levels by reintroduction of most of the settled activated sludge from a previous treatment run, hence the name of the process.

In situ aeration

For contaminated aquifers it is possible to supply oxygen through forced-aeration spargers. Air sparging, however, is inefficient and expensive (Davis-Hoover, 1991). Air cannot be delivered far into a contaminated soil and air must be supplied continuously. Jamison et al. (1975) used forced aeration to supply oxygen for hydrocarbon biodegradation in a groundwater supply which had been contaminated by gasoline. Nutrient addition without aeration failed to stimulate biodegradation, but when both nutrients and oxygen were supplied, it was estimated that up to 1000 barrels of gasoline were removed by stimulated microbial degradation. Such manipulations to supply oxygen probably are not feasible in open systems where natural forces such as wind and wave action will have to be relied upon for turbulent mixing and resupply of oxygen to support biodegradation of oil.

To overcome oxygen limitation it is also possible to add hydrogen peroxide in appropriate and stabilized formulations (American Petroleum Institute, 1987; Yaniga and Smith, 1984; Brown et al., 1984, 1985; Thomas et al., 1987; Berwanger and Barker, 1988). The decomposition of hydrogen peroxide releases oxygen which can support aerobic microbial metabolism. The practical concentration of hydrogen peroxide in injected water is around 100 ppm (Brown et al., 1984; Yaniga and Smith, 1984). Berwanger and Barker (1988) investigated in situ biorestoration involving stimulating aerobic biodegradation in a contaminated anaerobic, methane-saturated groundwater situation using hydrogen peroxide as an oxygen source. Hydrogen peroxide, added at a nontoxic level, provided oxygen which promoted the rapid biodegradation of benzene, toluene, ethyl benzene, and o-, m-, and p-xylene. Frankenberger et al. (1989) studied a 1000-gal spillage of diesel fuel from a leaking underground diesel fuel storage tank.

Too rapid hydrogen peroxide decomposition creates gas pockets that interfere with subsequent pumping operations. For this reason, hydrogen peroxide is applied in conjunction with stabilizers that slow down its decomposition (Brown et al., 1984). Stabilizer formulations are proprietary and their composition is not published. Some compounds with stabilizing properties such as phosphates, may do double duty as stabilizers and fertilizers.

In a field study, hydrogen peroxide was injected into a contaminated aquifer at a rate of 750 mg/L for half a year (Huling et al., 1991). The gradient of oxygen concentration from the injection regions clearly indicated that oxygen was delivered to the water-saturated zone of the soil. The levels of oxygen released were high enough to cause some inhibition of bacterial activities. Approximately half of the oxygen released from the hydrogen peroxide was transferred to the gas phase.

In another study of hydrogen peroxide addition to support bioremediation, Barenschee et al. (1991) found that hydrocarbons in diesel
fuels were degraded 4 to 7 times faster when hydrogen peroxide was
added than when nitrate was added as a terminal electron acceptor for
respiration. The addition of hydrogen peroxide resulted in a two order
of magnitude higher count of hydrocarbon-degrading microorganisms,
a fivefold increase in carbon dioxide production, and biodegradation of
70–80 percent of the contaminating hydrocarbons. Flathman et al.
(1991) similarly found that hydrogen peroxide addition to test soil
columns greatly increased the rates of biodegradation of hydrocarbons
in JP-5 fuel. Lee and Raymond (1991) reported great success in restor-
ing a gasoline-contaminated aquifer through the use of air stripping
and hydrogen peroxide–bioaugmented bioremediation.

The U.S. Coast Guard and the U.S. EPA performed a field evaluation
of bioreclamation on fuel spills at the USCG Air Station at Traverse
City, Michigan, in which peroxide was injected into a spill of aviation
gasoline (Wilson, 1991). Eleven gal/min of water with peroxide concentrations near 750 mg/L were injected into wells located across the area
contaminated with gasoline. Clean water from another part of the
aquifer was injected deep below the contaminated interval at 22
gal/min in order to raise the water table and flood the entire vertical interval contaminated with gasoline. After 18 months of operation, the
concentration of benzene in monitoring wells up to 100 ft from the injection wells was less than 0.1 µg/L. The concentrations of the other
alkylbenzenes were below 5 µg/L in wells out to 50 ft from the injection
wells. However, core material only 7 ft from the injection well still contained 700 mg/kg total petroleum hydrocarbons.

Nutrients and Fertilization

Nitrogen, phosphorus, and other mineral nutrients are necessary for
incorporation into biomass. Concentrations of available nitrogen and
phosphorus often limit rates of microbial degradative activities; for example, there are numerous reports that nitrogen and phosphorus concentrations in seawater limit rates of hydrocarbon degradation
following oil spills (Atlas and Bartha, 1972; Bartha and Atlas, 1973;
Floodgate, 1973, 1979; Gunkel 1967; LePetit and Barthelemy, 1968;
LePetit and N'Guyen, 1976; Leahy and Colwell, 1990). Based on
Kuwait crude oil at 14°C, the nitrogen demand is 4 nmol of nitrogen per
µg of oil (Floodgate, 1979). Colwell et al. (1978) concluded that oil from
the Metulla spill was degraded slowly in the marine environment, most
probably because of limitations imposed by the relatively low concentrations of nitrogen and phosphorus available in seawater. Much like

in aquatic environments, nitrogen and phosphorous availability may also limit hydrocarbon biodegradation in terrestrial situations (Dibble and Bartha, 1979*a*; Bossert and Bartha, 1984; Bartha, 1986).

Soil fertilization and land treatment for disposal of oily wastes

In *land treatment,* oily sludges are applied to soil. Scientific monitoring of spreading of such oily wastes over land masses began in the early 1970s (Kincannon, 1972; Francke and Clark, 1974). Later, the process was systematically optimized in field and laboratory experiments (Lehtomake and Niemela, 1975; Maunder and Waid, 1973, 1975; Raymond et al., 1976; Huddleston and Meyers, 1978; Dibble and Bartha, 1979*b*; Arora et al., 1982; Brown and Donelly, 1983; Sandvik et al., 1986; Shailubhai, 1986; Amaral, 1987; Tesan and Barbosa, 1987).

Practical recommendations concerning land treatment were summarized by the American Petroleum Institute (1980) in the United States and by CONCAWE (1980*b*) in Europe. Oil is applied at rates to acieve approximately 5 percent hydrocarbon concentration in the upper 15- to 20-cm layer of the soil. Hydrocarbon concentrations above 10 percent are definitely inhibitory to the biodegradation process. This limit translates to approximately 100,000 L hydrocarbon per ha, usually in 3 to 4 times as high sludge volume (Dibble and Bartha, 1979*a*). The soil pH is adjusted to a value between 7 and 8 or to the nearest practical value, using agricultural limestone. Nitrogen and phosphorus fertilizers are applied in ratios of hydrocarbon:N = 200:1 and hydrocarbon:P = 800:1. Undegraded hydrocarbons do not leach readily into the groundwater from the land treatment sites (Dibble and Bartha, 1979*b*), and the environmental impact of properly operated sites appears to be minimal (Arora et al., 1982).

Song et al. (1990) concluded that the process is most suitable for the medium fuel distillates. While gasoline responds to bioremediation in surface soils, biodegradation cannot keep up with evaporation rates and most of the product is lost to the atmosphere. In laboratory experiments, Song et al. (1990) had very limited success with No. 6 (residual) fuel oil. However, Jones and Greenfield (1991) reported quite promising results in an on-site bioremediation effort in Florida. Soil contaminated by an average of 10,000 ppm of No. 6 fuel oil was treated with fertilizer. The soil was turned, cultivated, and kept moist by sprinklers when necessary. In 300 days about 90 percent of the contaminant was eliminated, leaving approximately 1,000 ppm residue that included multiring PAHs.

Aquifer nutriation

When a polluted aquifer is pumped in order to keep the pollution from spreading, the recovered water is supplemented with mineral nutrients (nitrogen and phosphorus) and aerated. The combination of biodegradation and air stripping frees the recovered water from the dissolved pollutants. This water, containing now substantial numbers of hydrocarbon-degrading microorganisms, is reinjected into the aquifer around the perimeter of the polluted plume. This "pump-and-treat" cleanup operation is then aided by the in situ activity of the injected microorganisms (Lee and Ward, 1985; Brown et al., 1985; Thomas et al., 1987). To maximize in situ activity, the water may be supplemented prior to injection with additional mineral nutrients and materials serving as electron sinks for hydrocarbon oxidation.

Bioremediation of aquifers contaminated with halogenated aromatics, haloethanes, and halomethanes presents additional complex problems (Kuhn et al., 1985; Wilson et al., 1986; Berwanger and Barker, 1988). While some of these materials are dehalogenated anaerobically, others cannot serve as substrates under either aerobic or anaerobic conditions and are attacked only cometabolically.

A spill of JP-4 was remediated with nitrate and mineral nutrients (Wilson, 1991). A study area 30 ft by 30 ft was flooded with 200 gal/min of water from underneath the spill. The water took 8 h to move across the contaminated interval, then a week to move to the large production wells that supplied water to the infiltration gallery above the study area. Unamended water was circulated for 2 months to bring the water and oil to chemical equilibrium. Then 10 mg/L of nitrate as N was circulated for an additional 2 months. Benzene was brought below 0.1 mg/L before nitrate was added. After addition of nitrate, the other alkylbenzenes were brought below 5 µg/L.

Adding phosphates to aquifers can result in precipitation that may plug the aquifer (Aggarwal et al., 1991). Laboratory tests using hydrocarbon-contaminated soils indicated that a phosphate concentration in soil over the aquifer of 20 mg/L is sufficient to provide excess phosphate for microbial growth. Phosphate, however, will precipitate as a calcium salt if added to calcareous soils that have high calcium concentrations. In sandy quartz soils phosphate will not form excess precipitates if added as orthophosphates up to 20 mg/L. For higher concentrations the phosphate can be added as trimetaphosphate.

Oleophilic fertilizers

For marine oil spills, Atlas and Bartha (1973) developed an oleophilic nitrogen and phosphorus fertilizer that would remain in contact with oil at the oil-water interface where microbial hydrocarbon biodegradation occurs. The fertilizer designed by Atlas and Bartha (1973) contains paraffinized urea and octyl phosphate, but a range of other oleophilic nitrogen and phosphorus compounds could serve equally well (Atlas and Bartha, 1976). Atlas and Bartha tested the effectiveness of oleophilic fertilizers for stimulating oil biodegradation in nearshore areas off the coast of New Jersey (Atlas and Bartha, 1973), in Prudhoe Bay, and in several ponds near Barrow, Alaska (Atlas and Schofield, 1975; Atlas and Busdosh, 1976); tests included in situ as well as in vitro experiments in each case. Also, the fertilizer was tested in microcosms for potential Arctic applications (Horowitz and Atlas, 1977). In each case there was a naturally occurring microbial population that was capable of petroleum biodegradation when this oleophilic fertilizer was added to an oil slick, and in each case, addition of oleophilic fertilizer stimulated biodegradative losses.

Olivieri et al. (1976) described a slow-release fertilizer containing paraffin-supported magnesium ammonium phosphate as the active ingredient for stimulating petroleum biodegradation. After 21 days, 63 percent of the oil had disappeared when fertilizer was added, compared with 40 percent in a control area. Olivieri et al. (1978) also found a combination of soybean lecithin and ethyl allophane to provide good oleophilic sources of phosphorus and nitrogen, respectively. Bergstein and Vestal (1978) found that oleophilic fertilizer enhances biodegradation of crude oil in oligotrophic lakes and ponds.

The low availability of iron has been shown to limit hydrocarbon biodegradation, but the same limitation was not evident in sediment-rich nearshore seawater (Dibble and Bartha, 1976). No other mineral nutrients were found or are suspected to be limiting for oil biodegradation in seawater, but in some freshwater environments the sulfate concentration may be insufficient to support optimal oil biodegradation (Bartha, 1986). Dibble and Bartha (1976) found additional stimulation of crude oil biodegradation when oleophilic iron was added as ferric octoate along with nitrogen and phosphorus. Greater stimulation was observed only in sediment-free offshore seawater, not in sediment-rich nearshore seawater. Addition of oleophilic iron appears to be useful only in open ocean areas where iron concentrations are particularly low.

In the aftermath of the *Amoco Cadiz* oil spill of 1978, a commercial oleophilic fertilizer was developed by Elf Aquitaine (Paris, France) (Sirvins and Angeles, 1986; LaDousse et al., 1987; Sveum and LaDousse, 1989; Tramier and Sirvins, 1983; LaDousse and Tramier, 1991). The product, called Inipol® EAP 22, contains urea as a nitrogen source, lauryl phosphate as a phosphate source, and oleic acid as a carbon source to boost the populations of hydrocarbon-degrading microorganisms. It is formulated as a microemulsion. Laboratory experiments demonstrated significant enhancement of oil biodegradation; in some experiments 60 percent of added oil was biodegraded in fertilized flasks compared to 38 percent in unfertilized ones within 60 days (LaDousse et al., 1987). Even greater enhancement, 70 percent biodegradation with the fertilized compared to only 20 percent when unfertilized, was found in high-energy oxygen saturation tests (LaDousse et al., 1987). Field tests also showed enhanced rates of oil biodegradation when Inipol EAP 22 fertilizer was applied, even in cold Arctic tests (Sirvins and Angeles, 1986; Sveum and LaDousse, 1989).

Pritchard and Costa (1991) found that Inipol EAP 22 was not as effective as water-soluble nitrogen- and phosphorus-containing fertilizers for enhancement of biodegradation of subsurface oil. They reported that 50–60 percent of the ammonia and phosphate in Inipol EAP 22 was released within a few minutes of application, followed by slower release over the following several weeks. Safferman (1991) tested several slow-release nitrogen formulations for the treatment of shorelines contaminated in the *Exxon Valdez* spill. In his tests, isobutyraldehyde diurea briquettes gave the best results in terms of gradual ammonia release.

Despite sampling and interpretation complications resulting from the high variability in oil distribution on the beaches, it was possible to show statistically that oil biodegradation (as measured by changes in residue weights and oil chemistry) was significantly greater on the beach treated with the fertilizer solution than it was on the control beach (Prince et al., 1990; Pritchard and Costa, 1991). After 45 days, approximately 3 to 4 times more oil remained on the control test beach than on the fertilizer-solution-treated beach. This corresponded to an enhanced biodegradation rate of about two- to threefold. Results appeared to be similar on the Inipol-Customblen-treated beach, but for this beach statistically significant differences from the control were more difficult to establish. However, it appeared that accelerated biodegradation (approximately a two- to threefold increase) occurred early in the test when nutrient concentrations were highest. These results imply that fertilizer reapplication (maintaining nutrient concentrations at high levels for long periods) is important.

As a result of the EPA-Exxon project, bioremediation of oil-contaminated beaches was shown to be a safe cleanup technology; no adverse ecological effects were observed (Fox, 1990; Office of Technology Assessment, 1991). The addition of fertilizers caused no eutrophication, no acute toxicity to sensitive marine test species, and did not cause the release of undegraded oil residues from the beaches. The success of the field demonstration program has now set the stage for the consideration of bioremediation as a key component (but not the sole component) in any cleanup strategy developed for future oil spills.

References

Adriaens, P. H., P. E. Kohler, and D. Kohler-Staub. 1989. Bacterial dehalogenation of chlorobenzoates and coculture biodegradation of 4,4'-dichlorobiphenyl. *Appl. Environ. Microbiol.* 55:887–892.

Adrian, N. R., and J. M. Suflita. 1990. Reductive dehalogenation of a nitrogen hetero-cyclic herbicide in anoxic aquifer slurries. *Appl. Environ. Microbiol.* 56:292–294.

Aggarwal, P. K., J. L. Means, and R. E. Hinchee. 1991. Formulation of nutrient solutions for *in situ* biodegradation. In R. E. Hinchee and R. F. Olfenbuttel (eds.), *In Situ Bioreclamation: Applications and Investigations for Hydrocarbon and Contaminated Site Remediation.* Butterworth-Heinemann, Boston, pp. 51–66.

Amaral, S. P. 1987. Landfarming of oily wastes: Design and operation. *Water Sci. Technol.* 19:75–86.

American Petroleum Institute. 1980. *Manual on Disposal of Petroleum Wastes.* American Petroleum Institute, Washington, D.C.

American Petroleum Institute. 1987. *Field Study of Enhanced Subsurface Biodegradation of Hydrocarbons Using Hydrogen Peroxide as an Oxygen Source.* American Petroleum Institute Publ. 4448. American Petroleum Institute, Washington, D.C.

Arora, H. S., R. R. Cantor, and J. C. Nemeth. 1982. Land treatment: A viable and successful method of treating petroleum industry wastes. *Environ. Int.* 7:285–292.

Atlas, R. M. (ed.). 1984. *Petroleum Microbiology.* Macmillan, New York.

Atlas, R. M., and R. Bartha. 1972. Degradation and mineralization of petroleum in seawater: Limitation by nitrogen and phosphorus. *Biotechnol. Bioeng.* 14:309–317.

Atlas, R. M., and R. Bartha. 1973. Stimulated biodegradation of oil slicks using oleophilic fertilizers. *Environ. Sci. Technol.* 7:538–541.

Atlas, R. M., and R. Bartha. 1976. Biodegradation of oil on water surfaces. U.S. Patent 3,939,127.

Atlas, R. M., and M. Busdosh. 1976. Microbial degradation of petroleum in the Arctic. *In* J. M. Sharpley and A. M. Kaplan (eds.), *Proceedings of the Third International Biodegradation Symposium.* Applied Science Publ., London, pp. 79–86.

Atlas, R. M., and E. A. Schofield. 1975. Petroleum biodegradation in the Arctic. *In* A. W. Bourquin, D. G. Ahearn, and S. P. Meyers (eds.), *Impact of the Use of Microorganisms on the Aquatic Environment.* EPA 660-3-75-001. U.S. Environmental Protection Agency, Corvallis, OR.

Barenschee, E. R., P. Bochem, O. Helmling, and P. Weppen. 1991. Effectiveness and kinetics of hydrogen peroxide and nitrate-enhanced biodegradation of hydrocarbons. *In* R. E. Hinchee and R. F. Olfenbuttel (eds.), *In Situ Bioreclamation: Applications and Investigations for Hydrocarbon and Contaminated Site Remediation.* Butterworth-Heinemann, Boston, pp. 103–124.

Bartha, R. 1986. Biotechnology of petroleum pollutant biodegradation. *Microb. Ecol.* 12:155–172.

Bartha, R., and R. M. Atlas. 1973. Biodegradation of oil in seawater: Limiting factors and artificial stimulation. *In* D. G. Ahearn and S. P. Meyers (eds.), *The Microbial Degradation of Oil Pollutants.* Publ. no. LSU-SG-73-01. Center for Wetland Resources, Louisiana State University, Baton Rouge, pp. 147–152.

Beraud, J.-F., J. D. Ducreux, and C. Gatellier. 1989. Use of soil-aquifer treatment in oil pollution control of underground waters. *In Proceedings of the 1989 Oil Spill Conference.* American Petroleum Institute, Washington, D.C., pp. 53–59.

Bergstein, P. E., and J. R. Vestal. 1978. Crude oil biodegradation in Arctic tundra ponds. *Arctic* 31:158–169.

Berwanger, D. J., and J. F. Barker. 1988. Aerobic biodegradation of aromatic and chlorinated hydrocarbons commonly detected in landfill leachate. *Water Pollut. Res. J. Can.* 23(3):460–475.

Bossert, I., and R. Bartha. 1984. The fate of petroleum in soil ecosystems. *In* R. M. Atlas (ed.), *Petroleum Microbiology.* Macmillan, New York, pp. 473–476.

Brown, K. W., and K. S. Donnelly. 1983. Influence of soil environment on biodegradation of a refinery and a petrochemical sludge. *Environ. Pollut. Ser. B* 6:119–132.

Brown, R. A., R. D. Norris, and R. L. Raymond. 1984. Oxygen transport in contaminated aquifers. *In Proceedings of the Conference on Petroleum Hydrocarbons and Organic Chemicals in Ground Water—Prevention, Detection, and Restoration.* National Water Well Association, Worthington, OH, pp. 441–450.

Brown, R. A., R. D. Norris, and G. R. Brubaker. 1985. Aquifer restoration with enhanced bioreclamation. *Pollut. Eng.* 17:25–28.

Brunner, W., S. H. Southerland, and D. D. Focht. 1985. Enhanced biodegradation of poly-chlorinated biphenyls in soil by analog enrichment and bacterial inoculation. *J. Environ. Quality* 14:324–328.

Colwell, R. R., A. L. Mills, J. D. Walker, P. Garcia-Rello, and V. Campos-P. 1978. Microbial ecology studies of the Metula spill in the Straits of Magellan. *J. Fish. Res. Board Can.* 35:573–580.

CONCAWE. 1980a. *Disposal Techniques for Spilt Oil*. Rep. 9/80, CONCAWE, The Hague.

CONCAWE. 1980b. *Sludge Farming: A Technique for the Disposal of Oily Refinery Wastes*. Rep. 3/80, CONCAWE, The Hague.

Cooney, J. J. 1984. The fate of petroleum pollutants in fresh-water ecosystems. *In* R. M. Atlas (ed.), *Petroleum Microbiology*. Macmillan, New York.

Davis-Hoover, W. J., L. C. Murdoch, S. J. Vesper, H. R. Pahren, O. L. Sprockel, C. L. Chang, A. Hussain, and W. A. Ritschel. 1991. Hydraulic fracturing to improve nutri-ent and oxygen delivery for in situ bioreclamation. *In* R. E. Hinchee and R. F. Olfenbuttel (eds.), *In Situ Bioreclamation: Applications and Investigations for Hydrocarbon and Contaminated Site Remediation*. Butterworth-Heinemann, Boston, pp. 67–82.

Delaune, R. D., G. A. Hambrick, and W. H. Patrick. 1980. Degradation of hydrocarbons in oxidized and reduced sediments. *Mar. Pollut. Bull.* 11:103–106.

Dalton, H., and D. I. Stirling. 1982. Co-metabolism. *Philosoph. Trans. R. Soc. London Ser. B* 297:481–491.

Dibble, J. T., and R. Bartha. 1976. The effect of iron on the biodegradation of petroleum in sea-water. *Appl. Environ. Microbiol.* 31:544–550.

Dibble, J. T., and R. Bartha. 1979a. Effect of environmental parameters on the biodegra-dation of oil sludge. *Appl. Environ. Microbiol.* 37:729–739.

Dibble, J. T., and R. Bartha. 1979b. Leaching aspects of oil sludge biodegradation in soil. *Soil Sci.* 127:365–370.

Dolfing, J., and J. M. Tiedje. 1987. Growth yield increase linked to reductive dechlorina-tion in a defined 3-chlorobenzoate degrading methanogenic coculture. *Arch. Microbiol.* 149:102–105.

Fathepure, B. Z., J. M. Tiedje, and S. A. Boyd. 1988. Reductive dechlorination of hex-achlorobenzene to tri- and dichlorobenzenes in anaerobic sewage sludge. *Appl. Environ. Microbiol.* 54:327–330.

Flathman, P. E., K. A. Khan, D. M. Barnes, J. H. Carson, S. J. Whitehead, and J. S. Evans. 1991. Laboratory evaluation of the utilization of hydrogen peroxide for en-hanced biological treatment of petroleum hydrocarbon contaminants in soil. *In* R. E. Hinchee and R. F. Olfenbuttel (eds.), *In Situ Bioreclamation: Applications and Investigations for Hydrocarbon and Contaminated Site Remediation*. Butterworth-Heinemann, Boston, pp. 125–142.

Floodgate, G. D. 1973. A threnody concerning the biodegradation of oil in natural water. *In* D. G. Ahearn and S. P. Meyers (eds.), *The Microbial Degradation of Oil Pollutants*. Publ. no. LSU-SG-73-01. Center for Wetland Resources, Louisiana State University, Baton Rouge, pp. 17–24.

Floodgate, G. D. 1979. Nutrient limitation. *In* A. W. Bourquin and P. H. Pritchard (eds.), *Microbial Degradation of Pollutants in Marine Environments*. EPA-66019-79-012. Environmental Research Laboratory, Gulf Breeze, FL, pp. 107–119.

Floodgate, G. 1984. The fate of petroleum in marine ecosystems. *In* R. M. Atlas (ed.), *Petroleum Microbiology*. Macmillan, New York.

Fogel, M. M., A. R. Taddeo, and S. Fogel. 1986. Biodegradation of chlorinated ethanes by a methane-utilizing mixed culture. *Appl. Environ. Microbiol.* 51:720–724.

Fox, J. E. 1990. More confidence about degrading work. *Bio/Technology* 8:604.

Francke, H. C., and F. E. Clark. 1974. *Disposal of oil wastes by microbial assimilation*. Report Y-1934. U.S. Atomic Energy Commission, Washington, D.C.

Frankenberger, W. T., Jr., K. D. Emerson, and D. W. Turner. 1989. In situ bioremedia-tion of an underground diesel fuel spill: A case history. *Environ. Manage.* 13:325–332.

Galli, R., and P. L. McCarty. 1989. Biotransformation of 1,1,1-trichloroethane, trichloromethane, and tetrachloromethane by a *Clostridium* sp. *Appl. Environ. Microbiol.* 55:837–844.

Genthner, B. R. S., W. A Price, and P. H. Pritchard. 1989a. Anaerobic degradation of chloroaromatic compounds in aquatic sediments under a variety of enrichment conditions. *Appl. Environ. Microbiol.* 55:1466–1471.

Genthner, B. R. S., W. A. Price, and P. H. Pritchard. 1989b. Characterization of anaerobic dechlorinating consortia derived from aquatic sediments. *Appl. Environ. Microbiol.* 55:1472–1476.

Grbic-Gallic, D., and T. M. Vogel. 1987. Transformation of toluene and benzene by mixed methanogenic-cultures. *Appl. Environ. Microbiol.* 53:254–260.

Gunkel, W. 1967. Experimentell-okologische Untersuchungen uber die limitierenden Faktoren des mikrobiellen Olabbaues in marinen Milieu. *Helgol. Wiss. Meeresunters.* 15:210–224.

Hambrick, G. A., III, R. D. DeLaune, and W. H. Patrick, Jr. 1980. Effect of estuarine sediment pH and oxidation-reduction potential on microbial hydrocarbon degradation. *Appl. Environ. Microbiol.* 40:365–369.

Horowitz, A., and R. M. Atlas. 1977. Continuous open flow-through system as a model for oil degradation in the Arctic Ocean. *Appl. Environ. Microbiol.* 33:647–684.

Huddleston, R. L., and L. W. Cresswell. 1976. Environmental and nutritional constraints of microbial hydrocarbon utilization in the soil. *In Proceedings of the 1975 Engineering Foundation Conference: The Role of Microorganisms in the Recovery of Oil.* NSF/RANN, Washington, D.C., pp. 71–72.

Huddleston, R. L., and J. D. Meyers. 1978. Treatment of refinery oily wastes by land-farming. Paper presented at the 85th National Meeting of AICHE. Philadelphia, PA, American Institute of Chemical Engineers, New York.

Hurling, S. G., B. E. Bledsoe, and M. V. White. 1991. The feasibility of utilizing hydrogen peroxide as a source of oxygen in bioremediation. *In* R. E. Hinchee and R. F. Olfenbuttel (eds.), *In Situ Bioreclamation: Applications and Investigations for Hydrocarbon and Contaminated Site Remediation.* Butterworth-Heinemann, Boston, pp. 83–102.

Jamison, V. M., R. L. Raymond, and J. O. Hudson, Jr. 1975. Biodegradation of high-octane gasoline in groundwater. *Dev. Ind. Microbiol.* 16:305–312.

Jones, M., and J. H. Greenfield. 1991. In situ comparison of bioremediation methods for a Number 6 residual fuel oil spill in Lee County, Florida. *In Proceedings of the 1991 International Oil Spill Conference.* American Petroleum Institute, Washington, D.C., pp. 533–540.

Kincannon, C. B. 1972. *Oily waste disposal by soil cultivation process.* EPA-R2-72-100. U.S. Environmental Protection Agency, Washington, D.C.

Kuhn, E. P., P. J. Colberg, J. L. Schnoor, O. Wanner, A. J. B. Zehnder, and R. P. Schwartzenbach. 1985. Microbial transformations of substituted benzenes during infiltration of river water to groundwater: Laboratory column studies. *Environ. Sci. Technol.* 19:961–968.

Labrie, P., and B. Cyr. 1990. Biological remediation of shoreline oily waste from marine spills. *In Proceedings of the Thirteenth Annual Arctic and Marine Oil Spill Program Technical Seminar.* Environment Canada, Ottawa, Canada, pp. 339–387.

LaDousse, A., C. Tallec, and B. Tramier. 1987. Progress in enhanced oil degradation. Paper presented: *Proceedings of the 1987 Oil Spill Conference.* Abstract 142. American Petroleum Institute, Washington, D.C.

LaDousse, A., and B. Tramier. 1991. Results of 12 years of research in spilled oil bioremediation: Inipol EAP 22. *In Proceedings of the 1991 International Oil Spill Conference.* American Petroleum Institute, Washington, D.C., pp. 577–581.

Lee, K., and E. M. Levy. 1991. Bioremediation: Waxy crude oils stranded on low-energy shorelines. *In Proceedings of the 1991 International Oil Spill Conference.* American Petroleum Institute, Washington, D.C., pp. 541–547.

Lee, M. D., and R. L. Raymond, Sr. 1991. Case history of the application of hydrogen peroxide as an oxygen source for in situ bioreclamation. *In* R. E. Hinchee and R. F. Olfenbuttel (eds.), *In Situ Bioreclamation: Applications and Investigations for Hydrocarbon and Contaminated Site Remediation.* Butterworth-Heinemann, Boston, pp. 429–438.

Lee, M. D., and C. H. Ward. 1985. Biological methods for the restoration of contaminated aquifers. *Environ. Toxicol. Chem.* 4:743–750.

Leahy, J. G., and R. R. Colwell. 1990. Microbial degradation of hydrocarbons in the environment. *Microbiol. Rev.* 54:305–315.

LePetit, J., and M. H. Barthelemy. 1968. Les hydrocarbures en mer: Le probleme de l'epuration des zones littorales par les microorganismes. *Ann. Inst. Pasteur Paris* 114:149–158.

LePetit, J., and M.-H. N'Guyen. 1976. Besoins en phosphore des bacteries metabolisant les hydrocarbures en mer. *Can. J. Microbiol.* 22:1364–1373.

Lehtomake, M., and S. Niemela. 1975. Improving microbial degradation of oil in soil. *Ambio* 4:126–129.

Linkfield, T. G., J. M. Suflita, and J. M. Tiedje. 1989. Characterization of the acclimation period before anaerobic dehalogenation of halobenzoates. *Appl. Environ. Microbiol.* 55:2773–2778.

Little, C. D., A. V. Palumbo, S. E. Herbes, M. E. Lindstrom, R. L. Tyndall, and P. J. Gilmer. 1988. Trichloroethylene biodegradation by a methane-oxidizing bacterium. *Appl. Environ. Microbiol.* 54:951–956.

McCarty, P. L., L. Semprini, M. E. Dolan, T. C. Harmon, C. Tiedeman, and S. M. Gorclick. 1991. *In situ* methanotrophic bioremediation for contaminated groundwater at St. Joseph, Michigan. In R. E. Hinchee and R. F. Olfenbuttel (eds.), *On-Site Bioreclamation: Processes for Xenobiotic and Hydrocarbon Treatment.* Butterworth-Heinemann, Boston, pp. 16–40.

Maunder, B. R., and J. S. Waid. 1973. Disposal of waste oil by land spreading. In *Proceedings of the Pollution Research Conference,* 20–21 June 1973, Wairakei, New Zealand. Information Series No. 97. New Zealand Department of Scientific and Industrial Research, Wellington.

Maunder, B. R., and J. S. Waid. 1975. Disposal of waste oil by land spreading. *Paper presented at the Third International Biodeterioration Symposium,* 17–23 August, University of Rhode Island, Kingston.

Mohn, M. M., and J. M. Tiedje. 1990a. Strain DCB-1 conserves energy for growth from reductive dechlorination coupled to formate oxidation. *Arch. Microbiol.* 153:267–271.

Mohn, M. M., and J. M. Tiedje. 1990b. Catabolite thiosulfate disproportionation and carbon dioxide reduction in strain DCB-1, a reductively dechlorinating anaerobe. *J. Bacteriol.* 172:2065–2070.

Moore, A. T., A. Vira, and S. Fogel. 1989. Biodegradation of trans-1,2-dichloroethylene by methane-utilizing bacteria in an aquifer simulator. *Environ. Sci. Technol.* 23:403–406.

Morgan, P., and R. J. Watkinson. 1989. Hydrocarbon biodegradation in soils and methods for soil biotreatment. *CRC Crit. Rev. Biotechnol.* 8(4):305–333.

Novick, N. J., and M. Alexander. 1985. Cometabolism of low concentrations of propachlor, alachlor, and cycloate in sewage and lake water. *Appl. Environ. Microbiol.* 49:737–743.

Office of Technology Assessment. 1991. *Bioremediation for Marine Oil Spills.* United States Congress, Washington, D.C.

Olivieri, R. P., P. Bacchin, A. Robertiello, N. Oddo, L. Degen, and A. Tonolo. 1976. Microbial degradation of oil spills enhanced by a slow-release fertilizer. *Appl. Environ. Microbiol.* 31:629–634.

Olivieri, R., A. Robertiello, and L. Degen. 1978. Enhancement of microbial degradation of oil pollutants using lipophilic fertilizers. *Mar. Pollut. Bull.* 9:217–220.

Palumbo, A. V., W. Eng, P. A. Boerman, G. W. Strandberg, T. L. Donaldson, and S. E. Herbes. 1991. Effects of diverse organic contaminants of trichloroethylene degradation by methanotrophic bacteria and methane-utilizing consortia. In R. E. Hinchee and R. F. Olfenbuttel (eds.), *On-Site Bioreclamation: Processes for Xenobiotic and Hydrocarbon Treatment.* Butterworth-Heinemann, Boston, pp. 77–91.

Pavoni, J. L., J. E. Heer, Jr., and D. J. Hagerty. 1975. *Handbook of Solid Waste Disposal, Materials and Energy Recovery.* Van Nostrand Reinhold, New York.

Pfaender, F. K., and M. Alexander. 1972. Extensive microbial degradation of DDT in vitro and DDT metabolism by natural communities. *J. Agric. Food Chem.* 20:842–846.

Prince, R., J. R. Clark, and J. E. Lindstrom. 1990. *Bioremediation Monitoring Program.* Joint Report of Exxon, the U.S. EPA, and the Alaskan Dept. of Environmental Conservation, Anchorage, AK.

Pritchard, P. H., and C. F. Costa. 1991. EPA's Alaska oil spill bioremediation project. *Environ. Sci. Technol.* 25:372–379.

Raymond, R. L., J. O. Hudson, and J. W. Jamison. 1976. Oil degradation in soil. *Appl. Environ. Microbiol.* 31:522–535.

Safferman, S. I. 1991. Selection of nutrients to enhance biodegradation for remediation of oil spilled on beaches. *In Proceedings of the 1991 international Oil Spill Conference.* American Petroleum Institute, Washington, D.C., pp. 571–576.

Sandvik, S., A. Lode, and T. A. Pedersen. 1986. Biodegradation of oil sludge in Norwegian soil. *Appl. Microbiol. Biotechnol.* 23:297–301.

Semprini, L., G. D. Hopkins, P. V. Roberts, and P. L. McCarty. 1991. *In situ* biotransformation of carbon tetrachloride, Freon-113, Freon-11 and 1,1,1-TCA under anoxic conditions. *In* R. E. Hinchee and R. F. Olfenbuttel (eds.), *On-Site Bioreclamation: Processes for Xenobiotic and Hydrocarbon Treatment.* Butterworth-Heinemann, Boston, pp. 41–58.

Sirvins, A., and M. Angeles. 1986. *Biodegradation of Petroleum Hydrocarbons.* NATO ASI Series, Volume G9.

Shailubhai, K. 1986. Treatment of petroleum industry oil sludge in soil. *Trends Biotechnol.* 4:202–206.

Song, H.-G., and R. Bartha. 1990. Effects of jet fuel spills on the microbial community of soil. *Appl. Environ. Microbiol.* 56:646–651.

Song, H.-G., X. Wang, and R. Bartha. 1990. Bioremediation potential of terrestrial fuel spills. *Appl. Environ. Microbiol.* 56:652–656.

Stevens, T. O., and J. M. Tiedje. 1988. Carbon dioxide fixation and mixotrophic metabolism by strain DCB-1, a dehalogenating anaerobic bacterium. *Appl. Environ. Microbiol.* 54:2944–2948.

Stevens, T. O., T. G. Linkfield, and J. M. Tiedje. 1988. Physiological characterization of strain DCB-1, a unique dehalogenating sulfidogenic bacterium. *Appl. Environ. Microbiol.* 54:2938–2943.

Sveum, P., and A. LaDousse. 1989. Biodegradation of oil in the Arctic: Enhancement by oil-soluble fertilizer application. *In Proceedings of the 1989 Oil Spill Conference.* American Petroleum Institute, Washington, D.C., pp. 439–446.

Tesan, G., and D. Barbosa. 1987. Degradation of oil by land disposal. *Water Sci. Technol.* 19:99–106.

Thomas, J. M., M. D. Lee, P. B. Bedient, R. C. Borden, L. W. Carter, and C. H. Ward. 1987. *Leaking Underground Storage Tanks: Remediation with Emphasis on in situ Bioreclamation.* EPA/600/S2-87/008. U.S. Environmental Protection Agency, Ada, OK.

Tramier, B., and A. Sirvins. 1983. Enhanced oil biodegradation: A new operational tool to control oil spills. *In Proceedings of the 1983 Oil Spill Conference.* American Petroleum Institute, Washington, D.C.

Uchiyama, H., T. Nakajima, and O. Yagi. 1989. Aerobic degradation of trichloroethylene at high concentration by a methane-utilizing mixed culture. *Appl. Environ. Microbiol.* 55:1019–1024.

Vogel, T. M., and D. Grbic-Gallic. 1986. Incorporation of oxygen from water into toluene and benzene during anaerobic fermentative transformation. *Appl. Environ. Microbiol.* 52:200–202.

Vogel, T. M., and P. L. McCarty. 1985. Biotransformation of tetrachloroethylene to trichloroethylene, dichloroethylene, vinyl chloride and carbon dioxide under methanogenic conditions. *Appl. Environ. Microbiol.* 49:1080–1083.

von Wedel, R. J., J. F. Mosquera, C. D. Goldsmith, G. R. Hater, A. Wong, T. A. Fox, W. T. Hunt, M. S. Paules, J. M. Quiros, and J. W. Wiegand. 1988. Bacterial biodegradation and bioreclamation with enrichment isolates in California. *Water Sci. Technol.* 20:501–503.

Ward, D. M., and T. D. Brock. 1978. Anaerobic metabolism of hexadecane in marine sediments. *Geomicrobiol. J.* 1:1–9.

Ward, D., R. M. Atlas, P. D. Boehm, and J. A. Calder. 1980. Microbial biodegradation and the chemical evolution of Amoco Cadiz oil pollutants. *Ambio* 9:277–283.

Wilson, B. H., G. B. Smith, and J. F. Rees. 1986. Biotransformations of selected alkylbenzenes and halogenated aliphatic hydrocarbons in methanogenic aquifer material: a microcosm study. *Environ. Sci. Technol.* 20:997.

Wilson, J. 1991. Performance evaluations of in situ bioreclamation of fuel spills at Traverse City, Michigan. *In Proceedings of the In Situ and On-Site Bioreclamation: An International Symposium.* Butterworth, Stoneham, MA.

Wilson, J. T., and B. H. Wilson. 1985. Biotransformation of trichloroethylene in soil. *Appl. Environ. Microbiol.* 49:242–243.

You, I.-S., and R. Bartha. 1982. Stimulation of 3,4-dichloroaniline mineralization by aniline. *Appl. Environ. Microbiol.* 44:678–681.

Yaniga, P. M., and W. Smith. 1984. Aquifer restoration via accelerated in situ biodegradation of organic contaminants. *In Proceedings of the Conference on Petroleum Hydrocarbons and Organic Chemicals in Ground Water—Prevention, Detection, and Restoration.* National Water Well Association, Worthington, OH, pp. 451–470.

Zeyer, J., E. P. Kuhn, and P. R. Schwarzenbach. 1986. Rapid microbial mineralization of toluene and 1,3-dimethylbenzene in the absence of molecular oxygen. *Appl. Environ. Microbiol.* 52:944–947.

Zeyer, J., P. Eicher, J. Dolfing, and P. R. Schwarzenbach. 1990. Anaerobic degradation of aromatic hydrocarbons. *In* D. Kamely, A. Chakrabarty, and G. S. Omenn (eds.), *Biotechnology and Biodegradation.* Gulf Publishing, Houston, TX, pp. 33–40.

Chapter

3

Principles and Practices of Biotreatment Using Altered Microorganisms

B. D. Ensley

Envirogen
Lawrenceville, New Jersey

G. J. Zylstra

Rutgers University
New Brunswick, New Jersey

Any commercial application of microorganisms, including to the degradation of hazardous wastes, soon leads to speculation that desired activities of the microorganisms might be improved by genetic alterations. The means to introduce such genetic changes can range from simple dependence on natural mutation frequencies, to the use of mutagens, to sophisticated molecular biology techniques. Each of these approaches has its own strengths and weaknesses, discussed below.

Scientists seeking to improve microorganisms through mutation should bear in mind that bacteria and other microorganisms that degrade hazardous waste have already adapted to perform that job. Changes made to the genetic makeup of a microorganism must by necessity be made in ignorance of the history of the genes or their evolution. There may be excellent reasons why a microorganism doesn't degrade a compound faster or why a catabolic pathway does not have a broader substrate range. While there are numerous examples of successful genetic alterations that result in markedly improved performance, such endeavors must be started with an appreciation for the inherent characteristics of the original microorganism. The microorganism has had tens and perhaps hundreds of centuries of adaptation and selection to evolve a degradative pathway that is in balance with other physiological needs of the cell and thus permits the microorganism to compete in its natural environment. If a researcher succeeds in constructing a new microorganism that displays faster or more complete metabolism but can no longer compete in the environment with its natural cousins, this new microorganism will be useless in most applications of hazardous waste degradation.

Genetic approaches to strain improvement in degradative microorganisms are complicated by the fact that most hazardous wastes are not completely degraded by a single enzymatic reaction; rather, entire pathways requiring as many as five or more steps are generally needed for effective mineralization. Thus genetic alterations to an enzyme product must be compatible with the complicated machinery of a complete metabolic pathway. One super enzyme made by a lucky mutation isn't enough for improved performance; it also has to be integrated into a living system. In addition, most degradative pathways require cofactors and the right environmental conditions. This means that whole, living microorganisms are required for degradation of all but a very few hazardous wastes. A mutation that is severely deleterious or causes a genetic instability may produce all the desired effects in terms of reactivity or specificity, but still won't be useful (in or out of the laboratory).

This chapter describes a range of methods of increasing complexity that can be used to genetically alter microorganisms in hopes of improving their performance. "Improving" in this case means altering a cell's performance to that desired by researchers or other practitioners of remediation. It is helpful always to bear in mind that the organism itself may not regard these changes as an improvement. The means available today to introduce stable genetic changes into microorganisms include the use of mutagenesis and screening or selection, conjugation and/or recombination of desired pathways or pathway steps, transposons, gene cloning, protein engineering, and site-directed mutagenesis.

Mutagenesis and Screening

This discussion will emphasize screening of mutant strains since mutagenesis is by far the easier step. A number of protocols exist for use of commonly available mutagens,[1,2] but finding the desired mutation is always difficult. Mutagenic techniques include exposure to chemical mutagens, the use of ultraviolet radiation, freezing and thawing, or simply depending on the natural mutation frequency of microorganisms caused by mistakes during replication, exposure to cosmic radiation, or other environmental sources of mutation. Chemical mutagens that can cause changes in the genome fall into several classes. Agents such as base analogs, intercalative dyes, radiation, or reactive chemicals all can cause mutations. Base analogs such as 5-bromouracil or 2-aminopyrine are incorporated during replication cycles, thus causing GC to AT or AT to GC mutations. Intercalative dyes such as acridines or ethidium bromide can cause frame-shift mutations. Radiation such as ultraviolet or x-rays leads to pyrimidine dimer formation or free radical attack on the DNA, both of which can lead to error-prone repair or deletion. Reactive chemicals such as hydroxylamine, 4-nitroquinoline oxide, ethyl methanesulfonate, or methyl methanesulfonate function to chemically modify the DNA, leading to mispairing and error-prone DNA repair.

An extremely potent mutagen is N-methyl-N-nitro-N-nitroquinoline, which induces mutations by an error-prone DNA repair pathway. A procedure for inducing mutations in *Streptomyces* using nitroquinoline is as follows: A culture of cells adjusted to pH 8.5 is incubated with nitroquinoline at a concentration of 100–400 µg/ml for 20 min. The cells are then washed free of the mutagen by centrifugation and resuspension in fresh broth. Cultures are then grown in nutrient medium and plated to obtain various mutants.[3] What makes any of these methods effective is the design and implementation of a screening or selection strategy that favors the identification of the desired traits regardless of the mutagenic method used. This applies to the use of genetic manipulation or transposon mutagenesis as well. The success of any of these methods relies very heavily on a good, sound strategy for screening or selection.

One of the major disadvantages of simple mutagenic techniques is that most mutations result in a microorganism that is simply missing a certain enzymatic activity. This is useful if one is studying gene mapping or gene structure and function but losing an ability is usually not regarded as an improvement in a degradative microorganism. Simple mutations causing an increase in the overall rate of a reaction or those that will permit degradation of new compounds are much rarer and more difficult to select for after a mutagenic treatment. One example frequently used to illustrate that simple mutagenic and screening techniques can be used to improve biosynthetic microorganisms is the overproduction of antibiotics such as penicillin by fungal cultures. What is overlooked is the literally thousands of hours of brute force labor that went into individually screening isolates, one after the other, following a mutagenic treatment. The appeal of this simplistic approach to strain improvement quickly fades when one estimates that it would cost millions or tens of millions of dollars in laboratory expenses to repeat the process used in the isolation of current penicillin-synthesizing strains.

More sophisticated screening or selection techniques are needed if a desired improved strain is to be isolated with efficient utilization of personnel and in a reasonable amount of time.

There are several powerful methods available for the selection and enrichment of mutants defective in catabolic pathways. One of the most effective depends on a microorganism being killed if it displays the properties of a wild-type cell. Enrichments for desired mutants can be achieved by causing the wild-type cells to form a toxic intermediate during degradation of a substrate analog. An example of this is the use of halogenated—in particular, fluorinated—analogs of aromatic hydrocarbons to enrich for mutants defective in degradation of the natural non-halogenated molecules. This approach has been described as *lethal synthesis*, where toxic molecules such as fluorocitrate accumulate during the degradation of fluorinated analogs.[4] Selective enrichment has been used to isolate pseudomonads defective in catabolism of molecules such as cymene with fluorinated analogs.[5] This method is so powerful that after growth of *Pseudomonas putida* in the presence of 5 mM 5-fluorosalicylate without any previous mutagenesis, up to 100 percent of the surviving microorganisms were defective in their ability to metabolize salicylate. This method was also successful in generating defective mutants with halogenated benzoates, phthalates, hydroxybenzoates, and anisates.[5,6,7] In all cases, mutants were enriched from cultures that had not been exposed to any mutagens, again illustrating the ability of this method to enrich for the desired defective mutants at a high degree of efficiency. This type of selection, direct, powerful, and effective, is the most straightforward way of obtaining desired mutants that are defective in catabolic pathways.

Another selective method found to be highly effective is the treatment of cells to kill the remaining wild-type cells that can still grow and then rescuing the nongrowing survivors. This approach involves incubating cells in the presence of a target substrate against which defective organisms are desired, then treating the culture with a combination of penicillin or ampicillin plus cycloserine[8] (Fig. 3.1). Multiple rounds of exposing a culture to a target substrate, adding cycloserine and ampicillin to kill cells that are able to grow on this compound, and then rescuing and enriching mutants by washing the cells and plating on another growth compound result in a high proportion of mutants that can be isolated (up to several percent after three rounds of enrichment and rescuing). This technique is simple, effective, and very easy to use if the subject microorganisms are sensitive to ampicillin or penicillin and cycloserine. The simplest approach that gives desired results is usually the best, and this method should not be overlooked just because it is an older approach.

Although it is hard to imagine that a mutation causing the loss of an activity results in a microorganism with improved degradative proper-

- A mutation in a benzoate metabolic pathway is desired

- Wild type benzoate degrader can be mutagenized or used as-is by depending on natural mutation frequencies

- Incubate culture with benzoate until rapid growth commences

- Add ampicillin plus cycloserine and continue incubation to kill only growing cells

- Harvest cells by centrifuging, wash out antibiotics by resuspending pellet 2-3 times in fresh medium and re-centrifuging.

- Incubate cells in fresh growth medium with a non-selective carbon source such as succinate, glutamate, or casamino acids to "rescue" survivors (including desired mutants) of the antibiotic treatment.

- Harvest cells and incubate in fresh medium containing benzoate.

- Repeat the process of incubating the culture in benzoate, treating with antibiotic, and rescuing the survivors at least 2-3 more times.

- Plate the resulting cells on a selective medium in agar plates and screen for benzoate-defective mutants.

Figure 3.1   Enriching for a blocked mutant.

ties, loss of catabolic function by mutation is occasionally useful in the generation of microorganisms with broadened activity against hazardous waste. A microorganism containing a genetic lesion in the *meta*-cleavage pathway was used as a starting point in the production of a new strain containing a hybrid metabolic pathway.

Single point mutations have also been useful in broadening the range of substrates that can be attacked by a single microorganism. A spontaneous mutation resulting in the appearance of phenol-oxidizing activity has been utilized in the construction of a broad-substrate-specificity microorganism.[9] Populations of microorganisms with enhanced capabilities for the biodegradation of chlorinated organics have been generated using random UV mutagenesis.[10] A device including a continuous-flow UV radiation chamber connected to a bioreactor that continuously selected for microorganisms capable of growing at the expense of 4-chlorobenzoic acid, 2,4-dichlorobenzoic acid, and other complex chloroorganics has been successfully operated. The UV mutagenesis method was shown to enhance the degradation of chloroorganics by the continuous introduction of UV-induced mutations.

Mutagenesis and selection can also be used, at least theoretically, to broaden the substrate specificity of a particular enzyme or to alter the regulation of an enzyme or complete degradative pathway. One useful form of a mutation would be a microorganism harboring a degradative pathway that is constitutively synthesized rather than subject to induction by one or a few substrates. Many degradative pathways are active against substrates that do not induce the synthesis of the degradative pathway itself. Typically, a cosubstrate must be present as an inducer, but generation of constitutive mutants permits the entire spectrum of metabolic capabilities of a microorganism to act against a range of substrates whether an inducer is present or not. This is a useful tool in mutagenesis since constitutive synthesis can overcome a regulatory block that could prevent a microorganism from attacking and even growing at the expense of a much wider range of substrates.

Constitutive mutants of aromatic degradation pathways can easily be obtained by a method described by Parke and Ornston.[11] In this method, cells are alternatively grown on succinate (a noninducing carbon source) and on a compound found in the metabolic pathway that is not normally a growth substrate. During exposure to succinate in the medium, all cells will grow and spontaneous mutants will arise. Upon subculturing into medium containing the noninducing pathway substrate, only constitutive mutants will immediately grow. If no growth is observed, the culture is transferred again to succinate medium to allow additional growth and new spontaneous mutants to form. Alternately exposing the culture to these two substrates may eventually produce a constitutive mutant. This method could also be used to enrich for constitutive mutants following mutagenesis with chemical mutagens or transposons.

Clever schemes have also been developed for isolating mutants of regulatory pathways that have altered inducer specificity; the regulatory machinery is changed by mutagenesis so that new molecules cause induction of the desired enzyme pathway. The regulatory system of the benzoate pathway encoded on the TOL plasmid has been altered by chemical mutagenesis so that it is induced by 4-ethylbenzoate.[12] In this particular case, the regulatory machinery for the TOL pathway (the Pm promoter and the *xylS* gene) were transferred into the same cell by genetic engineering methods. The Pm promoter, instead of regulating the expression of the TOL genes, was placed in a position to control expression of a tetracycline resistance gene. *Escherichia coli* containing this genetic system would grow in the presence of tetracycline if a TOL pathway inducer such as benzoate was added to the growth medium, but would not grow with 4-ethylbenzoate (4EB) present since this compound was not an inducer of Pm. After plating cells on agar plates containing the mutagen ethylmethanesulfonate and 4-ethylbenzoate, a few tetracycline-resistant mutants were isolated. Some of these mutants displayed the property that 4-ethylbenzoate caused induction of the Pm promoter. Although the means used in this example for selection of a 4EB-inducible promoter system relied on genetic engineering techniques, the mutation itself was caused by simple chemical mutagenesis.

Transposon mutagenesis was used by Harker and coworkers to construct a constitutive 2,4-dichlorophenoxyacetic acid pathway utilizing the plasmid pJP4 and the transposon Tn *1721*.[13] In this case the transposon was inserted into the regulatory gene, causing constitutive expression. This construction was performed using natural plasmids and thus would not be considered the product of genetic engineering.

Natural Conjugation and Recombination

Conjugation is a method of moving desired genetic information from one organism into another. It is a natural process whereby one or more genes is one organism are transferred into another by allowing the two organisms to mate. Bacteria are promiscuous and will readily mate and exchange genetic information at a relatively high frequency with different species. While not all plasmids have the ability to mobilize themselves into a new host strain, they sometimes can be encouraged to do so by the presence of a second, "helper" plasmid that encodes the transfer of plasmid DNA from a donor microorganism into a recipient. This means of genetic transfer is very useful for introducing new DNA into a host that already contains certain desirable properties. A microorganism could be altered by this method so that it now degrades a broader range of substrates or now has new enzymatic activities in a host that was already adapted for survival or growth in a particular environment.

The process of conjugation in the laboratory is extremely simple. Typically, the recipient microorganism is chosen because of its desirable properties, and displays some characteristic allowing it to be selected away from the donor. One selection method is to make the recipient organism resistant to an antibiotic such as rifampicin; another is to mutate the donor organism to make it deficient in its ability to synthesize an essential amino acid or nucleotide. Once this has been accomplished, the donor organism containing the desired DNA and the recipient organism are mixed together on a filter, in broth, or on the surface of an agar plate and allowed to grow together. The resulting biological mass is suspended, washed, and replated on a medium that permits only the recipient organism to grow. If the plasmid DNA encodes a directly selectable function, such as growth on a new substrate, the ultimate recipient organisms can then be directly selected by growth on the target compound (Fig. 3.2).

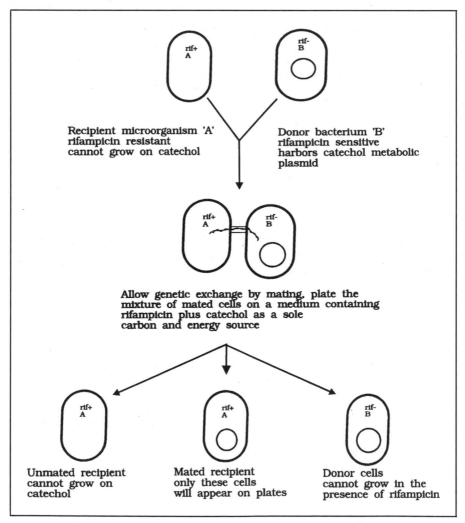

Figure 3.2 Conjugative transfer of genetic information.

The conjugative processes described here are effective for the genetic transfer of an extremely broad range of plasmid-borne genes. If a helper plasmid is required for successful transfer, then three separate strains are used in the mating procedure: The recipient, the strain carrying the desired plasmid, and the strain carrying the helper plasmid are all mated together. This genetic exchange process results in the successful transfer of the desired genetic information into the recipient. The new strain, produced by quasi-natural methods, is regarded as a natural microorganism, although it now contains properties that may be completely unique to this particular host. These properties confer new and hopefully desirable characteristics on the recipient and are a very effective means of introducing large amounts of new DNA into a desired host organism. Microbial strains can be naturally constructed using mobilization of chromosomal and plasmid genes by conjugative plasmids. This is the technique used by Kellog et al. and improved by Krockel and Focht.[14,15,16]

The use of natural genetic exchange processes by encouraging mating or other DNA exchange mechanisms can generate in the laboratory new hybrid degradative pathways that display an expanded range of substrates. This approach also can sometimes reveal new classes of compounds that serve as growth substrates for the recipient microorganism. While these processes depend upon recombination events or even the introduction of large (>100 kb) plasmids into microorganisms, this approach is not regarded as genetic engineering and is not strictly regulated by government agencies. Even though a particular strain may be unique to the laboratory in which it was produced, the use of this organism in the field to treat hazardous waste contamination is no more restricted than that of a naturally occurring, nonindigenous strain.

Some of the earliest examples of constructed hybrid metabolic pathways used the TOL plasmid introduced by conjugation. A strain of *Pseudomonas putida* which grows on salicylates via the formation of catechol and the *ortho*-cleavage pathway can also convert 3-methylsalicylate to 3-methylcatechol but cannot grow on 3-methylsalicylate because it has a very high specificity *ortho*-cleavage pathway. Transferring the TOL plasmid containing the nonspecific *meta*-cleavage pathway into this culture results in a microorganism that will grow at the expense of 3-methylsalicylate.[17]

Another example of a hybrid pathway generated by conjugation is described by Reineke and Knackmuss for a microorganism that can attack halogenated aromatics.[18,19] *Pseudomonas* strain B13 can degrade some chloroaromatics such as 3-chlorobenzoate via the chlorocatechol pathway. However, other substituted halobenzoates are not substrates for the first enzyme in this pathway, and attempts to mutate or select spontaneous mutants were not successful. These workers showed that a molecule such as 4-chlorobenzoate could serve as a growth substrate if the specificity of the first enzyme in the pathway could be broadened to attack this compound, since 4-chlorocatechol could be metabolized by the rest of the pathway and would serve as a growth substrate. This initial metabolic block was overcome by transferring the TOL plasmid, encoding a nonspecific benzoate dioxygenase, by conjugation into the recipient *Pseudomonas* strain B13. After selection, recipient cells that depended on the TOL-encoded benzoate dioxygenase for growth with 4-chlorobenzoate as a sole carbon and energy source were isolated. This is an excellent example of the use of conjugation and subsequent selection to improve the properties of a microorganism active against recalcitrant hazardous wastes such as halogenated organics.

Earlier work with *Pseudomonas* sp. strain B13 showed the structural instability of the TOL degradative plasmids after conjugation into *Pseudomonas putida*. Transfer of the plasmid by mating into a new biochemical background caused structural changes in the DNA that resulted in a new plasmid that would grow with *m*-toluate as a growth substrate and which produced derivatives that would grow at the expense of the novel substrate 4-chlorobenzoate.[20]

These conjugation experiments also caused large structural changes in the DNA encoding the TOL degradative pathway. Original plasmids underwent a large deletion that caused changes in the regulatory phenotype so that growth with 4-chlorobenzoate was possible. These experiments demonstrate the plastic nature of DNA; large changes in substrate specificity and even DNA structure can be caused by simply introducing a piece of DNA into a new microorganism. Since this process goes on in nature every day, it is not surprising that a huge number of related enzymatic pathways exist for the degradation of organic compounds.

Transposon Mutagenesis

Transposons are units of DNA of varying size and complexity that can "hop" to various places in a bacterial genome. Transposons appear to insert themselves into DNA sequences by recombination functions that are encoded by the transposons themselves.[21] Transposons and other insertion sequence elements have common distinct features. The termini of these DNA units carry inverted repeats of 10 to 40 bp thought to serve as recognition sites for transposition enzymes (such as transposases). These repeats flank a central region containing a number of genes that encode transposition functions and other selectable markers. Insertion into a bacterial genome is accompanied by duplication of target DNA. Transposons are able to integrate by recombination into any number of foreign DNA sequences (lacking any apparent DNA homology).

The advantage that transposons bring to the genetic alteration of hazardous-waste-degrading microorganisms is their property of rearranging DNA. Transposons mediate fusions, duplications, deletions, inversions, and the addition of new genes. All of these actions can alter and hopefully improve the performance of microorganisms used to degrade hazardous waste. Transposons have been very useful in studying gene function and structure, and some hydrocarbon metabolic pathways such as the TOL system are themselves carried on transposons.[22,23] The catabolic genes of the TOL plasmid have also been localized and characterized by transposon mutagenesis.

Other metabolic pathways carried on transposons include a chlorobenzoate catabolic pathway carried on transposon Tn5271 that permits growth of the host organism on 3- and 4-chlorobenzoate.[24] This transposon mobilizes into a range of different host bacteria during community adaptation to the presence of 4-chloroaniline, a common industrial and agricultural pollutant.[25] The chlorobenzene dioxygenase genes from *Pseudomonas* sp. strain P51 are also carried on a transposable element, Tn5280.[26] The transposon-borne nature of catabolic functions such as these may be a mechanism for their dissemination in the environment when an appropriate substrate such as chlorobenzene is present. A recent report indicates that the genes encoding naphthalene degradation are borne on a defective transposon, designated Tn4655.[27]

The insertion of a transposon is a catastrophic event for a structural gene that is the site of insertion. Since transposons are usually large pieces of DNA, their insertion into any structural gene causes the destruction of the integrity of that gene and the complete loss of its function. In addition, transposons may have polar effects on the rest of the operon. That is, a transposon insertion not only inactivates a structural gene but it also often blocks the transcription of any other downstream gene that depends on the same promoter element. This characteristic is very useful in the study of structure-function relationships and gene order in a series of genes downstream from a single promoter. This is illustrated by the following example. A promoter P controls the synthesis of five genes in the order A, B, C, D, and E. A transposon insertion into gene E would cause loss of function of only gene E (Fig. 3.3). A transposon insertion into gene B would cause the loss of function of genes B, C, D, and E. In this way, a collection of transposon mutants, one or more in each gene in the operon, would identify the gene order in this operon.

Transposons have other uses besides gene inactivation and the study of gene structure. The advantages of transposon mutagenesis include the fact that it occurs at a relatively high frequency of 10^{-3} to 10^{-6} per cell.[28] Transposon mutations are natural events, and the products of their use are not considered by regulatory agencies to be genetically en-

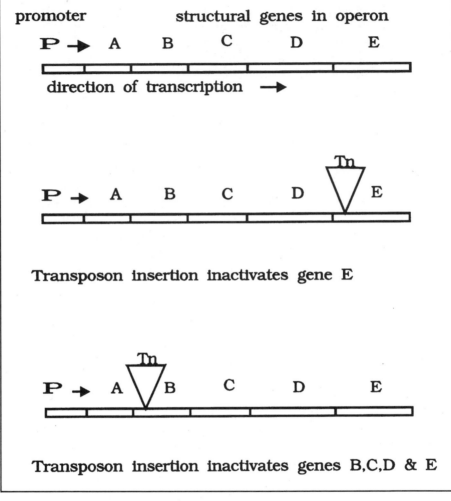

Figure 3.3 Transposon mapping of an operon.

gineered. A caveat here should serve as a warning, however. A frequently used method of introducing transposons into a cell is by transferring a plasmid carrying the transposon into the recipient cell. If the
plasmid itself (but not the transposon) has been the subject of any genetic engineering techniques such as gene cloning or site-directed mutagenesis, the resulting transposon mutant may be regarded at this time
by the U.S. EPA as genetically engineered, even if the plasmid itself cannot replicate in the mutagenized cell. Therefore, if one is planning to use
a transposon to generate a strain with improved properties for hazardous waste degradation and wishes to avoid regulatory oversight and

expense, a transposon and plasmid delivery system should be chosen that does not contain genetically engineered information of any sort.

A major advantage of transposon mutagenesis and the use of transposons as genetic transfer elements is the simplicity with which such experiments can be successfully carried out. Although occurring at a relatively high frequency, the transposition events are rare enough that devising a selection strategy for a transposed genome is still necessary. A characteristic of most transposons is that they contain some type of selectable marker such as antibiotic resistance, resistance to heavy metals, or a positive gene function. Typically, a transposon can be introduced into a recipient cell by conjugation using a carrier plasmid. Since it is much more efficient to examine only the microorganisms that have actually undergone a transpositional event, the plasmid bearing the transposon frequently can be manipulated by some means so that it is lost from the recipient cell. Therefore, any cells that contain the selectable marker carried on the transposon after selection must have undergone a transpositional event. One means of accomplishing this is to use a plasmid whose replication machinery is temperature-sensitive. The plasmid is then grown in the donor cells at a permissive temperature and transferred by conjugation into a recipient cell which is then grown at a temperature not permissive for replication of the plasmid. At the end of this selection cycle any cells still carrying the selectable marker carried on the transposon will have undergone a transpositional event.

A few of these cells will have a transposon inserted in or near the gene of interest unless that particular transposon only inserts in one or a few sites on the chromosome of the recipient cell. A typical experiment for mutation or transpositional insertion of DNA into a recipient pseudomonad or other hydrocarbon-degrading microorganism is performed as follows. An *E. coli* capable of mating with the intended recipient *Pseudomonas* is transformed with a plasmid bearing a transposon. The transposon itself carries a selectable marker such as resistance to tretacycline. The plasmid, although capable of being transferred by conjugation into the recipient *Pseudomonas,* cannot replicate in that host. The donor *E. coli* is then mated with the recipient *Pseudomonas.* After mating, the organisms are plated onto a medium that permits only the *Pseudomonas* to grow, such as a minimal medium containing an aromatic carbon source. Since the *E. coli* plasmid cannot replicate in *Pseudomonas,* selection for the marker carried on the transposon, such as resistance to tetracycline (this assumes that the *Pseudomonas* wasn't already tetracycline resistant), results in a collection of *Pseudomonas* colonies that now carry the transposon either somewhere in the chromosome or inserted into a plasmid that was already in the cell.

This type of experiment is a straightforward means of carrying out a complicated series of genetic events. A new, large fragment of DNA containing one or several genes has been inserted somewhere in the genome of a desired recipient. In the case of insertional inactivation of a gene, further cycles of selection are carried out. Screening or selection for loss of gene function at the same time that the selectable marker carried on the transposon is utilized can result in the isolation of one or more microorganisms carrying the transposon in the gene of interest.

Besides the catastrophic inactivation of structural genes, which is seldom useful for any commercial application, transposons can alter DNA in other ways that have considerably more potential for environmental applications. Since transposons can contain a number of structural genes encoding antibiotic resistance, transposon functions, and other uncharacterized activities, the promoters for these genes can exert an influence beyond the confines of the transposon DNA itself. Transposon-borne promoters may cause transcription of structural genes downstream from the point of insertion. Such an event may have occurred in a recently reported microorganism that normally degrades the common environmental contaminant trichloroethylene (TCE) only after growth in the presence of aromatic substrates such as toluene or phenol. This microorganism displays the constitutive synthesis of enzymes involved in TCE degradation after being subjected to transposon mutagenesis.[29]

Most natural functions of transposons, with the possible exception of activities such as those on the TOL plasmid, are not directly related to hazardous waste degradation, but some may improve the fitness of a particular host strain to survive in a new environment. Such transposon-borne gene functions include resistances to heavy metals. A host organism carrying such a resistance theoretically would be able to function in a new environment where heavy metals may be inhibitory or lethal to the original host strain. An example of this is the transposon Tn501.[30]

A drawback that transposon mutagenesis shares with other natural mechanisms of mutation is that it is by definition a random event. A desired mutation is relatively rare, and there is seldom a rational basis for using these types of mutagens to improve performance. If multiple mutations are required for an entire metabolic pathway to display the desired new characteristics, it may be impossible in a practical sense to create using natural mutagenic methods. A fundamental advantage of using transposons or other natural mutagens is that this approach is still extremely simple to carry out and the use of the products of such efforts is viewed favorably by regulatory agencies. This isn't much of a consideration in laboratory studies or with cultures that will be used in confined manufacturing processes. However, the treatment of hazardous waste is almost invariably an environmental use of microorganisms. Environmental releases of microorganisms that are not "naturally" altered still involves a high level of regulatory oversight and expense, sometimes making these less controlled genomic changes a more practical alternative.

In addition to disrupting gene function and occasionally causing constitutive synthesis of desired enzymes, another activity of transposons makes them highly attractive vehicles for altering bacterial genomes. A transposon by its nature carries any genes flanked by the insertion sequences along with it into the chromosome or resident plasmids of the recipient cell. Transposons can thus be used to deliver desired genetic information into a new host. In the context of hazardous waste degradation, most transposons have limited natural potential for enhancing the specificity of an enzyme pathway or adding desired new metabolic functions. Therefore, this attractive feature of transposons can only be utilized most effectively if one crosses the line separating the use of natural mutagenic methods from the use of genetic engineering technology. Any number of modern molecular biology techniques can be used to insert a new gene into a transposon, to alter the gene structure of the transposon itself, or even change the properties of the plasmid carrying the transposon into the cell. The resulting altered microorganism is regarded at this time as containing genetically engineered material and will be regulated more strictly than a "naturally" derived strain.

Once one has overcome or justified this practical compromise, transposons are extremely valuable genetic engineering tools. A new set of transposons have recently been described that can be used for insertional mutagenesis, for probing of promoter function, and for adding new genetic capabilities into the chromosome of a host microorganism.[31] These transposons have been altered by genetic engineering techniques to contain genes specifying resistance to antibiotics such as kanamycin or chloramphenicol and to contain a unique cloning site flanked by the inverted repeat sequences of Tn5. Some derivatives also contain promoterless structural genes such as *lacZ, luxAB,* or *xylE* so that they can be used to identify various promoter functions. These transposons are located on a plasmid that replicates only in particular strains of *E. coli,* but can be mated into a very large number of recipient microorganisms. The donor plasmid itself has also been engineered to provide the transposase functions necessary for transposon "hopping." This mobilizable plasmid delivers the transposon into the recipient, where the replication-deficient plasmid and its associated transposase function is lost. The transposon carries a selectable marker so that recipients grown under selection contain only the transposon, having lost both the plasmid and its transposase functions, so the transposon is now "stuck" in its new location.

These transposons have already been used to introduce new gene functions that broaden the specificity of the metabolic pathway in *Pseudomonas* species B13 so that additional substrates can be degraded by a single microorganism. A genetically engineered derivative of transposon Tn5 containing genes from the TOL-plasmid-encoded toluate 1,2-dioxygenase (*xylD*), dihydroxycyclohexadiene carboxylate dehydrogenase (*xylL*), and the positive regulator *xylS* was used to expand the degradation range of the recipient organism to include 4-chlorobenzoate.[32]

Molecular Biology Techniques

Over the last fifteen years there has arisen a collection of techniques involving the physical manipulation of DNA that permits one to rationally and purposely alter one or more genes on plasmids or chromosomes of microorganisms. These deliberate approaches have been widely used to improve the desired properties of industrial microorganisms, primarily for the manufacture of human therapeutic products. Other applications include altering microorganisms used in agriculture and pesticide manufacture as well as genetically engineered plants. Genetic engineering techniques are among the most powerful tools available for improving the characteristics of microorganisms.

While molecular biology processes are by definition less random, and generate more predictable results, the primary drawback of this approach is that the use of genetically engineered microorganisms (GEMs) in the environment is currently strictly regulated. Environmental releases, which include most hazardous waste degradation applications, are the subject of time-consuming (and expensive) oversight by the U.S. EPA, and by state and local regulatory agencies. Environmental uses of GEMs are also the subject of resistance at the local and national level by professional critics of genetic engineering and citizens who are concerned that these products could have adverse human health or environmental impacts. To date, the advantages of using methods that result in defined, rational, and predictable changes to a microorganism's genome for improved performance in hazardous waste degradation have not been demonstrated in the field. So far no field trials of genetically engineered organisms used for degradation of hazardous waste have been conducted. Since this situation is likely to change sometime in the future, a description of the usefulness of these techniques in constructing new and improved hazardous-waste-degrading microorganisms is included here.

Any genetic alteration for strain improvement begins with a change in the DNA encoding a single gene. This change may involve a single pair of nucleotides introduced by a random mutation, but the technique of site-directed mutagenesis available to practitioners of molecular biology brings a rational, designed approach to the alteration of genetic material. Mutations introduced into regulatory genes may result in constitutive synthesis of desired enzymes or cause a degradative pathway to be induced by a new stimulus. A mutation in a promoter sequence may also cause elevated levels of gene expression. Mutations in structural genes can change the rate of an enzymatic reaction or cause changes in the stability or half-life of an enzyme. Changes affecting the substrate-binding activities of catalytic proteins can also broaden substrate specificity and cause an enzyme in a biodegradative pathway to display activity against a broader range of substrates, thus improving its utility. All of these changes can be made (at least theoretically) by random mutagenesis, but if the DNA sequence of a gene encoding an enzyme or degradative pathway is known, then rational design changes can be introduced into these enzymes or pathways. These new properties may be extremely difficult or impractical to generate using more traditional techniques.

Site-directed mutagenesis has already been used to markedly increase the stability of a commercial enzyme (the protease subtilisin). Several mutations have been introduced into the structural gene encoding subtilisin, causing the new protein to display a tenfold longer lifetime at elevated temperatures and in the presence of detergents or oxidizing agents such as bleach.[33,34] Information about the active structure of this protein made it possible for the researchers to propose changes in the architecture that would cause an increase in stability. Their hypotheses were correct; some changes resulted in up to an order of magnitude increase in stability of the protein. Stability is a desired trait in both industrial enzymes and enzymes involved in the degradation of hazardous waste, since long enzyme lifetimes in harsh environments are the required standards of performance for environmental applications although enzyme normally is inside the cell where the environment is not so harsh.

Nucleotide-directed site-specific mutagenesis can be used to create base substitutions, deletions, and insertions. Methods for site-directed mutagenesis are described in detail in a recent *Methods in Enzymology* volume.[35] Methods have been developed which use genes cloned into a plasmid vector or even a phage such as M13.[36] Use of the phage M13 permits a simple process to be employed for the introduction of mutations. The mutagenesis occurs when a single strand of M13 DNA containing a cloned gene is mixed with a small oligonucleotide that hybridizes at the region of the desired mutation. The small oligonucleotide will have been synthesized to contain the desired mutation along with a homologous region. This mutant oligonucleotide will still hybridize to the wild-type, single-stranded DNA, and classic techniques are used to fill in the remaining DNA strand. This process results in one copy of the wild-type gene and one copy of a gene with a mutation in exactly the place it was designed for (Fig.3.4). Genes can also be carried on plasmids before the mutagenesis step, but double-stranded DNA must be made single-stranded before introducing the desired site-directed mutation. Methods have been developed so that high yields of mutants can be obtained using double-stranded plasmid DNA.[37]

An often noted disadvantage of site-directed mutagenesis is that after the mutagenesis event at best half of the DNA carries the desired mutation, and, in fact, it is seldom that even half of the DNA is mutated. Extremely low frequencies of mutation can result from the inef-

- **A gene to be mutated is cloned into a vector (plasmid or phage) that can be made single-stranded**

single-stranded vector DNA with cloned gene

- **A primer incorporating the desired mutation is hybridized to the DNA**

- **DNA polymerase and nucleotides are added to generate double-stranded DNA**

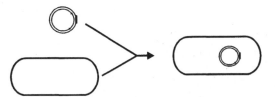

- **The mutagenized DNA is transformed into recipient cells and DNA containing the mutation is identified by screening**

Figure 3.4   Site-directed mutagenesis.

ficient in vitro reactions needed to incorporate a mutation and also from a heteroduplex expression phenomenon that favors the original sequence over the mutants.[38] The rarity of the desired mutation after site-specific mutagenesis requires subsequent colony-screening procedures similar to those employed for the selection and isolation of randomly generated mutants.

A rational approach to site-directed mutagenesis can be utilized to generate mutants at high frequencies, and it is even possible to isolate mutants that do not display a selectable phenotype. This highly efficient and mutant-favoring type of site-directed mutagenesis can be accomplished by the use of bacteria that synthesize uracil-containing DNA as a template. A gene to be mutated is cloned into plasmid DNA or M13 phage and transferred into an E. coli host that lacks the enzymes dUTPase (dut) and uracil glycosidase (ung). These mutations cause the cell's DNA to contain uracil rather than thymine. The uracil-containing template DNA is recovered from these cells and mixed with wild-type nucleotides, a primer containing the desired site-specific mutation, and DNA polymerase. This reaction causes a second strand of DNA to be synthesized, containing both thymine and the desired, specific mutation since the new strand is synthesized using a mutated primer. At the termination of the reaction, a mixture of two types of DNA will be present: the unmutated uracil-containing DNA that was used as a template at the beginning at the reaction and thymine-containing DNA that carries the site-specific mutation. The mixture of DNA is then transferred into a wild-type E. coli. Since this E. coli has the enzyme dUTPase, the uracil-containing DNA is destroyed but the mutated thymine-containing DNA is retained.[39]

Site-directed mutations are useful in applications beyond increasing the stability of an enzyme. Site-directed mutagenesis has also been used to alter substrate specificity. The substrate recognition of subtilisin has also been altered by site-directed mutagenesis.[40,41] By making changes in one or two amino acids in the active site of subtilisin, the recognition pattern for peptide-bond cleavage by this enzyme could be radically altered. This method resulted in the collection of different subtilisins, each reactive preferentially with its own particular set of polypeptide substrates. Some of these changes resulted in very rapid hydrolysis of substrates that are normally only poorly reactive with subtilisin. This site-directed mutagenesis approach demonstrates the practical value of this technique: precise changes can be introduced into a protein to cause it to react with the wider range or different substrates. Altering the substrate specificity of hazardous-waste-degrading enzymes would permit the same enzyme or degradative pathway to attack a broader range of substrates. This will always be important in hazardous waste degradation because contamination by pure compounds is rare; most sites contain mixtures of many different compounds. If a single microorganism can be used to degrade most or all of the hazardous waste at a particular site, this kind of alteration becomes extremely useful and reduces significantly the cost of generating and applying biomass for hazardous waste degradation. In the future, site-directed mutagenesis will become a widely used tool in the improvement of enzymes active against hazardous wastes.

With the advent of a new technique called *polymerase chain reaction* (PCR) for the in vitro replication of DNA, a new and highly directed method of introducing mutations into cloned genes has presented itself.[42,43] The ingredients for a PCR include a cloned gene that serves as the template DNA, the nucleotides that will be used to construct new DNA, oligonucleotide primers that serve to "prime" the reaction and are synthesized in such a way that at least one of the primers will hybridize at the site of mutation and actually carry the mutated sequences, and a thermosensitive DNA polymerase enzyme. The reaction participants are mixed together, the reaction is heated to melt apart the template DNA, and the oligonucleotide primers (including the ones containing the mutation) hybridize with the template DNA and prime the synthesis of a new DNA strand. This new DNA strand contains the desired mutation. Since these newly synthesized strands now become templates for the next round of synthesis, the mutated DNA rapidly comes to represent the major product of the PCR. At the end of the amplification cycles a very high proportion of mutated DNA is present for subsequent cloning, isolation, and characterization.

One of the most powerful applications of molecular biology techniques is its use to transfer one or more genes from one organism into another. This gene transfer can introduce new metabolic capabilities into the recipient organism. These cloning methods have become almost routine over the last fifteen years of use. There are descriptions of cloning experiments for single genes involved in the degradation of pesticides[44] and entire blocks of genes encoding the substantial degradation of complex molecules such as aromatic hydrocarbons or polychlorinated biphenyls (PCBs).[45] These methods are straightforward in use but require laboratory skills in the techniques of molecular biology.

Very often the introduction of one or a few genes into a microorganism can be used to markedly expand its degradative capabilities. Often a pathway with restricted utility in hazardous waste degradation can be markedly improved by introducing a single gene encoding a "gateway" enzyme with much broader specificity. This has been demonstrated in a series of experiments in the laboratory of K. N. Timmis, by the construction of microorganisms that degrade a broad range of chloroaromatics.[46] Gene-cloning techniques follow a relatively straightforward protocol, particularly if the gene in question confers a desired property such as growth on a new substrate. A microorganism containing the gene(s) of interest is grown in medium and the DNA recovered by one of a number of methods found in the Cold Spring Harbor manual.[47] The DNA is then "cut up" or digested with a variety of restriction enzymes. The restricted DNA fragments are cloned into a plasmid or phage using DNA ligase to ligate the digested DNA into a similarly digested vector DNA (either plasmid or phage DNA used to introduce the cloned gene into a new host). The appropriate host organism is now "transformed" with the recombinant DNA, and the individuals that have been successfully altered by this method are identified by isolation and screening or growth techniques.

These organisms may now contain new functional DNA conferring unique and desired properties upon the recipient microorganism and causing these cells to be labeled "genetically engineered." The introduced genetic material can encode a single gene or several in a complete pathway. The new properties may include the ability to degrade recalcitrant or difficult-to-treat molecules. One of the earliest patents protecting a genetically engineered microorganism described an organism that contained multiple pathways catalyzing the degradation of components in crude oil.[48]

Gene-cloning techniques offer some approaches that are not available by any other genetic alteration methods. By using cloning methods, entire biochemical pathways can be brought under the influence of promoters and regulatory systems that permit the pathway to be turned on or off by the addition of a specific inducer or a change in culture conditions. Altering regulation of degradative pathways can be a significant advantage if the hazardous waste is not an inducer of the degradative pathway or if the waste is still hazardous at concentrations below those necessary to cause induction of the pathway. Altered regulation could also aid in the degradation of compounds that do not serve as carbon or energy sources for growth of an organism. There are plasmids, referred to as *expression vectors,*[49] that carry a strong promoter sequence directing transcription into a region of one or more convenient cloning sites. A gene or series of genes from a biodegradative pathway can be transferred into a cloning site and become regulated by the strong promoter. Native promoter sequences carried on the cloned DNA can sometimes cause constitutive expression of degradative enzyme activity in the new host cell.

Both of the above approaches have been described in the scientific literature. The genes encoding the oxidation of aromatic compounds such as naphthalene give higher than wild-type enzyme activity in *E. coli* if the *lac* promoter on the pUC plasmid is used to direct expression.[50] The cluster of the genes encoding toluene monooxygenase has also been transferred into expression vectors. Cells carrying the recombinant vectors have been shown to degrade toluene and TCE after thermal or chemical induction of the hybrid promoter.[51] The genes encoding the pathway for degradation of certain PCB congeners have been transferred into a high-copy-number plasmid, and cells harboring this vector display constitutive synthesis of PCB degradative activity.[45]

The toluene dioxygenase genes from *Pseudomonas putida* F1 have been sequenced and cloned into expression vectors in *E. coli*[52] Upon induction it was shown that the expressed enzyme had a broad substrate range which included such disparate compounds as toluene, phenol, cresol, chlorobenzene, dichlorobenzene, xylene, naphthalene, biphenyl, and certain mono- and dichlorobiphenyls as well as TCE.[53]

In some instances the amount of an enzyme or enzymes present in the environment limits the rate of hazardous waste degradation. The genes encoding the degradative activity can be cloned downstream from a strong promoter carried on a plasmid with a high number of copies (30 or more per cell). This can cause the degradative genes to be overexpressed so that high levels of the degradative enzymes are present in the cells. Molecular biology and biochemical techniques have been refined to the point that up to 40 percent of the total protein in a cell can be a single molecule.[54] An organism containing this much enzyme may not be practical in environmental applications since so much of the cell's energy is dedicated to the synthesis of a protein that not much will be left over to degrade waste. More modest genetic amplification that results in the desired metabolic pathway representing a major enzymatic activity in the cell could radically accelerate the rate of hazardous waste degradation in those environments where the level of enzyme is the rate-limiting step.

The use of modern PCR methods in environmental microbiology brings new advantages of molecular biology techniques to this discipline. Applications of PCR DNA amplification in environmental microbiology have recently been outlined.[55] Specific environmental applications of PCR gene amplification are the use of this method to detect the presence of microorganisms in environmental samples and the use of PCR methods to track the spread and survival of genetically modified microorganisms.[56] PCR amplification can be used to detect indigenous microorganisms that may serve as indicators of chemical contamination or biological contamination from sources such as untreated sewage.

PCR can also aid in the cloning of partially characterized genes or sequences of genes. PCR DNA amplification has been used to isolate genes directly from natural environments,[57] and PCR methods can be used to amplify DNA with limited sequence information.[58]

These PCR methods emphasize the expanding role and technology base in molecular biology techniques. This technology is continually developing new and more powerful methods for gene cloning, manipulation, and expression. These rapidly changing and advancing technologies will alter the way bioremediation problems are approached, alter the quality of the biocatalyst available to researchers and practitioners of this activity, and improve the chances of cleaning up sites contaminated with recalcitrant hazardous waste.

References

1. Hopwood, D. A., 1970, The Isolation of Mutants. In J. R. Norris and D. W. Ribbons (eds.), Methods in Microbiology, Vol. 3A, Academic Press, New York, pp. 363–433.
2. Carlton, B. C., and B. J. Brown, 1981, Gene Mutation. In P. Gerhardt (ed.), Manual of Methods for General Bacteriology, American Society for Microbiology, Washington, D.C., pp. 222–242.
3. Baltz R. H., 1986, Mutagenesis in Streptomyces spp. In Manual of Industrial Microbiology and Biotechnology, A. L. Demain and N. A. Solemon (eds.), pp. 184–189.
4. Peters, R. A., 1952, Croonian Lecture. Lethal Synthesis. Proc. R. Soc. London Ser. B. 139:143–167.
5. Wigmore, G. J., and D. W. Ribbons, 1981, Selective Enrichment of Pseudomonas Spp. Defective in Catabolism after Exposure to Halogenated Substrates, J. Bacteriol. 146:920–927.
6. Wigmore, G. J., and D. W. Ribbons, 1980, p-Cymene Pathway in Pseudomonas putida Selective Enrichment of Defective Mutants by Using Halogenated Substrate Analogs, J. Bacteriol. 143:816–824.
7. Shira, K., 1986, Screening of Microorganisms for Catechol Production from Benzene, Agric. Biol. Chem. 50:2875–2880.
8. Ornston, L. N., Ornston, M. K., Chow, G., 1969, Isolation of Spontaneous Mutant Strains of Pseudomonas putida. Biochem. Biophys. Res. Commun. 36:179–184.
9. Ribbons, D. W., and P. A. Williams, 1982, Genetic Engineering of Microorganisms for Chemicals: Diversity of Genetic and Biochemical Traits of Pseudomonads. In Genetic Engineering of Microorganisms for Chemicals, A. Hollaender (ed.), Plenum Press, New York, pp. 211–232.
10. Kai, G., A. S. Weber, and W. C. Ying, 1991, Use of Continuous Flow UV-Induced Mutation Technique to Enhance Chlorinated Organic Biodegradation, J. Ind. Microbiol. 8:99–106.
11. Parke, D., and L. M. Ornston, 1976, Constitutive Synthesis of Enzymes in the Protocatechuate Pathway and of the ß-Ketoadipate Uptake System in Mutant Strains of Pseudomonas putida, J. Bacteriol. 126:272–281.
12. Ramos J. L., and K. N. Timmis, 1987, Experimental Evaluation of Catabolic Pathways of Bacteria, Microbiol. Sci. 4:228–237.
13. Harker, A., R. H. Olsen, and R. J. Seidler, 1989, Phenoxyacetic Acid Degradation by the 2,4-dichlorophenoxyacetic Acid (TFD) Pathway of Plasmid pJP4: Mapping and Characterization of the TFD Regulatory Gene, tfdR, J. Bacteriol. 171:314–320.
14. Kellog, S. T., D. K. Chatterjee, and A. M. Chakrabarty, 1981, Plasmid Assisted Molecular Breeding: New Technique for Enhanced Biodegradation of Persistent Toxic Chemicals, Science 214:1133–1135.
15. Krockel, L., and D. D. Focht, 1987, Construction of Chlorobenzene-Utilizing Recombinants by Progenitive Manifestation of a Rare Event, Appl. Environ. Microbiol. 53:2470–2475.
16. Adams, R. H., C. M. Huang, F. K. Higson, V. Brenner, and D. D. Focht, 1992, Construction of a 3-Chlorobiphenyl-Utilizing Recombinant from an Intergeneric Mating, Appl. Environ. Microbiol. 58:647–654.
17. Nakazawa, T., and T. Yokota, 1977, Isolation of a Motant TOL Plasmid with Increased Activity and Transmissibility from Pseudomonas putida, J. Bacteriol. 129:39–46.
18. Reineke, W., and H. J. Knackmuss, 1979, Construction of Haloaromatics Utilizing Bacteria, Nature 277:285–286.
19. Reineke, W., and H. J. Knackmuss, 1980, Hybrid Pathway for Chlorobenzoate Metabolism in Pseudomonas sp. B13 Derivatives, J. Bacteriol. 142:467–473.

20. Jeenes, D. J., W. Reineke, H.-J. Knackmuss, and P. A. Williams, 1982, TOL Plasmid pWWO in Constructed Halobenzoate-Degrading *Pseudomonas* Strains: Enzyme Regulation and DNA Structure, *J. Bacteriol.* 150:180–187.
21. Bukhari, A., J. A. Shapiro, and S. Adhyal, (eds.), 1977, DNA Insertion Elements, Plasmids and Episomes, Cold Spring Harbor Laboratory, New York, 782 pp.
22. Tsuda, N., and T. Iino, 1987, Genetic Analysis of a Transposon Carrying Toluene Degrading Genes on a TOL Plasmid pWWO, *Mol. Gen. Genet.* 210:270–276.
23. Nakazawa, T., S. Nouye, and A. Nakazawa, 1980, Physical and Functional Mapping of RP4-TOL Plasmid Recombinants: Analysis of Insertion and Deletion Mutants, *J. Bacteriol.* 144:222–231.
24. Nakatsu, C., J. Eng, R. Singh, N. Straus, and C. Wyndham, 1991, Chlorobenzoate Catabolic Transposon Tn5271 is a Composite Class 1 Element with Flanking Class 2 Insertion Sequences, *Proc. Natl. Acad. Sci. USA* 88:8312–8316.
25. Fulthorpe, R. R., and R. C. Wyndham, 1992, Involvement of a Chlorobenzoate-Catabolic Transposon, Tn5271, in Community Adaptation to Chlorobiphenyl, Chloroaniline and 2,4-dichlorophenoxy Acetic Acid in a Fresh Water Ecosystem, *Appl. Environ. Microbiol.* 58:314–325.
26. Dan Rur Mer, J. R., A. J. B. Zehender, and W. M. DeVoss, 1991, Identification of a Novel Composite Transposable Element, Tn5280, Carrying Chlorobenzene Dioxygenase Genes of *Pseudomonas* sp. Strain P51, *J. Bacteriol.* 173:7077–7083.
27. Tsuda, M., and T. Iino, 1990, Naphthalene Degrading Genes on Plasmid NAH7 are on a Defective Transposon, *Mol. Gen. Genet.* 223:33–39.
28. Franklin, N. C., 1978, Genetic Fusions for Operon Analysis, *Annu. Rev. Genet.* 12:193–221.
29. Shields, M. S., 1991, Construction of a *Pseudomonas cepacia* Strain Constitutive for the Degradation of Trichloroethylene and Its Evaluation for Field and Bioreactor Conditions, Abstracts, Annual Meeting, American Society for Microbiology, K-8, May 15–18, New Orleans, Louisiana, p. 576.
30. Stanisih, V. A., P. M. Bennett, and M. H. Richmond, 1977, Characterization of a Translocation Unit Encoding Resistance to Mercuric Ions That Occurs on a Nonconjugative Plasmid in *Pseudomonas aeruginosa, J. Bacteriol.* 129:1227–1233.
31. DeLorenzo, V., M. Herrero, U. Jakubzik, and K. M. Timmis, 1990, Mini-Tn5 Transposon Derivatives for Insertion Mutagenesis, Promotor Probing and Chromosomal Insertion of Cloned DNA in Gram-Negative Eubacteria, *J. Bacteriol.* 172:6568–6572.
32. Timmis, K. M., F. Rojo, and J. L. Ramos, 1988, Prospects for Laboratory Engineering of Bacteria to Degrade Pollutants. *In* Environmental Biotechnology, G. S. Omenn (ed.), Plenum Press, New York, pp. 61–79.
33. Narhi, L. O., Y. Stabinski, N. Levitt, L. Miller, R. Sachdev, S. Finley, S. Park, C. Kolvenbach, T. Aurakawa, and M. Zukowski, 1991, Enhanced Stability of Subtilisin by Three Point Mutations, *Biotechnol. Appl. Biochem.* 13:12–24.
34. Zukowski, M., Y. Stabinski, L. Narhi, J. Mauck, M. Stowers, and M. Fiske, 1990, An Engineered Subtilisin with Improved Stability: Applications in Human Diagnostics. *In* Genetics in Biotechnology of Bacilli, Vol. III, M. N. Zukowski, A. T. Ganesan, and J. A. Hoch (eds.), Academic Press, New York, pp. 162–193.
35. Wu, R., and L. Grossman, 1987, Methods of Enzymology, Vol. 154, Academic Press, New York.
36. Zoller, M. J., and M. Smith, 1982, Oligonucleotide-Directed Mutagenesis Using M13-Derived Vectors: An Efficient and General Procedure for the Production of Point Mutations in Any Fragment of DNA, *Nucl. Acid Res.* 10:6487–6500.
37. Y. Morinaga, T. Franceschini, S. Inouye, and M. Inouye, 1984, *Bio/Technology* 7:636–639.
38. Hemsley, A., N. Arnheim, M. D. Toney, G. Cortopassi, and D. J. Galas, 1989, A Simple Method for Site-Directed Motagenesis Using the Polymerase Chain Reaction, *Nucleic Acids Research*, 17:6545–6551.
39. Kunkel, T. A., 1985, Rapid and Efficient Site-Specific Mutagenesis without Phenotypic Selection, *Proc. Natl. Acad. Sci. USA* 82:488–492.

40. Wells, J. A., B. C. Cunningham, T. P. Graycar, and D. A. Estelle, 1987, Recruitment of Substrate Specificity Properties from One Enzyme into a Related One by Protein Engineering, *Proc. Natl. Acad. Sci. USA* 84:5157–5174.

41. Wells, J. A., D. B. Bowers, R. R. Bott, T. P. Graycar, and D. A. Estelle, 1987, Designing Substrate Specificity by Protein Engineering of Electrostatic interactions, *Proc. Natl. Acad. Sci. USA* 84:1219–1223.

42. Mullis, K. B., 1990, The Unusual Origin of the Polymerase Chain Reaction, *Sci. Am.* 262(4):56–65.

43. Mullis, K. B., and F. A. Faloona, 1987, Specific Synthesis of DNA *in Vitro* via a Polymerase-Catalyzed Chain Reaction, *Methods Enzymol.* 155:335–351.

44. Serdar, C. M., D. C. Murdock, and M. F. Rohde, 1989, Parathyonhydrolase Gene from *Pseudomonas diminuta* MG: Subcloning, Complete Nucleotide Sequence, and the Expression of the Mature Portion of the Enzyme in *Escherichia coli, Bio/Technology* 7:1151–1155.

45. Mondello, F. J., 1989, Cloning and Expression in *Escherichia coli* of *Pseudomonas* Strain LB400 Genes Encoding Polychlorinated Biphenyl Degradation, *J. Bacteriol.* 171:1725–1732.

46. Timmis, K. N., F. Rojo, and R. J. Ramos, 1988, Prospects for Laboratory Engineering of Bacteria to Degrade Pollutant. *In* Environmental Biotechnology: Reducing Risks from Environmental Chemicals through Biotechnology, G. S. Omenn (ed.), Plenum Press, New York, pp. 61–79.

47. Maniatis T., E. F. Frisch, and J. Sanbrook, 1982, Molecular Cloning—A Laboratory Manual, Cold Spring Harbor Laboratory, Cold Spring Harbor, New York.

48. Chakrabarty, A. M., 1981, Microorganisms Having Multiple Compatible Degradative Energy Generating Plasmids and Preparation Thereof. U.S. Patent 4,259,444.

49. Burnette, W. N., V. L. Marr, and W. Cieplak, 1988, Direct Expression of Bordetella Pertussis Toxin Sub-Units to High Levels in *Escherichia coli, Bio/Technology* 6:699–706.

50. Ensley, B. D., T. D. Osslund, M. Joyce, and M. J. Simond, 1988, Expression and Complementation of Naphthalene Dioxygenase Activity in *Escherichia coli. In* Microbial Metabolism and the Carbon Cycle, S. R. Hagedorn, R. S. Hanson, and D. A. Kunz (eds.), Harwood Academic, New York, pp. 437–455.

51. Winter, R. B., K. M. Yen, and B. D. Ensley, 1989, Efficient Degradation of Trichloroethylene by a Recombinant *Escherichia coli, Bio/Technology* 7:282–285.

52. Zylstra, G. J., and D. T. Gibson, 1989, Toluene Degradation by *Pseudomonas putida* F1: Nucleotide Sequence of the *tol*C1C2BADE Genes and Their Expression in *Escherichia coli, J. Biol. Chem.* 264:14940–14946.

53. Zylstra, G. J., L. P. Wackett, and D. T. Gibson, 1989, Trichloroethylene Degradation by *Escherichia coli* Containing the Cloned *Pseudomonas putida* F1 Toluene Dioxygenase Genes, *Appl. Environ. Microbiol.* 55:3162–3166; Zylstra, G. J., and D. T. Gibson, 1991, Aromatic Hydrocarbon Degradation: A Molecular Approach. *In* J. K. Setlow (ed.), Genetic Engineering: Principles and Methods, Vol. 13, Plenum Press, New York, pp. 183–203.

54. Fiescko, J., and T. Rich, 1986, Production of Human Alpha Consensus Interferon in Recombinant *Escherichia coli, Chem. Eng. Commun.* 45:229–240.

55. Steffan, R. J., and R. M. Atlas, 1991, Polymerase Chain Reaction: Applications in Environmental Microbiology, *Annu. Rev. Microbiol.* 45:137–161.

56. Chaudhry, G. R., G. A. Torazos, and A. R. Bhatti, 1989, Novel Method for Monitoring Genetically Engineered Microorganisms in the Environment, *Appl. Environ. Microbiol.* 55:1301–1304.

57. Paul, J. H., L. Cazares, and J. Thurmond, 1990, Amplification of the *rbcL* Gene from Dissolved and Particulate DNA from Aquatic Environments, *Appl. Environ. Microbiol.* 56:1963–1966.

58. Kalman, M., E. T. Kalman, and M. Cashel, 1990, Polymerase Chain Reaction (PCR) with a Single Specific Primer, *Biochem. Biophys. Res. Commun.* 167:504–506.

4

Bioreactors

Piero M. Armenante

Department of Chemical Engineering, Chemistry, and Environmental Science
New Jersey Institute of Technology
Newark, New Jersey

A *bioreactor* can be defined as a vessel in which biological reactions are carried out by microorganisms or enzymes contained within the reactor itself. In hazardous, municipal, or industrial waste treatment, bioreactors are used primarily to reduce the concentration of contaminants in incoming wastewaters to acceptably low levels. In particular, biological treatment appears to be especially versatile and cost effective when the concentration of pollutants in the wastewater is relatively low and the volumes to treat are large, thus making the use of other treatment alternatives (such as incineration) unattractive.

Wastewaters typically contain a number of contaminants that should be removed, or at least significantly reduced in concentration. These contaminants can be classified as follows:

- Immiscible floating materials (e.g., oils, floating solids)
- Suspended solids
- Soluble nonhazardous organic materials
- Soluble hazardous materials
- Soluble inorganic materials (e.g., ammonia and nitrates, phosphorous)
- Volatile materials

The choice of the possible sequence of specific treatments depends on the type and concentration of contaminants. In general, it is common practice to classify wastewater treatment processes in three categories, i.e., primary, secondary, and tertiary treatment.[5] *Primary treatment* pertains to the removal of easily separable materials such as oils, floating solids, or quickly settling solids, and the preparation of the wastewater (e.g., pH adjustment) for subsequent treatments. Primary treatment involves operations such as equalization, neutralization, sedimentation, oil separation, and floatation. *Secondary treatment* is typically the most important part of the process, and is used primarily to remove the bulk of the suspended solids, organic materials (both hazardous and nonhazardous), and other soluble materials. *Tertiary treatment,* involving processes such as sand filtration, reverse osmosis, adsorption, and electrodialysis, is used (if necessary) to remove any residual contaminants not eliminated during the previous treatment processes.

Biological treatment constitutes the process of choice during secondary treatment of wastewater. The microorganisms utilized during the treatment are typically capable of significantly reducing the content of pollutants by utilizing them as energy and nutrient sources, or electron acceptors during respiration. A number of measures indicating the level of concentration of pollutants have been devised and are commonly used in industrial practice. The most common are *biological oxygen demand* (BOD), originally devised by the British Royal Commission on Sewage Disposal in 1898, and *chemical oxygen demand* (COD).[14] These estimates are rather crude, but convey an idea of the biologically degradable and oxidizable material in the water. Wastwaters are also typically tested for a number of other compounds; measures include total phosphorous, nitrogen, and suspended solids. Biological treatment is routinely used to reduce the BOD from 100–250 to 5–15 mg/L, COD from 200–700 to 15–75 mg/L, phosphorous from 6–10 to 0.2–0.6 mg/L, nitrogen from 20–30 to 2–5 mg/L, and suspended solids from 100–400 to 10–25 mg/L.[5] In addition, it is becoming common practice to test the water for its content of EPA priority pollutants.[50] These pollutants are significantly more difficult to treat biologically, at least in conventional treatment plants, and it is likely that a significant fraction of the amounts apparently removed in such plants is actually removed by air stripping and adsorption rather than biooxidation.[14] On the other hand, over the last few years a number of specific microorganisms have been shown to be able to successfully attack and mineralize many hazardous materials that were previously thought not to be biologically degradable.[33,46] This may also have an impact on reactor design, since each class of microorganisms may have different requirements to be satisfied to successfully carry out its degradation activity.

In general, the main objective for the use of bioreactors is to generate an optimal environment for the biological activity to take place on a large scale, thus satisfying the requirements imposed on the quality (in terms of residual pollutant concentration) and quantity (in terms of flow rates) of the water to be treated and discharged. Therefore, bioreactors for wastewater treatment can be appropriately designed only if the kinetic parameters associated with (at least) the main biological reactions involved in the process, such as rate of pollutant removal per unit biomass, growth rate of the microorganisms, biomass yield per unit of substrate consumed, and nutritional requirements of the microorganisms, are known. In addition, the designer must have quantitative information on the mass transfer rate of nutrients (especially oxygen) to the microorganisms to produce a design that satisfies all the material and energy balances for the system. All these data are typically obtained from laboratory or pilot plant experiments, or from the experience accumulated through the operation of existing plants.

Special care must be used in the interpretation and utilization of data from small-scale tests, since scale-up effects may become significant during the design of large-scale installations. As an example, the mixing time required to completely homogenize the content of a 5-L fermenter with a rotating impeller is orders of magnitude lower than the corresponding time required to completely mix a 500 m³ reactor (assuming constant impeller tip speed), even if the two systems have exactly the same geometric proportions. In numerical terms, if the mixing time in the 5-L tank is 50 s, it can be expected to be about 40 min in the 500 m³ vessel. It is evident that this may have a significant effect, among other things, on the distribution and availability of key nutrients to all the microorganisms, especially if the nutrients are fed at a single point in the reactor.

This example illustrates that different aspects of the process (in this case, mixing time and size of the reactor) scale up according to different rules. If this were not the case it would be sufficient to test a process at a small scale in the lab and then simply "blow it up" to the required scale. In addition, engineering, as a discipline, would not exist. In fact, successful scale-up requires knowledge of the scale-up rules appropriate for the process, and the involvement of engineers in all aspects of the design of the plant.

Classification of Biological Reactor Systems

Before examining the characteristics of any specific bioreactor it will be convenient to examine the broad categories in which reactors can be classified. This will aid in understanding the advantages and disadvantages of each class of reactors.

Anaerobic vs. aerobic reactors

Anaerobic reactors differ from aerobic reactors primarily because the former must be closed in order to exclude oxygen from the system, since this could interfere with anaerobic metabolism. A noticeable exception is constituted by anaerobic ponds and the bottom of facultative ponds, in which anaerobic conditions are established as a result of stratification and oxygen depletion in the lower part of the pond. An additional reason to require closed anaerobic reactors is the odors associated with anaerobic fermentation. An anaerobic reactor must also be provided with an appropriate vent or collection system to remove the gases (mainly methane and carbon dioxide) produced during anaerobiosis.

Conversely, aerobic reactors containing suspended biomass almost invariably require the use of an air-sparging or bubbling system to provide the microorganisms with oxygen. One of the main drawbacks of oxygen as a key substrate is its low solubility in water (about 10 ppm at room temperature) as opposed to most other substrates (e.g., glucose, nitrates), which have much higher saturation concentrations. In addition, because of the low oxygen concentration gradient (equal, under the most favorable circumstances, to the difference between the saturation concentration and the actual concentration in the water), the driving force for the mass transfer of oxygen from the air bubbles to the water is quite small. Therefore, a large air-water interface must be generated in order to supply enough oxygen to the system. Typically, this is accomplished by the use of one or more impellers which break up large air bubbles and disperse them in the liquid, with significant expenditure of energy.

The vast majority of existing biological treatment plants are aerobic. The reasons for this preference over anaerobic systems are the greater range of wastewaters that can be treated, easier control and greater stability of the process, and more significant degree of removal of BOD, nitrogen, and phosphorus. Because of the slower metabolism, anaerobic systems require a longer residence time of the waste in the reactor. This translates into a larger reactor volume to treat the same amount of waste. The slow metabolism also implies that a longer period of time is required for anaerobes to colonize the reactor. This, in turn, means that startup time can be significant, and that a longer period of time is required to bring the reactor back to full operation in case the bacterial population is lost because of a process upset. Anaerobes also require a more precise control of the operating parameters such as temperature or pH.

Whereas aerobic degradation is typically carried out by many organisms operating by and large independently and in parallel, anaerobes live in consortia in which different classes of organisms are responsible for carrying out single steps of the degradation process. This makes anaerobic reactors more prone to failures. The reasons for this can be traced back to hydraulic, organic, or toxic overloading of the reactor.[5] Hydraulic overloading in continuous reactors is produced when the microbial population is washed out of the reactor as a result of too high a flow rate. This occurs especially when the population reproduces slowly, as in the case of anaerobes. This problem can be minimized by the use of immobilization, as described below in greater detail. Organic overload is produced when the wastewater contains a high concentration of organic compounds. This results in the rapid production of a significant amount of volatile acids by one of the intermediate classes of anaerobic organisms in the consortium (the acetogenic bacteria), and in the inhibition of the methanogens (the last organisms to act in a methanogenic consortium), with consequent failure of the reactor. Toxic compounds can also inhibit the activity of the methanogens or cause their washout, with resulting reactor failure.

However, anaerobic processes have advantages of their own. They are typically capable of tolerating higher loading rates, do not require high mechanical energy input for air dispersion (as in the aerobic case), and produce less biomass per unit of waste degraded. Furthermore, because of the lack of a significant gas phase, anaerobic reactors are more suited to treat streams contaminated with volatile hazardous materials that could otherwise be stripped out of the liquid by the gas.

One of the main drawbacks of the use of aerobic processes, such as the activated-sludge process, is the high volumetric production of biomass sludge. This sludge is typically treated anaerobically to reduce its volume and increase its stability. Under the appropriate conditions anaerobic metabolism typically results in the production of methane, which can be used as a fuel. In addition, anaerobes have recently been shown to have unique degradative capabilities, such as being able to reductively dehalogenate highly recalcitrant compounds that can then be treated aerobically. This appears to offer great potential for the treatment of hazardous EPA priority pollutants, a large number of which are chlorinated.[8,46]

Anaerobic respiration (in which an inorganic ion, such as sulfate or nitrate, is used as an electron acceptor instead of oxygen) can also be conveniently used to carry out specific degradation reactions, such as denitrification (conversion of nitrate to nitrogen gas), if the appropriate conditions in the reactor are maintained.

Continuous vs. batch reactors

Most large-scale wastewater treatment systems are operated in a *continuous mode,* in which a waste stream is continuously fed to the plant and a clarified stream is continuously removed. This is a common requirement especially if the waste is generated at a continuous rate. An important concept associated with continuous reactors is that of *residence time,* defined as the average amount of time spent by a fluid element in the reactor.[29] Numerically, the residence time τ is defined as the ratio of the reactor volume V to the volumetric flow rate Q:

$$\tau = \frac{V}{Q}$$

The greater the residence time, the longer the wastewater will reside, on average, in the reactor. This also implies that, for a given flow rate, the reactor volume will increase in direct proportion to the residence time required to treat the waste.

In the absence of feed fluctuation or process upsets, a continuous process can be expected to treat all the incoming waste for which it was designed. In addition, the concentration of pollutants, biomass, or nutrients can be expected to be constant at any specific location within the system, and not to change as a function of time. In other terms, the system operates at steady state. In practice, fluctuations are always present because of the intrinsic variability of any biological process, and because of changes in both operating conditions (such as variability in feed composition and flow rate) and operating parameters (such as daily and seasonal temperature fluctuations). The main drawback of continuous systems is that if a process upset occurs that cannot be satisfactorily corrected, the result is the discharge of a stream which does not meet the desired requirements.

Conversely, in *batch systems* the waste is charged to the reactor and the process allowed to proceed to completion. This gives the user the advantage of being able to follow the evolution of the process as a function of time and decide when a satisfactory level of decontamination of the batch material has been achieved before discharging it. In addition, batch systems are generally simpler, require minimal support equipment, and are well suited to treat small amounts of waste. Batch reactors are also used, by necessity, when the residence time required for the decontamination reaction is exceedingly large or when solids are treated. This is the case of composting, for example, in which waste decomposition is achieved in piles of solid waste (e.g., soil) mixed with bulking agents and nutrients. Liquid/solids treatment, a process similar to the continuous, large-scale activated-sludge process, is also operated in batch.[9] However, batch processes are rather labor-intensive, and require the presence of storage facilities to temporarily store the incoming waste material while the treatment process is proceeding in the reactor. This makes the use of batch processes impractical for large-scale operation.

An intermediate mode of operation, between the batch and the continuous processes, is the *semi-batch process,* in which the waste material is continuously fed to an otherwise batch-operated reactor. Once the detoxification process is complete, the reactor is emptied and the process is started anew. In wastewater treatment this approach is extensively used in time-stepped processes utilizing *sequencing batch reactors* (SBRs), as more discussed below.

Well-mixed vs. plug-flow reactors

Continuous bioreactors can be designed and operated to run as well-mixed reactors, plug-flow reactors, or a combination of these two extreme conditions.

In a *well-mixed reactor,* the reactor content is made completely homogeneous by the use of a mixing device, as shown in Fig. 4.1. Upon entering the reactor, the feed is assumed to be immediately homogenized within the liquid bulk. The stream leaving the reactor has the same composition as the liquid in the reactor itself. A well-mixed reactor is often referred to as a *continuous stirred-tank reactor* (CSTR).

In a *plug-flow system,* the feed material moves throughout the reactor without (ideally) interacting with the material that was fed before or after. Therefore, the composition of the fluid changes from point to point as each fluid element moves along the reactor and the reaction proceeds. A typical representation of such a reactor is that of a tube in which any slug of fluid fed at the beginning of the reactor will appear at the other end after a time interval equal to that required by the slug to cross the entire length of the reactor.

Both reactor types are used to describe the behavior of a number of bioreactor systems used in wastewater treatment, such as the activated-sludge process, described below in greater detail. In addition, a number of intermediate combinations are also possible.

Plug-flow reactors are typically more efficient in most wastewater treatment applications since the biological reactions responsible for waste decontamination are proportional to the concentration of the waste compounds. As a result, a well-mixed reactor, in which the bulk concentration of the contaminants must necessarily be low (since this is also the concentration of the effluent), produces a lower waste conversion efficiency per unit of reactor volume than a plug-flow reactor, in which the effluent material is not backmixed with the incoming waste. For the same reason, plug-flow reactors are more sensitive to system upsets, such as shock loading, since the excess waste load is not diluted in the entire reactor volume as in the well-mixed case. In any case, ideal plug flow is rather difficult to achieve in practice since a certain degree of backmixing is unavoidable due to longitudinal disper-

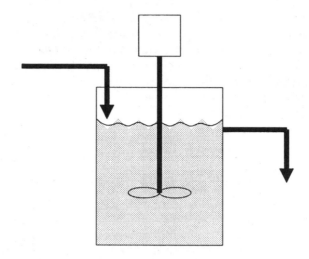

(a) Continuous Stirred-Tank Reactor (CSTR)

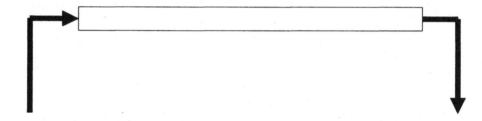

(b) Plug Flow Reactor

Figure 4.1 Well-mixed reactor (CSTR) and plug-flow reactor.

sion. Therefore, combined systems comprising elements of both reactor types are often encountered in practice.

Recycle streams

In many situations it is advantageous to provide the reactor with a *recycle stream,* i.e., a side stream from the reactor effluent that is partially recycled back to the reactor itself (Fig. 4.2). Even if the reactor is of the plug-flow type, it is evident that such a recycle stream introduces some level of backmixing in the system.[29] This is true especially if the ratio of the recycle stream flow rate to the feed flow rate is high. Recycle streams can therefore be used to increase the level of flexibility of a system.

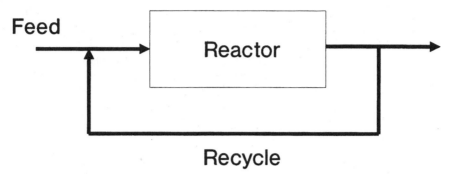

Figure 4.2 Reactor provided with recycle.

Furthermore, recycle streams increase the residence time of the material to be treated in the reactor.

Even more important is the use of recycle streams in which only a desired and separable component of the reactor content is recycled. This is the case in many anaerobic and aerobic waste treatment reactor systems. In these systems the biomass constitutes the "biocatalyst" responsible for the degradation reactions. Therefore, it is often convenient to maximize the biomass concentration in the bioreactor in order to improve the degradation efficiency of the reactor itself. This can be achieved by using a clarifier after the reactor. In the clarifier the biomass is partially separated from the liquid and then recycled to the reactor. In these systems the clarifier often becomes the key component of the entire system, and most system failures can be attributed to its inability to separate the biomass during upset conditions.

Immobilized vs. suspended biomass systems

Another way to maximize the retention of biomass within the reactor is by preventing the biomass from leaving the reactor with the effluent stream in the first place. This can be achieved by immobilizing the biomass on a solid support that cannot be removed with the effluent. For this reason a number of reactor configurations exist in which the reactor is provided with a packing material to which the microorganisms can adhere. This immobilization process is often enhanced by the preference of the microbial population to grow attached on a solid support rather than in suspension. In other cases, the solids on which the microorganisms are attached are small particles (such as sand) that can be easily kept in suspension (i.e., *fluidized*) by the waste stream entering the reactor, but that are not light enough to be entrained by the effluent stream. Finally, the organisms can also be immobilized on a solid support on which the waste effluent is then dripped (as in the case of trickle-bed reactors) or which is periodically immersed in the waste solution (as in case of rotating-disc contactors).

In all these systems, as the biomass grows thicker on the surface on which it is attached, it becomes increasingly difficult for the innermost organisms to receive enough nutrients. Eventually, this leads to the loss of the biomass mat contained on a given particle or surface area (the so-called "sloughing" process). This phenomenon helps to maintain a constant level of biomass in the reactor, which would otherwise eventually plug up.

Basic Bioreactor Configurations

A very large number of different bioreactor types are in use in industry. They differ not only in their physical configuration but also because of the different requirements imposed on the process and on the microbial metabolic activity. For example, aerobic reactors will almost invariably have some provision for air supply and distribution, a feature absent in anaerobic reactors, and since air can be supplied in a variety of ways, a corresponding number of aerobic reactor types can be expected. In addition, provisions for microbial immobilization, continuous or batch operation, and mixing, among others, constitute the essential features of a particular reactor type.

In general, bioreactors for wastewater treatment fall into two broad categories:

- *Bioreactors with suspended biomass.* In these reactors the biomass responsible for the degradation action is freely suspended, and therefore it must be partially recovered and recycled back to the reactor for the process to be self-sustainable. The activated-sludge process and all its variations are based on the use of these types of reactors. Some continuous anaerobic processes also utilize this approach. Agitated reactors and surface-aerated reactors are some examples of this category.

- *Bioreactors with immobilized biomass.* In these reactors the biomass is immobilized on some kind of support, and is not lost with the effluent (except in desired amounts). These reactors are used in a number of applications in aerobic treatment such as packed-bed and trickle-bed reactors, rotating-disc contactors, and fluidized-bed reactors. In addition, this is the approach utilized in the design of most continuous anaerobic treatment processes.

In the remainder of this section a brief overview of the most common reactor configurations used in biological wastewater treatment and their characteristics is provided.

Agitated bioreactors

Agitated reactors are metal or concrete vessels provided with a mechanical agitation system, typically consisting of motors driving one or more impellers. If the reactors are small and closed they will generally be cylindrical, made of metal, and have a centrally mounted shaft provided with impellers and driven by a motor. Vertical baffles mounted near the vessel wall are used to eliminate swirling of the liquid mass. Large, open facilities commonly encountered in aerobic treatment plants are typically rectangular basins with several agitation points equally distributed across the surface of the basin. An example of an agitated and aerated bioreactor is given in Fig. 4.3.

Agitation in these systems serves two purposes. The first is to achieve homogeneous mixing and to disperse dissolved nutrients and

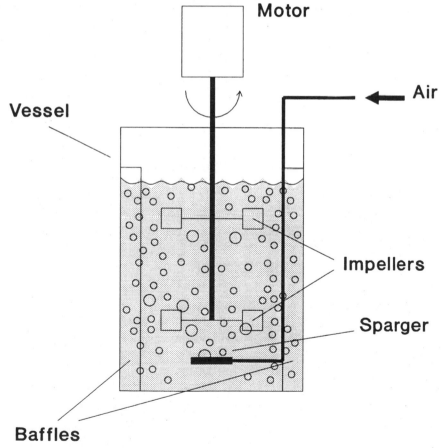

Figure 4.3 Agitated and aerated reactor.

biomass throughout the reactor. For each impeller the power consumption P can be calculated from the equation

$$P = N_p \rho N^3 D^5$$

where N is the agitation speed (in revolutions per unit time), D is the impeller diameter, and N_p is the power number (also called *Newton number*), i.e., a nondimensional constant which is a function of the impeller type. For example, for six-bladed disc turbines N_p is equal to 5. The values for N_p associated with other types of impellers can be found in the literature.[7,35,38]

If the reactor is aerobic, the second objective of the agitation system is to disperse the air by breaking up the air bubbles, thus increasing the gas-liquid interfacial area and promoting oxygen transfer. Disc turbines are very effective impellers for gas-liquid dispersion, but consume a significant amount of agitation power. A number of impeller types are now available that can provide comparable results (in terms of air dispersion efficiency and pumping capability) with less power consumption.[35] An air pipeline hooked to a compressor is used to pump the air to the sparging point within the reactor. Air is typically sparged just under the lower impeller, so that it has to pass through it and be dispersed into fine bubbles. The impeller power consumption under aerated conditions P_g is only a fraction of that under unaerated conditions. The relationship between P and P_g is given by

$$\frac{P_g}{P} = f(N, Q_g)$$

where the function $f(N, Q_g)$ depends on the agitation speed and the aeration rate Q_g. However, for most situations the value of $\rho(N, Q_g)$ is a constant in the range 0.4–0.6.[35] The rate of oxygen transfer per unit liquid volume N_{O_2}, is given by the expression

$$N_{O_2} = k_L a \, (C_s - C)$$

where C_s and C are the oxygen concentrations at saturation and in the water, respectively and the value of $k_L a$ (equal to the mass transfer coefficient k_L times the interfacial area per unit liquid volume a) is given by[31]

$$k_L a = \beta (P_g/V)^{0.7} v^{0.6}$$

In this equation v is the superficial velocity of the gas, and the numerical value of the constant β is between 1.2 and 2.3, if all the variables are in SI units. Typical oxygen transfer efficiencies are in the range 1 to 1.8 kilograms of O_2 transferred per kilowatthour of power consumed.[14]

These expressions show that the rate of oxygen transfer does not increase linearly with either the power consumption or the agitation speed. This can become important in the optimization of an aerobic system, for which the power consumption can become a significant part of the operating costs.

Surface-aerated bioreactors

Surface aerators are typically found in large aeration basins. They consist of an impeller rotating very close to the liquid surface and entraining air at the air-liquid interface, as shown in Fig. 4.4. The air-liquid mixture is then dispersed throughout the basin by the impeller's rotating action. If large basins are aerated, a number of such devices are typically used. The main advantage of this air-sparging method is that no external air compressors are used. The motor/impeller assembly can be anchored to a fixed beam hanging over the basin, or can be allowed to float over the waste liquor. The latter arrangement is especially convenient to ensure that the impeller is always placed at the same depth independently of level fluctuations.

Different impeller configurations are available. They typically con-

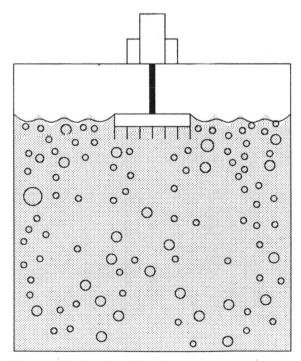

Figure 4.4 Surface-aerated reactor.

sist of open or closed turbines of different shapes and blade types.[45] The power consumption of these impellers is typically lower than that of submerged impellers. Power numbers of the order of 1 to 2 are common.[53] Typical oxygen transfer efficiencies are between 1.5 and 2.5 kg O_2/kWh. In order to ensure that the biomass is completely suspended by the impeller, the depth of the aeration tanks is typically less than about 4 m.[14]

Oxygen transfer rates comparable to those for surface aerators are also obtained with brush mechanical aerators, in which high-speed horizontally rotating cylindrical brushes, partially immersed in the liquid, are used to disperse the air and homogenize the liquid. This type of aeration equipment is primarily used in oxidation ditches.

Rotating-disk bioreactors

These reactors are exclusively used in aerobic treatment processes. They consist of a number of discs mounted on a horizontal shaft, partially immersed in the liquid, and slowly rotating, as shown in Fig. 4.5. The biomass is attached to both surfaces of the disks. The rotating action alternately exposes it to the liquid (providing the dissolved nutrients) and the air (providing the oxygen). Additional details on the operation and properties of this type of reactor are given in the section titled "Aerobic Reactor Systems."

Packed-bed bioreactors

Packed-bed reactors are one of the most common separation devices and reactor configurations used in the chemical industry. In wastewater treatment applications, microorganisms immobilized on their internals are used to degrade the contaminants in the wastewater.

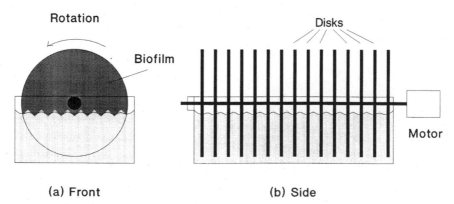

Figure 4.5 Rotating-disc contactor.

Packed-bed reactors consist of a vessel enclosure (which can be open or closed depending on the duty) provided internally with a packing material. The packing is designed so that it has a large surface where the liquid phase, the gas phase (if any), and the immobilized organisms can interact. Typically, the packing for biological wastewater treatment is made of loose materials, such as pebbles, lava rock, or plastic particles.

Packed-bed reactors can be used for both anaerobic and aerobic processes. Anaerobic packed beds (also called *anaerobic filter reactors*) are closed vessels in which only the liquor to be treated is circulated, typically in the upflow direction, over the bed (Fig. 4.6). The same configuration can also be used for aerobic treatment, even when filamentous organisms are used.[4,30] However, this approach requires that the stream itself be oxygenated separately so that the microorganisms can still be supplied with oxygen. This process can be especially advantageous if the compound to be treated is volatile and can be stripped out by bubbling air directly into the column. Recently, this approach has been used to treat a stream contaminated with methylene chloride by passing the stream through a column containing an immobilized bed of a single aerobic organism.[40]

In the vast majority of cases, however, aerobic packed-bed reactors require the additional presence of a gas phase. In this case the liquid phase is added from the top of the reactor, from which it trickles down by gravity, and the air is sparged at the bottom. If the reactor is open,

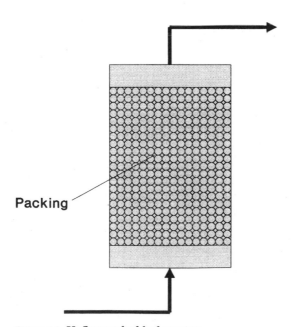

Figure 4.6 Upflow packed-bed reactor.

air moves through the bed by natural convection. The most widely used application of this technology in wastewater treatment is in trickling filters, which will be described below in greater detail.

Fluidized-bed bioreactors

Another way to confine microorganisms within a reactor is to immobilize them on particles of a material heavier than the liquid, and then suspend them using the liquid itself. This is the approach used in *fluidized-bed reactors* (Fig. 4.7). Different materials such as sand, activated carbon, or resin can be used as immobilization particles. Suspension is achieved when the frictional drag exerted by the fluid on the particle is equal to the force of gravity. Experimentally it has been found that, for particles smaller than 1 mm, the fluid velocity at which fluidization is achieved is given by the following equation:[34]

$$v_{fl} = v_d E_L^{\,n} 10^{-(d/D)}$$

where v_{fl} is the fluidization velocity of the fluid, v_d is the terminal velocity of the particles in free fall in the liquid, E_L is the bed porosity, n is an exponent which is a function of the flow regime, d is the particle diameter, and D is the diameter or the reactor.

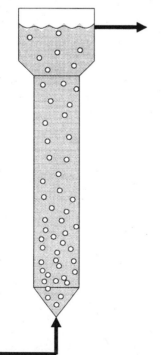

Figure 4.7 Fluidized-bed reactor.

Because of the high surface area per unit of reactor volume the concentration of biomass in fluidized-bed reactors can be 10 times greater than that achievable without immobilization.[9]

Diffused aeration tanks and airlift bioreactors

In a number of aerobic bioreactors, the energy supplied to the gas in the form of pressure is also used to homogenize or even circulate the liquid. In the aeration basins of the traditional activated-sludge process described below, one alternative way to supply air (instead of using mechanical agitation) is by using *diffusers,* devices placed deep inside the tank and capable of producing air bubbles. Many diffuser types are available, and they can be divided into two categories depending on the way they produce the bubbles.[45] The first method is by passing air through a porous material. Fine bubbles are obtained this way. The second method is by forcing the air to pass through orifices such as those found in bubble cups to obtain larger bubbles. Jet nozzles can also be employed, as described under "Jet Bioreactors" below. The diffusers are typically mounted on a network of air pipes placed on the bottom of the basin to provide uniform oxygenation of the liquor. The bubble swarms generated by the diffusers also produce an induced liquid flow which mixes the liquid vertically and functions to suspend the biomass and the suspended solids. In these aeration basins the air distribution is uniform and therefore no significant net liquid recirculation throughout the basin is induced as a result.

However, if the air is supplied only at specific points in the basin or the reactor, a significant liquid recirculation flow can be produced. This is the principle behind *airlift reactors* (Fig. 4.8). This concept is mainly used in deep cylindrical reactors provided with draft tubes, i.e., cylindrical tubes, open at both ends, and placed just above the air-sparging point. As the air bubbles rise through the tube, they induce a liquid flow upwards inside the tube, and a downward flow in the region outside the tube. This outer region is also devoid of most of the bubbles, which disengage and leave the reactor as they emerge from the draft tube.

This principle is used in deep-shaft bioreactors which can be as much as 100–300 m deep, have very low power consumption (0.17 kW/m^3), and have excellent oxygen transfer efficiencies (6 kg O_2/kWh).[21,45]

Jet bioreactors

Jet bioreactors utilize liquid jets to recirculate the reactor content and to disperse any gas that is present (Fig. 4.9). The liquid feeding the jet is typically taken from the bioreactor itself through an external loop provided with a pump. Any incoming reactor feed stream may also be

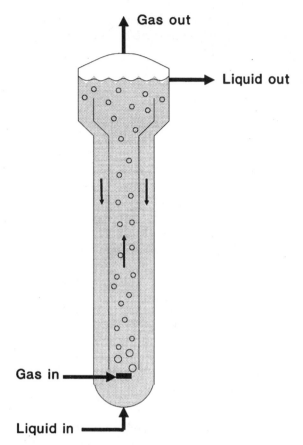

Figure 4.8 Airlift reactor.

input to the jet. In most cases the jet is submerged and the nozzle is oriented in such a way as to recirculate the material in the reactor. For example, if the jet is placed at the bottom of the tank, the liquid feeding it will be taken from the middle or the top of the reactor. In order to establish a strong flow circulation pattern within the reactor, the jet nozzle may be placed within draft tubes or in an asymmetric position within the reactor.

Liquid jets may also be fed with the gas to be dispersed. A large number of two-phase nozzle designs exists for this purpose. For example, in free jet nozzles the liquid power jet and the gas are input through separate lines, the gas line outlet being at the very end of the liquid jet nozzle to maximize gas dispersion.[41,47] In another design, a venturi liquid ejection nozzle sucks in the gas to be dispersed.[24] Finally, jets may be combined with mechanical dispersion devices to enhance the gas dispersion efficiency. This is the approach used in the design of the radial-

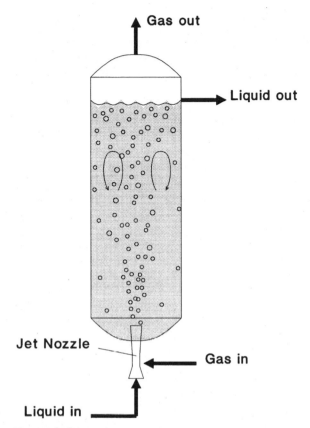

Figure 4.9 Jet reactor.

flow nozzles found in some large-scale commercial applications for wastewater treatment, such as Hoechst's BIOHOCH® reactors.[32,45] In this design, an upward vertical liquid jet impinges on a disc from whose center air is sparged. To enhance the liquid recirculation, these radial nozzles are then placed within draft tubes within the reactor.

When jets are used, the jet placement and vessel configuration become an important part of the overall design. For example, BASF utilizes a large number of two-phase jets (40 to 56) to disperse air in aeration tanks 125 m long, 20 m wide, and 5 m deep, with oxygen dispersion efficiencies of 1.5–1.6 kg O_2/kWh.[41,45] Other arrangements are also used in which the nozzles are grouped in clusters of four on the aeration tank bottom, each nozzle discharging at a 45° angle.[45] In tower reactors for wastewater treatment the reactor is typically deeper and can have different shapes, such as a cylinder or an upside-down cone frustum, in order to promote internal recirculation. The volume of these reactors can be as small as 700 m^3 or as large as 15,000 m^3 with diameters and heights up to 46 m and 20 m, respectively.[45] The sparging devices can be two-phase nozzles of different types (as in the Bayer tower rectors) or radial-flow nozzles with draft tubes (as in Hoechst's BIO-HOCH® reactors). These reactors can have high oxygen transfer efficiencies (3.8 kg O_2/kWh).[45]

An additional application of liquid jets can be found in plunging-jet bioreactors. In this design the jet nozzle is placed above the liquid, discharging the jet straight or at an angle into the liquid mass. Oxygenation of the liquid in the reactor occurs in three stages associated respectively with oxygen transfer from the air to the free jet, from the air to the surface of the liquid pool in the reactor, and from the air bubbles entrained by the plunging jet to the liquid bulk. The oxygenation rate is a function of a number of parameters, such as jet speed, nozzle size, type, angle of incidence, and position, and is typically within the range 0.92–3.90 kg O_2/kWh.[47] However, efficiencies as high as 8 kg O_2/kWh have been reported in some wastewater treatment plants.[45]

Loop bioreactors

Loop bioreactors are simply plug-flow reactors with recycle. However, the term *loop reactor* is generally used for reactors in which the rate of recirculation is much higher than the feed rate. In these reactors the liquid moves primarily in a loop, typically as a result of a pump placed in the recycle line. In wastewater treatment, oxidation ditches are good example of loop reactors (see below under "Oxidation Ditches").

Other reactor types

A large number of reactors can be found in use in industry. However, the vast majority of these reactors can be classified as one of the types described above, or, at least, as a combination of some elements of them. Additional information on reactor classification can be found in specialized texts. In addition, industrial reactors may present a number of different features (such as bubble diffusers for air dispersion, airlift systems for liquid recirculation, impellers for homogenization and particle suspension) to satisfy the specific requirements imposed on the reactor.

Aerobic Reactor Systems

Aerobic stabilization ponds

Aerobic ponds are one of the oldest methods of wastewater purification. They typically consist of water reservoirs ranging in size from a few hundred square meters to many square kilometers and surrounded by natural or artificial containment barriers, where wastewater flows in and in which a variety of different microorganisms carry out their metabolic activities.

Whereas in most biological treatment processes heterothophic organisms are primarily responsible for the decontamination activity, in aerobic ponds autotrophic organisms, such as algae, play a fundamental role.[13,32] These organisms' photosynthesis produces oxygen which is used by other microorganisms to attack and degrade the pollutants. In turn, the microbial population responsible for the pollutant degradation produces carbon dioxide and other waste products that can be used by algae. Algae require light. Therefore, aerobic ponds are typically shallow (0.15 to 1 m) to allow the light to reach the entire depth of the pond and maintain it under aerobic conditions. In addition, the shallow depth increases the air-water interfacial area per unit volume of pond, thus increasing the oxygen transfer rate. Other higher organisms, such as protozoa, are also present in this ecosystem.

Because of the high area/volume ratio, aerobic ponds can undergo significant temperature changes, both seasonal and daily, as a result of insolation, temperature variations, and evaporation rate. These temperature changes have a significant effect on the microbial population, and ultimately on the performance of the pond.

Stabilization ponds can be operated as impoundment basins, in which there is no liquid outflow, but only evaporation, mineralization of the pollutants to volatile compounds such as carbon dioxide, and settling of the undegradable material. Alternatively, stabilization ponds can be operated as continuous, flow-through basins with a given retention time.[14]

Organic loadings in aerobic ponds are lighter than in other types of ponds (such as facultative or anaerobic ponds) since higher loadings would result in the promotion of anaerobic conditions, especially at the bottom of the pond. Typical loadings are of the order of 0.01 kg $BOD/(m^3 \cdot day)$ with BODs up to a few hundreds of milligrams per liter. The retention time of aerobic ponds ranges from a few days up to about 100 days, with BOD removal rates in excess of 80 to 90 percent.[14]

Aerated lagoons and lagoon systems

Aerated lagoons are similar to the aerobic ponds described above, with the major difference that the former are provided with surface aerators to promote oxygen transfer and maintain aerobic conditions throughout the depth of the lagoon. Historically, aerated lagoons have in fact evolved from nonaerated lagoons. Aerated lagoons have depths between 2 and 5 m. These values are significantly larger than for nonaerated lagoons, since the agitation system is designed to transfer enough oxygen and provide enough mixing power to ensure aerobic conditions even at the bottom of the lagoon, and ensure complete suspension of the solids off the lagoon bottom.[13,14]

The mechanical power required by the agitation system to achieve these objectives is quite significant (typically of the order of 3 to 4 kW/m^3). However, there are aerated lagoons whose agitators are specifically designed to deliver much less power. As a result these lagoons are not well mixed. Instead, they operate aerobically throughout most of their depth but maintain anaerobic conditions at the bottom, where the aerobic biomass (sludge) settles and is anaerobically decomposed. Such lagoons are called *facultative lagoons,* and will be examined later in the section on mixed anaerobic-aerobic systems.

From a microbiological point of view, well-aerated, fully aerobic lagoons operate similarly to the activated-sludge process described in the next section, where aerobic organisms feed on the organic material to be disposed of and create microbial flocs. However, unlike the activated-sludge process, many aerobic lagoons typically operate as a single-pass, flow-through system without recycle. This means that the biomass density is typically low (0.05 grams of dry biomass per liter),[32] and that the effluent from the lagoon contains microbial flocs which need to be separated before discharge of the treated wastewater. For these reasons aerobic lagoons are typically a part of a *lagoon system.* In such systems, the aerobic lagoon is followed by a facultative lagoon and/or by a settling lagoon in which the aerobic sludge is partly decomposed or allowed to settle. Some aerated lagoons are even operated with a recycle stream in which the aerobic sludge is partially recycled from a settling facility placed after the lagoon.[9] Therefore, such systems are operated identically to an activated-sludge process.

Like the nonaerated types, aerobic lagoons can be significantly affected by daily and seasonal temperature variations, which may have various effects on the microbial population and its pollutant removal efficiency.

Activated-sludge process reactors

The activated-sludge process is one of the oldest wastewater treatment processes, and the basic concept is now implemented in a number of different ways. Historically, the practice of blowing air through wastewater to reduce its pollutant concentration has been used since ca. 1890. In so doing, the growth of aerobic organisms which can feed on the pollutant is promoted, thus preventing the growth of anaerobes and the generation of noxious odors. However, it was not until the 1910s that a significant improvement in the process was made by recycling part of the biomass generated during the aeration process (the so-called *activated sludge*) back to the aeration basin. The introduction of this recycling step led to an increase in the biomass concentration within the aeration tank, thus speeding up the rate at which the degradation process would proceed.

All activated-sludge processes follow this general scheme. The wastewater is first introduced into a bioreactor (the aeration basin) in which air is sparged and aerobic growth is promoted (Fig. 4.10). Under optimal conditions, the microbial consortia present in the reactor produce a polysaccharide gel which is responsible for producing the agglomeration of these microorganisms into the microbial flocs called activated sludge. The formation of such agglomerates results in the removal of the suspended solids present in the wastewater by incorporating them into the floc. Colloidal materials and other organic material can also be absorbed onto the floc. The microorganisms then proceed to attack all this material and to transform it primarily into biomass and carbon dioxide, with consequent decontamination of the wastewater.

The effluent from the aerated bioreactor, containing a significant amount of biomass, is fed to a settling tank or *clarifier* (sludge separator), a device capable of separating the clear supernatant from the bulk of the biomass. The supernatant is fed to a polishing treatment process (if necessary) or discharged. The biomass (sludge) is partially recycled to the aeration bioreactor. The excess is disposed of or treated separately to reduce its volume and water content, and to improve its stability. Excess sludge constitutes a significant waste product of the activated-sludge process, and several alternatives are used for its treatment and final disposal.

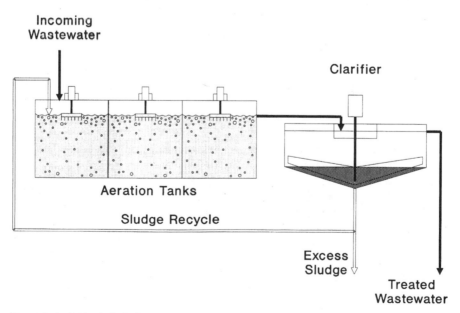

Figure 4.10 Activated-sludge process.

Under unfavorable conditions such as toxic loadings, temperature fluctuations, or pH changes, filamentous organisms dominate the microbial populations in the aeration tank. When this happens the process "bulks" and the resulting microbial flocs tend to remain suspended instead of settling. As a result, the clarifier is not capable of separating and recycling the biomass, and the entire process fails. Bulking can also be affected by the reactor design, as will be seen below.

Activated-sludge processes differ not in their basic operating principle (which is always the same) but in the configurations of the two primary components of the process, namely the aerated bioreactor, and the sludge/supernatant separator (clarifier). A large number of reactor designs now exist,and the most significant of these will be examined in the following sections.

Well-mixed activated-sludge reactors. The basic configuration of this type of reactor is a completely mixed and aerated tank in which the composition of the liquor is constant everywhere, as shown in Fig. 4.11a. This can be achieved quite easily in a small tank provided with an appropriately designed agitation system. If the tank volume is significant, multiple agitators or internal recycles and jets are used. These reactors typically consist of concrete basins in which aeration is provided by surface aeration or sparging with air diffusers. Details on the agitation, power, and aeration requirements were given above, under "Basic Bioreactor Configurations."

Well-mixed systems have the advantage of minimizing nutrient depletion (including oxygen depletion) in any part of the basin. In addition, they are quite tolerant of shock loadings since any fluctuation in the composition of the feed is dampened by the dilution produced when the feed is mixed with the reactor content.

If the wastewater to be treated contains a high concentration of easily degraded carbohydrates, the composition of the microbial population will be dominated by filamentous organisms, and bulking will be likely. Under these circumstances, the use of a well-mixed system is not recommended since this would tend to maintain optimal bulking conditions in the reactor. However, well-mixed aerobic tanks are commonly employed to treat wastewater with complex and more recalcitrant pollutants, such as those contained in the effluents of chemical plants.

From a reactor design point of view, continuous, well-mixed systems are equivalent to CSTRs. Therefore, they can be designed by writing steady-state mass balances for the biomass and the substrate (i.e., the organic loading of the feed), once the constraints for the system are imposed (such as the flow rate of wastewater to be treated, the degree of

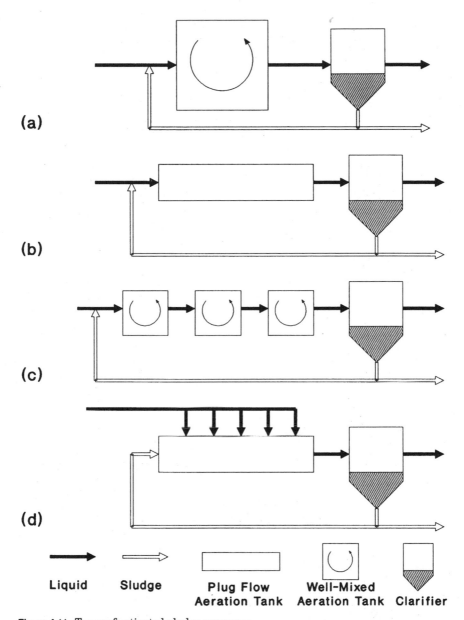

(a)

(b)

(c)

(d)

Liquid   Sludge   Plug Flow   Well-Mixed
                    Aeration Tank   Aeration Tank   Clarifier

Figure 4.11 Types of activated-sludge processes.

organic removal desired) and the kinetic rate constants for the degradation process are known. The mass balance for the substrate can be written as[5]

$$\frac{1}{\Theta_s} = \mu_{max} \frac{s_a}{s_a + K_s} - k_e$$

where Θ_s is the average residence time of the activated sludge in the system (the so-called *sludge age*), defined as the ratio of the sludge volume to the sludge removal rate; μ_{max} and K_s are the average Monod constants for the microbial population; k_e is the endogenous respiration rate constant; and C_a is the concentration of the pollutant in the reactor and in its effluent. According to its definition Θ_s is given by

$$\Theta_s = \frac{Vm_a}{(1 - \text{ß})Qm_e + \text{ß}Qm_r}$$

where V is the reactor volume; m_a, m_e, and m_r are the concentrations of biomass in the reactor and its effluent, in the effluent from the clarifier, and in the recycle stream, respectively; Q is the feed flow rate; and $(1 - \text{ß})Q$ and $\text{ß}Q$ are the flow rates of the sludge removed from the clarifier and the treated wastewater from the clarifier, respectively. Furthermore, from a mass balance for the substrate one gets

$$Vm_a = \frac{(C_0 - C_a)YQ\Theta_s}{1 + k_e \Theta_s}$$

where C_0 is the concentration of pollutants in the incoming feed and Y is the yield coefficient (amount of biomass produced per amount of pollutant removed). Finally, the aeration reactor volume can be calculated from

$$V = Q\Theta_s \left(1 + \alpha + \alpha \frac{m_r}{m_a}\right)$$

once the recycle ratio α (equal to the ratio of flow rate of the recycle stream fed back to the reactor to the incoming feed flow rate) has been fixed.

Some typical values for the kinetic parameters at 20°C are $Y = 0.5$; $\mu_{max} = 2$–6 days^{-1}; $K_s = 30$–300 mg/L; $k_e = 0.05$ days^{-1}. The sludge age is typically from 6 to 15 days.[5]

Plug-flow activated-sludge reactors. Plug-flow systems are typically more efficient to treat most wastewaters. In such systems the wastewater flows through long and narrow aeration tanks, as shown in Fig. 4.11b, thus minimizing any backmixing effect. In practice, it is difficult to obtain a true plug flow in any real system. Furthermore, the presence of air bubbles tends to increase the turbulence of the system and promote axial dispersion.

Plug flow typically results in the growth of a good-quality sludge with good settling characteristics. This is especially important in those cases in which the wastewater composition would promote the growth of filamentous organisms, as in the case in which it contains high levels of easily degradable materials that promote bulking. In this situation, the use of plug flow typically results in a rapid drop in pollutant concentration in the first part of the reactor, and minimizes the possibility that the concentration in the rest of the reactor will be such as to promote the growth of filamentous organisms. Hence, the sludge leaving the reactor has excellent settling properties and can be effectively separated in the clarifier.

One major drawback of plug-flow aeration tanks is their sensitivity to shock loading, since any increase in pollutant concentration in the wastewater is not distributed throughout the entire vessel but is confined to the slug of wastewater containing it. For this reason plug-flow reactors are often preceded by a smaller reactor operating as a CSTR, in which concentration fluctuations are partially dampened.

Activated-sludge cascade reactors.   In order to eliminate the sensitivity of plug-flow activated-sludge reactors to shock loading and still retain some of their positive flow characteristics several methods can be used. One is to "break down" the reactor into a series of smaller well-mixed reactors feeding each other in a cascade, as shown in Fig. 4.11.c. Other combinations include the use of a distribution system feeding the incoming wastewater to many points along the entire length of the plug-flow reactor (Fig. 4.11d), or the use of intermediate settling tanks for the partial separation and recycling of the activated sludge.

Oxidation ditches.   Oxidation ditches are essentially large-scale loop reactors in which aerobic degradation is carried out (Fig. 4.12). They typically consist of long (30 to 200 m), in-ground concrete basins having the shape of a running track, in which the wastewater moves in a circle. The feeding point of the incoming wastewater is just before one of the straight runs of the ditch. The wastewater is pumped horizontally and longitudinally in a plug-flow motion by submerged axial impellers. The water velocity is such (0.3–0.5 m/s) that most solids are kept in suspension and no anaerobic conditions are produced at the bottom of the ditch.[14] From a reactor design point of view, oxidation ditches can be thought of as plug-flow reactors with a significant recycle stream.

Air is sparged at a number of locations along the loop, in such a way as to form bubble curtains crossed by the water flow. As the bubbles rise, they move in cross-flow with respect to the water. By carefully spacing the air injection points, alternating oxygen-rich and oxygen-de-

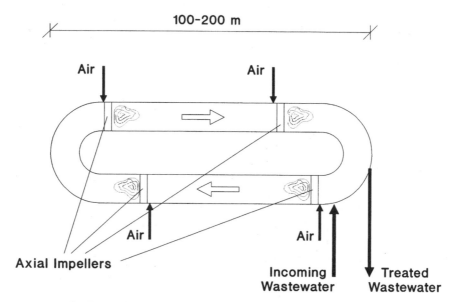

Figure 4.12 Oxidation ditch.

pleted regions can be created along the loop. Where oxygen is abundant, the oxidation of ammonium to nitrates (nitrification) can take place (in addition to a reduction of the BOD, as usual). Conversely, where oxygen is deficient, anoxic biological reactions are favored, such as the conversion of the nitrates formed during the nitrification process to elemental nitrogen (denitrification), thus completing the removal of the initial organic nitrogen loading. This constitutes an attractive feature of oxidation ditches. The effluent from the ditch is continuously removed at a point along the loop preceding the feeding point and sent to a clarifier where the biomass is partially separated and recycled, as in any other activated-sludge process.

These systems tend to be relatively insensitive to shock loading since such a load is more or less equally distributed throughout the loop as a result of the high ratio of water flow in the loop to incoming feed flow.

Other activated-sludge reactor configurations. In the traditional activated-sludge reactor configurations, aeration is carried out throughout the entire reactor volume. In addition, the air bubbles generated in these reactors have a relatively short residence time because of the small depth of the reactor. As a result, a significant portion of the expenses to operate aerobic reactors goes in power consumption for aeration and mixing. A number of alternatives have emerged over the years to eliminate this problem. The most obvious is to increase the reactor depth. This has two main advantages, namely an increase in the residence time of the air bubbles in the liquid, and an increase in the partial pressure of oxygen due to the increased hydrostatic pressure at the point where air is sparged (i.e., the bottom of the reactor). In addition, by using jets or gas spargers inserted in draft tubes it is possible to circulate the liquid without the need for any external mechanical agitation.

These improvements have resulted in the design of airlift bioreactors such as the *deep-shaft reactor,* in which a well drilled in the ground (up to 300 m deep) is used as a bubble column.[32,45] Because of the presence of a draft tube only a part of the reactor is used to aerate the wastewater. This, in turn, results in development of a strong liquid circulation pattern without using any impeller.

As previously mentioned, deep-shaft reactors have resulted in excellent oxygen transfer rates with reduced power consumption. Other alternatives, such as the BIOHOCH® reactor described above utilize similar principles to produce liquid circulation, air dispersion, and improved oxygen transfer rates.[32]

Enclosed, oxygen-rich activated-sludge reactors. Another way to increase the oxygen partial pressure (and hence the transfer rate) without building tall reactors is to increase the composition of oxygen in the gas mixture sparged in the reactor. Oxygen-rich mixtures are typically obtained from air (as raw material) using cryogenic distillation or pressure swing adsorption. In most cases it is convenient to build such an air separation unit on site.

Sparging can be achieved in a number of ways, but primarily by mechanical agitation (either submerged or surface). In order to maximize the retention time of the oxygen gas in the system, the reactors are closed. In addition, the reactors can be arranged in cascade to maximize the consumption of oxygen prior to gas venting. The gas phase (also containing oxygen) must be continuously purged to remove the carbon dioxide formed as a result of aerobic respiration. Oxygen-rich systems have found use in the treatment of wastewater from the pulp and paper, chemical, and related industries. In addition, this approach is used in the treatment of wastewater with potential odor problems (e.g., from fish-processing plants) because of the low volumetric rate of gas released.

Clarifiers and thickeners. Although clarifiers and thickeners are not bioreactors, they constitute an essential part of the activated-sludge process. Any of the activated-sludge processes described above relies on the separation of the biomass from the treated wastewater in order to (1) produce a clear effluent and (2) recycle part of the biomass to keep the process viable. These two essential steps are accomplished in the clarifier or thickener following the aeration tank. In fact, when an activated-sludge process fails to meet its effluent specifications, this typically results from the failure of the clarifier to perform as designed.

The difference between the terms *clarifier* and *thickener* is rather marginal—clarifiers being nearly identical to thickeners except that they are typically lighter in construction and used when the sludge volume is smaller.[38] A clarifier is used in most wastewater treatment applications.

The production of good-quality, rapidly settling sludge is essential for the good performance of a clarifier. Sludge sedimentation typically occurs as zone settling, in which the biomass flocs adhere to each other and settle as a blanket.[14] A clear interface can be observed between the sedimenting biomass and the clarified supernatant. Three zones can be identified during a batch sedimentation process: a top clear liquid, a bottom sedimented biomass which is undergoing compression as time goes by, and an intermediate zone where the biomass decreases its sedimentation velocity as it approaches the sedimented bottom layer. The sedimentation process initially proceeds at a constant velocity, before slowing down as the settled biomass begins to be compacted. The final compression-sedimentation process proceeds at a much slower pace. In the actual clarifier the same sedimentation process observed in batch can be expected.

Standard tests have been developed to determine the settling quality of the sludge. After 30 min of settling a good sludge should have a volume about 40 times that of the total suspended solids. If the sludge volume is instead much larger, then bulking is likely to occur in the clarifier.[5]

Clarifiers can be classified according to their shape, which is either rectangular or circular, and the position of the feeding point and the withdrawal points of the sludge and the clarified effluent. Rectangular clarifiers are elongated rectangular basins of rectangular cross section (2 to 30 m in width) with a length 3 to 5 times their width,[38] fed along one end. The liquid continuously moves through in the clarifier in plug flow at slow velocity, and in the process the biomass flocs sediment to the bottom. The clarified liquid is removed at the other end of the clarifier by passing over a weir. The sludge is moved by a series of scrapers attached to an endless chain which continuously rake the bottom of the clarifier and push the sludge in a direction opposite to that of the liquid. The sludge is collected at the bottom of the clarifier, in a sump placed at the bottom of the point where the incoming liquid is fed to the clarifier.[12]

Circular clarifiers are large cylindrical basins (3 to 150 m in diameter), with a bottom sloped toward the center of the basin, as shown in Fig. 4.13. They work on the same principle as the rectangular clarifiers. The feed is typically in the center. The water then moves radially and

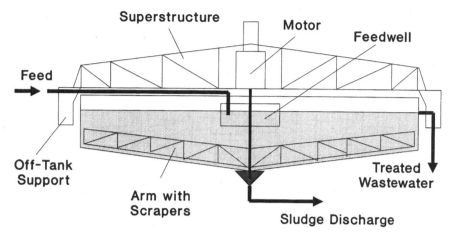

Figure 4.13 Clarifier.

is finally collected at the rim, where it flows over a circular weir placed around the clarifier. A ranking mechanism is used to scrap the clarifier bottom and push the sludge that has sedimented toward a central sump where it is collected.[14,38]

Sequencing batch reactors

A *sequencing batch reactor* (SBR) is a reactor in which an activated-sludge process is carried out in a time-oriented, sequential manner using a single vessel for all the phases of the process. The same steps involved in a conventional, continuous activated sludge process (such as aeration, pollutant oxidation, sludge settling, and recycling) are now conducted in batch one after the other.

In an SBR process, each cycle starts with the reactor nearly empty except for a layer of acclimated sludge on the bottom (Fig. 4.14). The reactor is then filled up with the wastewater and the aeration and agitation are started. The biological degradation process begins during the filling step and proceeds, once the reactor has been filled up, until a satisfactory level of degradation of the pollutant is achieved. Then the aeration and agitation are stopped, and the sludge begins to settle. Depending on the time allowed for the sedimentation, anaerobic reaction can occur, which may reduce the organic content of the sludge. Once the sludge has settled, the clear top layer of treated wastewater is discharged, and a new cycle can begin. Anaerobic sludge digestion may also be included as one of the steps in the cycle.[18]

From a reactor design point of view, an SBR is a conventional aerated basin, provided with aeration, agitation, and decanting systems that can accommodate the large changes in liquid level accompanying each cycle of the process.[49] This can be accomplished using floating

Wastewater
Feed

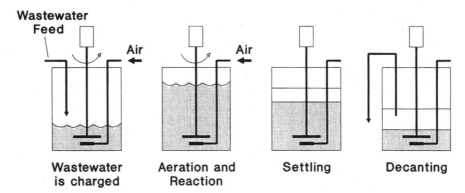

| Wastewater is charged | Aeration and Reaction | Settling | Decanting |

Figure 4.14 Operation of an SBR.

mixing and aeration equipment, and a floating decanting device connected to a pump to empty the reactor.[23]

The main advantage of SBRs is that they can accommodate large fluctuations in the incoming wastewater flow and composition without failing. The same may not be true in conventional activated-sludge processes, in which an increase in the incoming flow rate results in a lower residence time of the wastewater in the aeration tank and of the sludge in the clarifier, with potential failure of one of them or both. In addition, toxic shocks or significant changes in organic loading may produce alterations in the makeup of microbial populations of conventional activated-sludge processes, with consequent bulking or process failure. Instead, the wastewater residence time in SBRs can be extended until the microbial population has recovered and completed the degradation process. Similarly, the settling time can be varied to allow complete settling before discharging. In other words, SBR processes, like all batch processes, are more flexible.[50] On the other hand, the use of SBRs to treat a continuous wastewater flow requires the simultaneous use of multiple reactors and/or the presence of holding facilities to store the wastewater until an SBR becomes available. SBRs have been used also in denitrifying applications.[6]

Trickle-bed reactors

The *trickle-bed reactor,* often referred to as a *trickling filter* (in spite of the fact that it has nothing to do with mechanical filtration) has been used for wastewater treatment since its development by Corbett in England in the late 1890s.[32,48] Trickling filters can be classified as packed-bed reactors in which the wastewater trickles down the packing and is decontaminated by the biomass growing on the packing itself. A

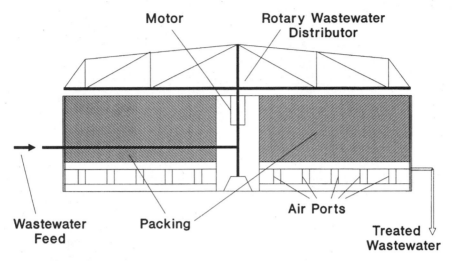

Figure 4.15 Trickling filter.

trickling filter such as that depicted in Fig. 4.15 usually consists of a large, open cylinder, 1 to 12 m high, loosely packed with a coarse material, 40 to 80 mm in size, such as lava rocks, slag, bricks, or plastic (typical void volume, 50 percent; typical surface area, 90–100 square meters per cubic meter of packing).[32] The reactor is provided with a water distribution system, such as a series of rotating arms placed above the packing, that evenly distributes the wastewater on the surface of the top layer of packing. Typical hydraulic loadings are in the range 1–4 m³/(day • m²) for low-rate filters to 10–50 m³/(day • m²) for high-rate filters. Organic loadings are between 0.1 and 5 kg BOD/(m³ • day).[44]

During the operation of the filter a biofilm is formed on the surface of the entire packing in the reactor. As the wastewater moves down the filter, it gets progressively decontaminated by the action of the biomass. The biomass that directly or indirectly lives on the organic nutrients contained in the wastewater forms a true ecosystem comprising bacteria, fungi, protozoa, and even worms and insect larvae, as well as blue-green algae on the illuminated surfaces.[19,32] Bacteria and fungi are primarily responsible for the removal of the organic loading in the wastewater. These organisms form a slime film on the packing, about 0.25 mm thick, which is aerobic from the air/film interface down to a depth where oxygen can penetrate (about 0.1–0.2 mm),[50] and is anaerobic below that point down to the surface of the packing. Superior organisms, such as insect larvae, prey on the lower organisms in the film, and play an important role in the growth and control of the biofilm.

In order to avoid plugging, the filter must be operated at an equilibrium point at which the rate of biomass growth is equal to the rate at which the biomass is rinsed away. In weakly loaded trickling filters, the flow rate to the filter is low and so is the rate of mechanical removal of biomass by the wastewater. Therefore, the BOD in the wastewater fed to these filters must be very low. On the other hand, in highly loaded filters the wastewater is mixed with a recycled stream from the outlet of the filter. The resulting flow is so high that a significant amount of biomass is mechanically removed from the filter. Therefore, these filters can take wastewaters containing a higher level of BOD without plugging.

The lower level of the reactor is provided with openings through which the air required to provide the microorganisms with oxygen can pass. This air then moves through the filter by natural convection (upward in the winter, downward in the summer) as a result of the temperature difference between the packing and the external air.

As in the activated-sludge process, a final settling tank is provided to separate the treated wastewater from the biomass sludge rinsed away from the filter. Trickling filters tend to be superior to conventional activated-sludge processes in terms of operating costs, clarity of effluent, and sensitivity to shock loading. However, they require a higher capital investment, are more sensitive to external temperature changes, and may present significant insect and odor problems.

Rotating biological contactors

Rotating biological contactors (RBCs) are another class of reactors that, like trickling filters, utilize a biofilm exposed to the wastewater to remove the pollutants. In this case, the biofilm is established on the surface of large plastic discs (1.7–3.7 m in diameter) mounted on a horizontal shaft slowly rotating at about 2 to 5 rotations per minute, with its axis placed horizontally just above the surface of the liquid to be treated (Fig. 4.5). As a result, about 40 percent of the disc area is always immersed in the wastewater. Because of the rotation the different sectors of the discs are alternatively exposed to wastewater and air, thus allowing the biofilm to be oxygenated without air compression and sparging. The biofilm formed on the discs is about 0.3 to 4 mm thick and is in contact with the wastewater not only when it is immersed in it but also when it is in the air, since a thin film of liquid is always carried along.[44,50]

In continuous treatment plants, the wastewater moves through longitudinal basins in a direction parallel to the rotating shaft and perpendicular to the discs. Typically, more than one of such units (stages) are used in series. The discs are tightly packed on the shaft so as to maximize the surface area per unit shaft length, and hence per unit wastewater volume (typical area/volume ratio: 100 m^2 to 1 m^3).[44]

A relationship has been developed to determine the contamination removal accomplished by RBCs:[15]

$$\frac{C_{out}}{C_{in}} = 14.2 \left[\frac{(Q/A)^{0.558}}{\exp(0.320N_s)\,C_{in}^{0.684}\,T^{0.248}} \right]$$

where C_{out} and C_{in} are the concentration of the pollutants (in g/m^3 BOD), Q is the mass rate of pollutant in g/day, A is the disc area in ft^2, N_s is the number of stages, and T is the temperature in °C.

The biofilm is periodically sloughed off as a result of it growth, which reduces the amount of nutrients reaching the microorganisms attached to the disc, and makes them enter the endogenous respiration phase. This results in their inability to cling to the disc surface. A clarifier typically follows an RBC unit to remove the treated wastewater from the excess biomass sloughed off.

RBCs are typically associated with low energy consumption and good tolerance to shock loading.[36] However, they are quite sensitive to temperature fluctuations.

Aerobic digestors

The sludge from primary treatment or the excess sludge produced during any of the secondary aerobic wastewater treatments can be processed in a number of ways to reduce its organic content prior to its final disposal. One method is *aerobic digestion,* in which the cellular protoplasm of the biomass in the sludge is oxidized through the process of endogenous respiration. This occurs when the biomass has exhausted its external supply of oxidizable matter and begins to use its own cellular material as a source of nutrients. About 75 to 80 percent of the cell mass can be oxidized this way to carbon dioxide, ammonia, salts, and water, the remainder being made of stable, inert material not easily degraded any further.

Aerobic digestors are typically batch reactors charged with aerobic sludge and aerated for an extended period of time until the desired amount of organic removal has been achieved. The process can be operated with air or an oxygen-rich gas mixture in a closed vessel. The processing time is quite long, typically in excess of 30 days. Thermophilic bacteria, operating at a temperature between 25 and 50°C, can also be used. The increased temperature has a positive effect on the kinetics of the process, which can be carried out in a much shorter time (3 to 4 days).[9] Alternatively, the digestor can be operated continuously. In this case a clarifier is added to the aeration tank, similar to the activated-sludge process.

The digestion process can be modeled by a first-order kinetics. For batch digestion the process can be expressed as[1]

$$\frac{X_{end}}{X_{beg}} = e^{-kt}$$

where X_{beg} and X_{end} are the concentrations of the biodegradable fraction of suspended solids at the beginning and end of the process, respectively; k is the kinetic constant of the digestion process; and t is the degradation time. A continuous process in a well-mixed system can be represented by

$$\frac{X_{out}}{X_{in}} = \frac{1}{1 + kt_{res}}$$

where X_{in} and X_{out} are the concentrations of the biodegradable fraction of suspended solids in the incoming and outgoing streams, respectively; and t_{res} is the residence time of the material in the reactor.

The oxygen consumed in the process can be estimated as 1.4 kilograms of oxygen per kilogram of biomass degraded. Given the long residence time required by the process, this typically implies that low-power agitators (0.01–0.02 kW/m^3) and surface aeration can be used.

Other aerobic reactors

In addition to the reactor configurations described above, all the reactor types described in the previous section (fluidized-bed reactors or jet reactors) can be used. Specialized applications may require the adaptation of the reactor to the constraints of the process. This may especially be the case for the treatment of hazardous waste materials using specialized organisms. For example, in one instance, a fluidized-bed reactor with separate aeration has been successfully used to treat a stream heavily contaminated with methylene chloride using a pure culture of *Hyphomicrobium*.[40] The use of external aeration was dictated by the need to minimize air stripping of the toxic, volatile compound. Other similar applications are likely to appear in the future as bioremediation of streams contaminated with toxic or hazardous material becomes more common.

Anaerobic Reactor Systems

Anaerobic ponds

Anaerobic ponds and lagoons are water basins similar to the aerobic lagoons and ponds described above, but in which anaerobic conditions are established. This can be achieved if the organic loading of the incoming wastewater is especially high, if the lagoon is not agitated, and if it is deep enough for the establishment of anaerobic conditions throughout its depth (except for the very top layer near the surface).

These ponds can be quite large, from a few dozen square meters to several square kilometers. Typical depths for these ponds are from 1 to 5 m. Loadings to anaerobic ponds have been reported to be as high as 2.95 kg BOD/(m³ • day) with an initial BOD of about 3000 mg/L. Residence times also vary significantly from 3 to 4 days to up to 245 days.[14]

In some cases the addition of nutrients, such as nitrogen and phosphate, to the wastewater may be necessary to the proper maintenance of the nutritional balance required by anaerobic microorganisms. Odor control can be a problem, but odors can be minimized by adding sodium nitrate.[14]

Anaerobic ponds can also be used as settling ponds in which settling material is removed by sedimentation to the bottom of the pond and anaerobically digested before the wastewater is passed to an aerobic lagoon.

Anaerobic bioreactors for wastewater treatment

Anaerobic organisms have been used since the late 1880s to treat wastewaters containing large amounts of suspended solids.[48] Because they present many advantages over aerobic processes, such as low sludge production, low power consumption, and the generation of methane as a valuable by-product, anaerobic process are now used to treat a variety of highly contaminated wastewaters. Nevertheless, the use of anaerobic systems has not been as widespread as one might expect, most likely because of the problems commonly associated with their operation, such as sensitivity to toxic pollutants. Indeed, the sequential operation of anaerobic consortia makes them (and especially the methanogens which are the last organisms in the anaerobic chain) more vulnerable to process upsets. However, the poor reputation that some anaerobic processes still have may have also partially resulted from the failure of operators and designers to take advantage of the knowledge accumulated over the last twenty years on the microbiology of anaerobic consortia. It is likely that as this knowledge is incorporated into the design and operation of new plants (as has begun to happen in the last few years), anaerobic treatment processes will become an increasing popular alternative in wastewater treatment.

Like their aerobic counterparts, anaerobic reactors for wastewater treatment can be classified according to the way they retain the biomass within the reactor system. The different types of reactors will now be examined in some detail.

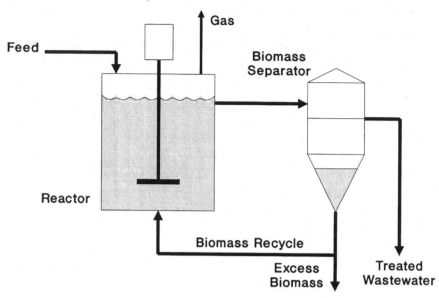

Figure 4.16 Anaerobic contact process.

Anaerobic contact process reactors. The *contact process* is the anaerobic equivalent of the aerobic activated-sludge process. The basic reactor configuration consists of a closed vessel, typically provided with agitation, in which the biomass is suspended in flocs and to which the wastewater is continuously fed. The wastewater retention time is much smaller than the average biomass retention time because the reactor is followed by a settling tank where the biomass is separated from the supernatant and partially recycled to the reactor. A schematic of this process is given in Fig. 4.16.

The kinetics of anaerobic digestion is quite sensitive to temperature. Therefore, the reactor may be provided with a heated jacket or a heating element to deliver the required energy to keep the temperature higher than ambient. This may have a significant impact on the operation of the reactor. If the reactor is not heated, the system hydraulic retention time can be of the order of 30 to 60 days. Under mesophilic conditions (32 to 37°C) the retention time can be reduced to 10 to 15 days, and at even higher temperatures (60°C, corresponding to thermophilic conditions) can be lowered to 4 days.[39]

The pH in the reactor must be maintained within acceptable limits for the conversion to proceed—i.e., between 6.7 and 7.8. If required, pH control can be achieved by lime additions to the feed.

A typical reactor has a cylindrical shape with conical or dish bottom and is provided with agitation to homogenize the mixture and suspend the biomass. The gas produced during the process (primarily methane and carbon dioxide) is removed from the top of the reactor. Significantly more complex can be the design of the settling tank or biomass separation device. Anaerobic flocs tend to have low density and contain entrapped gas bubbles produced by the anaerobic metabolism. Therefore, they settle very slowly. In addition to traditional settling tanks with conical bottoms other separation devices can be used, such as lamellar separators, in which the slurry from the reactor is passed through slanted channels which promote the coalescence of the suspended solids against the lamellas forming the channels and their sedimentation to the bottom of the reactor. Additional methods to promote solid separation, such as vacuum degassing or centrifugation, have also been attempted.[28]

High biomass content (5 g/L) and significant BOD removal (80 percent of an initial BOD of 5600 mg/L), with a reactor capacity of 3 to 5 kg COD/(m^3 • day) can be achieved with this type of reactor.[28]

Anaerobic filter reactors.   A schematic for the *anaerobic filter reactor* is provided in Fig. 4.17. Despite the name, the reactor can be classified as a packed-bed reactor. This reactor has some similarity to the aerobic

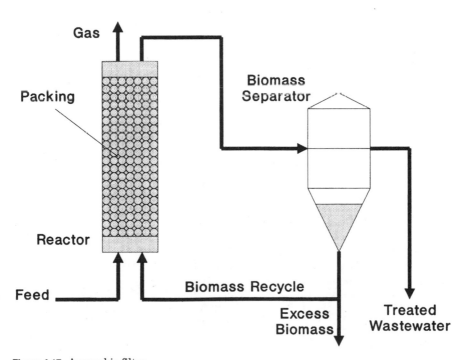

Figure 4.17 Anaerobic filter.

trickling filter, in that the biomass is attached to a solid support to prevent washout.[52] However, the anaerobic filter is completely filled up with liquid flowing either upward or downward (except for the gas formed during the process). Different packing materials can be used, such as Rashig rings, gravel, and plastic packing of different types and shapes.[17] The packing is generally rather coarse (2 to 6 cm in size) since anaerobic filters have a tendency to clog. This can occur because the anaerobic organisms can form large flocs instead of a thin film attached to the surface of the packing, filling the interspace between the packing, and producing channeling effects and biomass sloughing.[32] Because of the low radial diffusion, packed beds typically operate in plug flow. As previously noted, this makes these systems more vulnerable to toxic shocks and the effects of feed composition changes.

However, anaerobic filters have a number of advantages, namely high efficiency at low organic loading, high loading capacity, overall stability in the presence of toxic substances, and low energy requirements. COD reductions of 4 to 10 $kg/(m^3 \cdot day)$ with residence times of the order of 4 to 18 h have been reported.[45]

UASB process reactors.   The acronym UASB stands for *upflow anaerobic sludge blanket.* The process, which operates using a column reactor without any packing material, can be described as a system with an internal biomass recycle based on gravity settling.

As shown in Fig. 4.18, the wastewater feed injected at the bottom of the column first meets a thick layer (1.5–2.5 m) of biomass granules

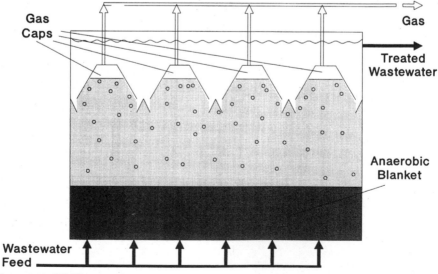

Figure 4.18 UASB reactor.

produced by the anaerobic organisms under the appropriate conditions. These granules, forming a "blanket" of high biomass content (from 60–70 g/L[28] up to 100–150 g/L[14]), are primarily responsible for the removal of the pollutants. The granules have a high enough density to enable them to be retained in the lower portion of the column without being entrained by the wastewater rising through the reactor. Above this biomass layer there is a 2- to 4-m region containing individual organisms, smaller flocs, and small granules, having a biomass concentration of about 15–25 g/L. This biomass is kept in suspension and moved upwards by the liquid flow and by the rising gas bubbles formed during the anaerobic process. This region is capped by bell-shaped gas-collecting devices that allow only the liquid to pass to the uppermost part of the reactor. Above the gas-collecting devices there is a 2-m calming region that enables a large fraction of the solids to sediment back into the lower regions where they can continue their metabolic activity.

This type of reactor has been used in industrial applications with removal efficiencies of about 4–10 kg COD/(m^3 • day). However, even higher rates [96 kg COD/(m^3 • day)][14] have been reported. Retention times are of the order of 3 to 8 h.

Anaerobic fluidized-bed reactors. In anaerobic fluidized-bed reactors the biomass retention within the reactor is achieved by immobilizing the microorganisms on small support particles which can still be suspended by an upward moving fluid, but which are dense enough to avoid being swept away with the effluent. Solids such as carbon particles or sand can be used for this purpose.

Fluidized-bed reactors have only relatively recently been used in full-scale anaerobic wastewater treatment.[25] In a particular application fluidized-bed reactors have been used to degrade the effluent from a brewer's yeast plant. These reactors had a 4.6-m diameter, a height of 21 m (including a 13-m packed-bed height) and 215 m^3 of fluidized-bed volume. These reactors operated at a very high biomass concentration (40 g/L), superficial liquid velocity (10–30 m/h), and COD removal efficiency [20–27 kg/(m^3 • day)].[20] Good performance of this type of reactor has also been reported by other sources.[14]

Other anaerobic reactors. In principle, a number of reactor configurations can be used to design anaerobic reactors. The primary difference will consist in the way the biomass is retained within the reactor, separated from the clarified stream and recycled, and/or immobilized on a solid support material. Examples of other types of reactors can be found in the literature.[2,27,45]

Anaerobic sludge digestion

All the aerobic wastewater treatment processes described previously produce a significant amount of excess sludge that must be disposed of. One common practice is to use anaerobic sludge digestion to reduce the sludge volume and its organic content. Since anaerobes, and especially methanogens, are less efficient users of the organic material contained in the sludge, the anaerobically digested sludge contains a smaller amount of cellular material. This results in a digested sludge which is typically very stable, nonputrescible, very low in pathogen content, and can be landfilled or even used as fertilizer. In addition, anaerobic digestion is also applied to the sludge produced during primary treatment. This sludge is usually less homogeneous and more difficult to degrade than activated sludge.

Anaerobic sludge digestion is carried out either in batch or continuous processes. In its simplest and oldest form the sludge digestor is simply a closed basin (sometimes even a pit) into which the sludge is transferred and then allowed to undergo digestion for a long period of time. The process can even be operated continuously in reactors of variable sizes (500 to 10,000 m^3).[32] If the basin is unheated and unmixed, the process can be expected to take from 30 to 60 days.[9] More technologically advanced reactors include provisions for feed preheating, slurry mixing and sludge dispersion, inoculation, temperature control, pH control, gas collection and removal. Sludge dispersion can be accomplished in a number of ways such as mechanically agitating the mixture with impellers (with or without a draft tube), recycling the gas in the head space to the bottom of the reactor and dispersing it with nozzles or through draft tubes to homogenize the sludge/liquid slurry, or using an external recycle to pump the liquid from the upper part of the reactor to the bottom, where it is ejected with a jet nozzle. Some of these reactor configurations were examined above, under "Basic Reactor Configurations."

Good mixing and temperature control can significantly speed up the process. By operating in the mesophilic range (32 to 37°C), the retention time can be lowered to 15 days, with biomass concentrations of 1–4 g/L. In addition, since the methane produced in the process has a significant calorific value (23,100 kJ/m^3), it can be used to maintain the proper temperature in the reactor.

Mixed Systems

Mixed systems are defined here as those waste treatment processes that utilize both aerobic and anaerobic organisms to achieve the desired objective of producing an environmentally accepted and stable final waste product. As more knowledge becomes available on the microbiology of each of the two classes of microorganisms, they are likely to be selectively used to solve more difficult waste treatment problems by exploiting the specific degradation potentials of each group. In turn, this will require the design of appropriate reactor configurations capable of maintaining the desired conditions for the microbial activity to take place. Some of these systems are now examined.

Facultative ponds

The facultative pond is probably one of the oldest applications of mixed systems. *Facultative ponds* are nonagitated, nonaerated, extended water basins in which aerobic conditions are established in the top layer of the pond and anaerobic conditions prevail in the bottom layers. The establishment of such conditions is a function of the organic loading to the pond, its depth, and its temperature. Such ponds have a typical depth of 1 to 2 m, which is roughly intermediate between the depth of aerobic and anaerobic ponds. Embankments are typically built with slopes ranging between 15 and 25 percent,[14] with materials such as gravel and stones that can withstand erosion, and with provisions for ample fluctuations in volumetric capacity.

Facultative ponds have a rich variety of microbial populations, ranging from algae and aerobic bacteria in the top layers, to facultative organisms, to strict anaerobic consortia in the bottom layers and in the sediments. The thickness of the layer in which aerobic organisms will prevail will depend on the depth to which oxygen will penetrate or will be produced by algal photosynthesis. Therefore, the thickness of this layer will change significantly from day to night, and from season to season, as a result of the change in intensity of the incoming solar radiation. Furthermore, high organic loadings will increase the thickness of the anaerobic layer. A relationship between these factors and the depth of oxygen penetration has been established.[37] If the organic loadings become excessive [about 4000 kg BOD/(km^2 • day) for turbid wastewaters], the oxygen layer will all but vanish and anaerobic conditions will be established throughout the pond. This is likely to result in odor problems.

Facultative lagoons

Facultative lagoons are similar to facultative ponds in that they are stratified in anaerobic and aerobic layers. However, they are typically also equipped with surface agitators designed to mix and provide oxygen only to the top, aerobic layer. Therefore, the power delivered by such agitators is typically less than 1 kW/m^3, well below the power level in aerated lagoons. Because of the depth of the lagoon (comparable to that of aerated lagoons) this agitation intensity is not sufficient to suspend the sediment settling to the bottom of the lagoon, thus ensuring that the lower layer is anaerobic.

Facultative lagoons can be used as first receiving basins (for example, in the treatment of highly colored wastewaters that will not allow light to penetrate too deeply within the liquid bulk) or in lagoon systems (as described above), in which they typically receive wastewaters already pretreated in aerated lagoons.[13]

Activated-sludge/anaerobic-sludge digestion process

The individual components of these two processes were examined in the previous sections. Here, it is important to note that the *complete* treatment of wastewater using this approach (which is the most commonly used) necessitates a rapid aerobic oxidation process in which the organic loading is oxidized to carbon dioxide and partially converted to biomass (activated-sludge process), followed by the slower anaerobic reductive transformation of the waste product of this treatment process (i.e., biomass) to carbon dioxide and methane (anaerobic sludge digestion).

Combined anaerobic-aerobic processes for the treatment of high-strength effluents

As previously noted, the amount of biomass produced in any type of treatment can be considered proportional to the amount of waste removed. However, aerobic organisms are more efficient and rapid utilizers of the organic nutrients typically available in wastewater. Therefore, aerobic processes are more cost effective in the treatment of low-strength wastewaters, when the advantages associated with the efficiency of the waste removal offset the cost associated with the handling and treatment of the excess sludge generated in the process. In addition, anaerobic treatment alone does not always result in a reduction of the BOD levels low enough for effluent discharge.

However, as the concentration of pollutant becomes significant, the low ratio of biomass to BOD removal provided by anaerobes becomes attractive. In such cases, it can be cost effective to subject the wastewater to a first anaerobic treatment followed by an aerobic polishing stage. This was recently shown by Eckenfelder et al.,[14] whose economic analysis pointed out that if the wastewater has a BOD in excess of 1000 mg/L a combined anaerobic-aerobic process can be advantageous.

This approach has been used in different applications,[10,22] including a recent one involving the combined use of powered activated carbon and both anaerobic (first) and aerobic (second) stages.[11] These applica-

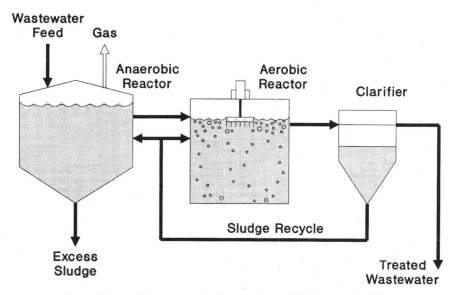

Figure 4.19 Anaerobic-aerobic process for the treatment of high-strength effluents.

tions were developed primarily to treat high-strength wastewater. In all these cases the reactors used for each stage were of the type described above for each class of organisms. An example of an anaerobic-aerobic process is shown in Fig. 4.19.

Combined anaerobic-aerobic processes for the treatment of hazardous waste

In addition to the advantages mentioned above, anaerobes can have an additional feature that makes them attractive in waste treatment. Anaerobic organisms have recently been shown to be responsible for a number of reductive reaction processes that could have a significant impact on the treatment of certain classes of hazardous compounds. In particular, anaerobic organisms have been shown to be capable of reductively dehalogenating a number of toxic compounds, such as chlorinated aromatics, that are very recalcitrant to aerobic degradation.[33,42,43] However, as the process proceeds and the molecule becomes more dechlorinated, the ability of the anaerobes to further dehalogenate it decreases. Fortunately, the resulting dehalogenated compounds can be degraded aerobically. Therefore, a possible alternative for the treatment of such compounds is their sequential exposure to specialized anaerobic and aerobic cultures, as shown in Fig. 4.20. If the process is operated continuously, it requires the sequential use of two reactors maintained under anaerobic and aerobic conditions, respectively. Because of

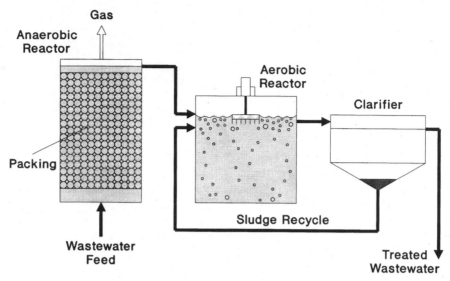

Figure 4.20 Anaerobic-aerobic process for the treatment of hazardous waste.

the relatively low concentration of the toxic compounds reactors with immobilized biomass appear to be favored. The validity of this approach
has recently been shown.[3,16,26] However, the process is still at the research stage.

References

1. Adams, C. E., W. W. Eckenfelder, and R. C. Stien. 1974. Modification to Aerobic
   Digestor Design. *Water Res.* 8:213–225.
2. Aivasidis, A., and C. Wandrey. 1988. Recent Developments in Process and Reactor
   Design for Anaerobic Wastewater Treatment. *Water Sci. Technol.* 20:211–218.
3. Armenante, P. M., D. Kafkewitz, G. Lewandowski, and C. M. Kung. 1992. Integrated
   Anaerobic-Aerobic Process for the Biodegradation of Chlorinated Aromatic
   Compounds. *Environ. Progr.* 11:113–122.
4. Armenante, P. M., G. Lewandowski, and I. U. Hag. 1992. Mineralization of 2-
   Chlorophenol by *P. Chrysosporium* Using Different Reactor Configurations,
   Hazardous Waste Hazardous Material. 9(3):213–229.
5. Bailey, J. E., and D. F. Ollis. 1986. Biochemical Engineering Fundamentals, 2nd ed.
   McGraw-Hill, New York.
6. Baltzis, B., G. Lewandowski, and S. Sanyal. 1991. Sequencing Batch Reactor Design
   in a Denitrifying Application. *In* D. W. Tedder and F. Pohland (eds.), Emerging
   Technologies in Hazardous Waste Management II. ACS Symposium Series.
   American Chemical Society, Washington, D.C.
7. Bates, R. L., P. L. Fondy, and J. G. Fenic. 1966. Impeller Characteristics and Power.
   In V. W. Uhl and J. B. Gray (eds.), Mixing, Vol. 1. Academic Press, New York.
8. Chaudhry, G. R., and S. Chapalamadagu. 1991. Biodegradation of Halogenated
   Organic Compounds. *Microbiol. Rev.* 55:59–79.
9. Cheremisinoff, P. N. 1990. Biological Treatment of Hazardous Waste, Sludges and
   Wastewater. *Pollut. Eng.* 22(5):87–94.

10. Collivignarelli, C., G. Urbini, A. Farneti, A. Bassetti, and U. Barbaresi. 1990. Anaerobic-Aerobic Treatment of Municipal Wastewaters with Full-Scale Upflow Anaerobic Sludge Blanket and Attached Biofilm Reactors. *Water Technol.* 22:475–482.
11. Copa, W. M., and T. J., Vollstedt. April 24, 1990. Two-Stage Anaerobic/Aerobic Treatment Process. U.S. Patent 4,919,815.
12. Daigger, G. T., and R. E. Roper. 1985. The relationship between SVI and the Activated Sludge Settling Characteristics. *J. Water Pollut. Contr. Fed.* 57:859–866.
13. Dienemann, E. A., J. F. Magee II, D. S. Kosson, and R. C. Ahlert. 1987. Rapid Renovation of a Sludge Lagoon. *Environ. Progr.* 6:158–165
14. Eckenfelder, W. W., Jr. 1989. Industrial Water Pollution Control. McGraw-Hill, New York.
15. Eckenfelder, W. W., Jr., Y. Argaman, and E. Miller. 1989. Process Selection Criteria for the Biological Treatment of Industrial Wastewaters. *Environ. Progr.* 8(1):40–45.
16. Fathepure, B. Z., and T. M. Vogel. 1991. Complete Degradation of Polychlorinated Hydrocarbons by a Two-Stage Biofilm Reactor. *Appl. Environ. Microbiol.* 57:3418–3422.
17. Genung, R. K., C. W. Hancher, A. L. Rivera, and M. T. Harris. 1982. Pp. 365–380. *In* E. L. Gaden, Jr. (ed.), Biotechnology and Bioengineering Symposium No. 12. Wiley, New York.
18. Goronszy, M. 1979. Intermittent Operations of the Expanded Aeration Process for Small Systems. *J. Water Pollut. Contr. Fed.* 41:279–287.
19. Grady, C. P. L., Jr., and H. C. Lim. 1980. Biological Wastewater Treatment. Marcel Dekker, New York.
20. Heijnen, J. J. 1984. Technik der Anaeroben Abwasserreinigung. *Chem.-Ing.-Tech.* 56:526–532.
21. Hines, D. A., M. Bailey, J. C. Oursby, and F. C. Roesler. 1975. Novel Aeration System Bows. *Water Wastes Eng.* 12(12):59–64.
22. Hoffman, C. A., W. M. Copa, and M. R. Mayer, Dec. 2, 1986. Method for Anaerobic Treatment of High Strength Liquors. U.S. Patent 4,626,354.
23. Irvine, R. L., and A. W. Busch. 1979. Sequencing Batch Reactors: An Overview. *J. Water Pollut. Contr. Fed.* 51:235–243.
24. Jackson, M. L. 1964. Aeration in Bernoulli Types of Devices. *AIChE J.* 10:836–842.
25. Jewoll, W. J, M S Switzenbaum, and J. M. Morris. 1981. Municipal Wastewater Treatment with the Anaerobic Attached Microbial Film Expanded Bed Process. *J. Water Pollut. Contr. Fed.* 53:482–490.
26. Kafkewitz, D., P. M. Armenante, G. Lewandowski, and C. M. Kung. 1992. Dehalogenation and Mineralization of 2,4,6-Trichlorophenol by the Sequential Activity of Anaerobic and Aerobic Microbial Populations. *Biotechnol. Lett.* 14:143–148.
27. Kobayashi, H. A., E. Conway de Macario, R. S. Williams, and A. J. L. Macario. 1988. Direct Characterization of Methanogens in Two High-Rate Anaerobic Biological Reactors. *Appl. Environ. Microbiol.* 54:693–698.
28. Lettinga, G., A. F. M. van Velsen, S. W. Hobma, W. De Leeuw, and A. Klapwijk. 1980. Use of the Upflow Sludge Blanket (USB) Reactor Concept for Biological Wastewater Treatment, Especially for Anaerobic Treatment. *Biotechnol. Bioeng.* 22:699–734.
29. Levenspiel, O. 1972. Chemical Reaction Engineering, 2nd ed. Wiley, New York.
30. Lewandowski, G., P. M. Armenante, and D. Pak. 1990. Reactor Design for Hazardous Waste Treatment Using a White Rot Fungus. *Water Res.* 24:75–82.
31. Middleton, J. C. 1985. Chapter 17: Gas-Liquid Dispersion and Mixing. pp. 322–355. *In* N. Harnby, M. F. Edwards, and A. W. Nienow (eds.), Mixing in the Process Industries. Butterworths, London.
32. Mudrack, K., H. Sahm, and W. Sittig. 1987. Environmental Biotechnology. pp. 623–660. *In* P. Praeve, U. Faust, W. Sittig, D. A. Sukatasch (eds.), Fundamentals of Biotechnology—1987. VCH, Weinheim, Federal Republic of Germany.
33. Neilson, A. H., A. Allard, P. Hynning, and M. Remberger. 1988. Transformations of Halogenated Aromatic Aldehydes by Metabolically Stable Anaerobic Enrichment Cultures. *Appl. Environ. Microbiol.* 54(9):2226–2232.

34. Ngian, K. F., and W. R. B. Martin. 1980. Bed Expansion Characteristics of Liquid Fluidized Particles with Attached Microbial Growth. *Biotechnol. Bioeng.* 22:1843–1856.
35. Oldshue, J. Y. 1983. Fluid Mixing Technology. McGraw-Hill, New York.
36. Opatken, E. J., H. J. Howard, and J. J. Bond. 1988. Biological Treatment of Leachate from a Superfund Site. *Environ. Progr.* 7:12–18.
37. Oswald, W. J. 1968. Advances in Anaerobic Pond System Design. *Water Resour. Symp.* 1:409–426.
38. Perry, R. H., and D. W. Green. 1984. Perry's Chemical Engineers' Handbook, 6th ed. McGraw-Hill, New York.
39. Pfeffer, J. T., and J. C. Liebman. 1976. Energy from Refuse by Bioconversion, Fermentation and Residue Disposal Processes. *Resour. Recovery Conserv.* 1:295–313.
40. Pierce, G. E. (Celgene Corp.). 1992. Personal communication.
41. Prokop, A., P. Janik, M. Sobotka, and V. Krumphanzl. 1983. Hydrodyanamics, Mass Transfer and Yeast Culture Performance of a Column Bioreactor with Ejector. *Biotechnol. Bioeng.* 25:1147–1160.
42. Quensen, J. F., J. M. Tiedje, and S. A. Boyd. 1988. Reductive Dehalogenation of Polychlorinated Biphenyls by Anaerobic Microorganisms from Sediments. *Science.* 242:752–756.
43. Reineke, W., and H. J. Knackmuss. 1980. Hybrid Pathway for Chlorobenzoate Metabolism in *Pseudomonas* sp. B13 Derivatives. *J. Bacteriol.* 142:467–473.
44. Rich, L. G. 1973. Environmental Systems Engineering. McGraw-Hill, New York.
45. Schuegerl, K. 1987. Bioreaction Engineering, vol. 2. Wiley, New York.
46. Suflita, J. M., A. Horowitz, D. R. Shelton, and J. M. Tiedje. 1982. Dehalogenation: A Novel Pathway for the Anaerobic Biodegradation of Haloaromatic Compounds. *Science.* 218:1115–1117.
47. Tojo, K., and K. Miyanami. 1982. Oxygen Transfer in Jet Mixers. *Chem. Eng. J.* 24:89–97.
48. Verstraete, W. H., and D. Schowanek. 1987. Aerobic versus Anaerobic Wastewater Treatment. *In* Proc. 4th Europ. Congress on Biotechnology, vol. 4, pp. 49–63.
49. Weber, A. S., and Matsumoto. 1987. Feasibility of Intermittent Biological Treatment for Hazardous Waste. *Environ. Progr.* 6:166–171.
50. Wentz, C. 1989. Hazardous Waste Management. McGraw-Hill, New York.
51. Wu, A. C., E. D. Smith, and Y. T. Hung. 1980. Modeling of Rotating Biological Contactor Systems. *Biotechnol. Bioeng.* 22:2055–2064.
52. Young, J. C., and M. F. Dahab. 1982. Operational Characteristics of Anaerobic Packed-Bed Reactors. pp. 303–316. *In* E. L. Gaden, Jr. (ed.), Biotechnology and Bioengineering Symposium No. 12. Wiley, New York.
53. Zlokarnik, M. 1979. Scale-Up of Surface Aerators for Waste Water Treatment. pp. 157–180. *In* T. K. Ghose, A. Fichter, and N. Blackebrough (eds.), Advances in Biochemical Engineering, vol. 11. Springer Verlag, Berlin.

5

Modeling Biological Processes Involved in Degradation of Hazardous Organic Substrates

Bruce E. Rittmann

Professor of Environmental Engineering
John Evans Northwestern University
Evanston, Illinois

Pablo B. Sáez

Associate Professor of Environmental Engineering
Pontificia Universidad Catolica de Chile
Santiago, Chile

This chapter develops the conceptual and quantitative foundations for modeling biological processes in which biodegradation of hazardous organic compounds occur. All models of biological processes must account for the growth and maintenance of active biomass, which requires utilization of primary electron-donor and electron-acceptor substrates to gain energy for cell synthesis and maintenance. Therefore, a model must begin with mass balances on active biomass and the primary substrates. Although the hazardous compounds may sometimes be primary substrates, they often cannot serve this role and must be removed by secondary utilization. Therefore, a separate mass balance must be provided for each secondary substrate. Utilizations of primary and secondary substrates are linked, because the primary substrates support growth of the biomass able to utilize the secondary substrates. The final key feature is to account for direct substrate interactions, such as inhibition and cosubstrate requirements. Secondary substrates can alter the growth of biomass by inhibiting primary-substrate utilization and cell yield, while primary substrates can directly inhibit the transformation of secondary substrates. These inhibitory effects are modeled through kinetic parameters, whose values are changed by the presence of the other (inhibitory) component. Cosubstrate effects occur when the degradative enzyme requires the target substrate and other cosubstrate. Several examples of cosubstrate kinetics are provided. An example illustrates model formulation and demonstrates how substrate interactions can significantly affect process performance.

Introduction

The core of modeling for biotreatment applications involves formulating and solving mass-balance equations for the important biological and chemical species. Because biotreatment systems can be too complex to comprehend by purely qualitative understanding, mathematical modeling is the essential tool for developing and applying biotreatment.

The process of model formulation provides the intellectual glue that connects the many materials and phenomena acting in a biotreatment system. Formulation identifies what factors are truly relevant and integrates these important factors. The quantitative framework of a model provides an objective criterion of importance: phenomena or species are important when their inclusion affects the mass balance. Thus, modeling is an intellectual tool that lets us integrate and prioritize many processes.

Solving the model then provides us with a practical tool. The solutions can be used to design a new process to meet treatment goals or analyze why an existing process does or does not achieve its goal. The quantitative nature of the model makes it useful for engineering applications that involve real volumes, flows, and times.

The best strategy is to make the model as simple as possible. First, only the important factors should be included. Adding unimportant factors clutters the model, obscures the essential connections, and creates increased opportunities for error. Once the proper factors are chosen, they must be quantified in ways that simply, but adequately, represent reality. Simple mathematical expressions make a model easier to understand and to solve mathematically. But note that oversimplification—neglect or too superficial treatment of important phenomena— must also be avoided. A good model balances simplicity and completeness.

The focus of this chapter is on what factors need to be considered when a model is formulated for biotreatment applications. It identifies phenomena that can be important, defines under what conditions the phenomena need to be included in a model, indicates how they interact, and provides guidance on how they can be quantified. These topics are essential for persons building models and for those who are trying to select models built by others. Emphasis is on situations in which biological processes are active in detoxifying hazardous organic chemicals. The chapter answers the questions "What factors should be included in simple, but adequate, models?" and "How can these important factors be included?"

Foundation of Biological Process Modeling

Goals of microbial metabolism

Bacteria and, to a smaller degree, fungi are responsible for the biodegradation reactions we use in biotreatment. Although most degradation reactions are part of these cells' normal metabolism, the metabolic goal of the microorganisms' metabolism is not elimination of environmental contaminants. Instead, the basic goal of microbial metabolism is to grow and sustain more of themselves. Therefore, model formulation must begin with active biomass and whatever factors permit that biomass to grow and sustain itself.

Microorganisms grow and sustain themselves by extracting nutrients, electrons, and energy from their environments. The nutrients are the C, N, P, S, and other minor elements that comprise the building blocks of cells' constituents: carbohydrates, amino acids, lipids, and nucleic acids. Electrons are required to reduce several of the nutrients to the chemical form used in cellular constituents and to generate the energy needed for synthesizing and maintaining the biomass.

The most basic process in microbial metabolism is the transfer of electrons from an electron-donor substrate to an electron-acceptor substrate. Figure 5.1 illustrates that the oxidation of the donor (D) releases electrons that are carried by a reduced internal cosubstrate (ICH_2). Some of the electrons carried by ICH_2 are transferred to an electron-acceptor substrate (A), and this transfer generates energy in the form of the high-energy storage compound adenosine triphosphate (ATP). The rest of the electrons and some of the ATP are used to generate new biomass, while the remaining ATP satisfies the cells' maintenance needs.

Because the transfer of electrons between the electron-donor and electron-acceptor substrates is essential for creating and maintaining biomass, these materials are called *primary substrates*. In order to have biomass active in any reaction, primary substrates must be available. For the bacteria most commonly active in detoxification, the *primary*

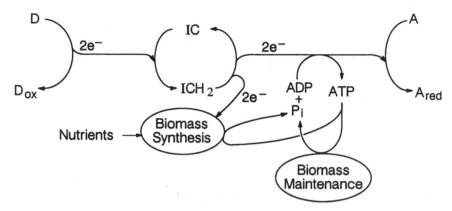

Figure 5.1 Typical electron and energy flows in a bacterial cell. D = primary electron-donor substrate, D_{ox} = oxidized electron-donor substrate, A = primary electron-acceptor substrate, A_{red} = reduced electron-acceptor substrate, ICH_2 = reduced internal cosubstrate, IC = oxidized internal cofactor, e^- = an electron, ATP = adenosine triphosphate, ADP = adenosine diphosphate, and P_i = inorganic phosphate.

electron donor is one of several organic compounds. The primary electron donor may or may not be one of the compounds targeted for detoxification. The *primary electron acceptor* usually is O_2, NO_3^-, NO_2^-, SO_4^{2-}, or CO_2, although organic compounds also can be used in fermentations.

Biotransformation phenomena

Once active biomass is present, any biotransformation reaction can occur, as long as the microorganisms possess enzymes for catalyzing the reaction. However, three questions should be answered before the model can be formulated correctly for the biodegradation of a target contaminant:

1. Is the target contaminant a primary or a secondary substrate?

2. Are other substrates modulating the biodegradation kinetics for the target contaminant?

3. Is all or part of the active biomass able to detoxify the target contaminant?

Primary or secondary substrate.   A target compound need not be the primary substrate for the bacteria that biodegrade it. When other materials serve as the primary substrate, a target compound can be biodegraded as a *secondary substrate,* a substrate whose oxidation (or sometimes reduction) yields too little electron and energy flow to support the biomass that degrades it (Kobayashi and Rittmann, 1982; Stratton et al., 1983).

Secondary substrates can be divided into two classes. The first class, called *low-concentration secondary substrates,* includes substrates that can contribute electron and energy flows to cell metabolism, but the rates of these flows are less than the minimum flow to maintain the biomass. The usual cause for the flows being very low is that the secondary substrate is present at a low concentration. Rittmann and McCarty (1980) derived S_{min}, the minimum concentration required to support steady-state biomass. Low-concentration secondary substrates could support biomass, but their concentration is less than S_{min}.

Another class of secondary substrate is called a *classical cometabolite,* a term used to describe a compound whose transformation cannot generate energy and electron flows (Alexander, 1981; Dalton and Sterling, 1982). Typically, a cometabolite is transformed by incidental metabolism, in which an enzyme that normally reacts with a different, but related, compound catalyzes a single transformation step for the cometabolite. Since further transformation steps do not occur, transformation of the cometabolite generates no electron and energy flows. Thus, a classical cometabolite cannot be a primary substrate, even if its concentration is large.

The term *cometabolism* is sometimes used too freely to indicate any situation in which another substrate enhances the biodegradation of a target compound. This overuse of the term is unfortunate, because it obscures its clear meaning: an obligate requirement for a primary substrate, because the cometabolite's transformation yields no electron and energy flows.

Modulators. The traditional view of biodegradation kinetics is that the rate of degradation is controlled by the concentrations of the target compound and the active biomass. While these traditional notions about kinetic controls are correct, other compounds also can affect the transformation kinetics. The general term for other compounds that affect the biotransformation is *modulator.* Three distinctly different types of modulators can act.

Cosubstrates react directly with the target compound (or perhaps a metabolic intermediate) at the biodegradative enzyme's reaction site. Although most cosubstrates are electron acceptors or donors, the cosubstrate reaction is not part of biomass-supporting electron flow. Instead, cosubstrates react directly with the target substrate.

One of the most widely recognized cosubstrates is molecular oxygen (O_2), which reacts in monooxygenase and dioxygenase reactions, such as the following monooxygenase reaction with trichloroethylene (TCE) (Little et al., 1988):

$$Cl_3HC_2 + ICH_2 + O_2 = Cl_2H_2O_2C_2 + IC + HCl$$

Hence, the dissolved oxygen concentration can directly affect the rate of TCE degradation.

The monooxygenase reaction also illustrates the second main type of cosubstrate: a *reduced internal cosubstrate,* ICH_2, supplies two electrons. Reduced internal cosubstrates are also required for reductive dehalogenation reactions (Vogel et al., 1987), such as reduction of TCE to dichloroethene:

$$Cl_3HC_2 + ICH_2 = Cl_2H_2C_2 + IC + HCl$$

Although the electrons carried by ICH_2 ultimately come from the primary electron-donor substrate, the internal concentration of ICH_2 directly controls the reaction kinetics for the monooxygenase and reductive dehalogenation reactions. Therefore, ICH_2 is an internal cosubstrate.

In the two previous cosubstrate examples, TCE is an example of a cometabolite, because the monooxygenase enzymes have methane as their normal substrate. Thus, methane is a primary substrate that requires two cosubstrates (ICH_2 and O_2), but TCE is a cometabolite whose transformation also requires the same two cosubstrates.

Inhibitors of substrate utilization slow enzyme-catalyzed reactions by reacting directly with the enzyme. One type of inhibition is called *competitive,* because the inhibitor competes for access to the reactive site on the enzyme. A second is *noncompetitive,* in which the inhibitor complexes with the enzyme at a nonreactive location, but in such a way that the enzyme's reactivity toward the target substrate is hindered. Competitive inhibitors usually are analogs of the target substrate; a good example is competition between TCE and methane, the normal substrate for the methane monooxygenase that transforms TCE (Tsien et al., 1989).

Inhibition can affect the biodegradation of a secondary substrate directly or indirectly. In the direct case, the inhibitor affects the enzyme degrading the secondary substrate. In the indirect case, the inhibitor retards the electron and energy flows of primary-substrate utilization. This inhibition of primary-substrate utilization can slow transformation of the secondary substrate by decreasing the amount of active biomass and/or by altering the availability of internal cosubstrates, such as ICH_2. Figure 5.2 illustrates places in which inhibition affects the primary electron and energy flows.

A special case of inhibition is called *self-inhibition,* in which high concentrations of a substrate inhibit its own degradation (Godrej and Sherrard, 1988; Sáez and Rittmann, 1989). Research on self-inhibitory compounds, such as phenolics, normally has involved target com-

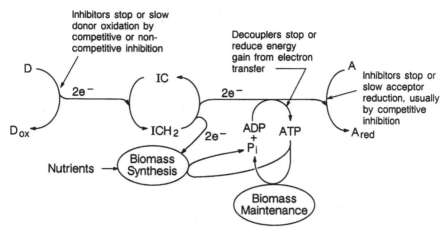

Figure 5.2 Illustration of how inhibitors can affect the primary flows of electrons and energy.

pounds that are the primary electron-donor substrate. Thus, self-inhibitory substrates probably hinder energy and electron flows at several locations and do not only inhibit their own enzyme-catalyzed transformation.

The final type of modulator is an *inducer*, which controls the formation of the enzymes that carry out the transformation reactions. A microorganism that has the genetic capability to transform a compound can do so only when the genes for that capability are expressed, which means that the code contained in DNA is transcribed to messenger RNA, which is translated to the enzyme. Although some genes are expressed all of the time, many degradative genes are expressed only when an inducer compound is present to initiate transcription of the DNA. Inducers usually are the substrate or a structurally similar analog.

Biomass active in transformation.   In the simplest case, all the biomass is equally capable of transforming all compounds. However, we know that biomass in real biotreatment systems is not uniform and should be differentiated. Four levels of *biomass differentiation* can be considered in model formulation: metabolically active versus inert biomass, suspended versus attached biomass, ecological diversity, and plasmid distribution.

Metabolically active bacteria, which can be active in biotransformation reactions, normally comprise only a fraction of what is measured by the traditional assays of "biomass," volatile suspended solids and dry weight. Much of what appears to be biomass is metabolically inert organic solids produced through the decay and death of active bacteria. Only the active biomass is capable of performing transformation reactions.

Some biomass is freely suspended, while other biomass is attached to solid surfaces as biofilms. The key distinction is that suspended biomass moves with the water flow, while the water flows past attached biomass. In addition, transport mechanisms that bring the substrates into contact with the bacteria often become limiting in biofilm systems (Rittmann and McCarty, 1980).

The active bacteria in biotreatment systems generally include many different strains. Although some ecological redundancy exists, great diversity is common because different strains are using different primary substrates present in the incoming water. Use of the different primary substrates creates ecological niches. Some niches are functionally very different, while others have only subtle differences. For example, bacteria that use phenol as the primary electron donor are totally distinct from bacteria that uses ammonium, but are relatively similar to those that use benzene.

The concept that a separate strain is required for each different substrate is seriously flawed. On the other hand, no strain uses all possible substrates. Thus, the reality lies between the extremes. Modeling of ecological selection must be based on primary-substrate utilization, because this is what controls the different strains' ability to grow and compete. A big unknown is how the presence of a secondary substrate affects ecological selection, since secondary substrates contribute little or nothing to cell growth.

The final issue in biomass differentiation is plasmid distribution. Plasmids are small strands of DNA that are not required for normal cell function, contain interesting degradative genes, and can be transferred between bacteria (Rittmann et al., 1990). Because plasmids are not essential, bacteria can exist quite well without them. However, if a cell acquires a plasmid, that cell can gain a degradative capability. If the fraction of the biomass containing the plasmid increases, then the concentration of biomass active in the degradative reaction increases proportionally. Likewise, the rate of degradation should increase in the same proportion.

The plasmid content in a biomass depends on a balance between plasmid transfer and plasmid loss. Plasmid transfer is aided by high cell concentrations, the availability of an energy source (e.g., primary substrates) to drive the conjugative reaction, and input of plasmid-containing donor cells (Smets et al., 1990; MacDonald et al., 1992). Loss rates are not well understood, but probably increase as the bacteria grow faster.

Mass balances

A mathematical model is composed of mass-balance equations for all of the critical chemical and microbial species. This section develops the mass balances that normally are required for a complete model. The rate terms that make up the mass-balance equations are discussed in the next section. Therefore, this section indicates where and how rate expressions are incorporated into a model.

Because biomass ultimately must utilize primary substrates to create and sustain itself, a mass balance on one or more primary substrates is essential. Most microbial types require two primary substrates: the electron donor and acceptor. The mass-balance equations for each substrate are given here for a completely mixed reactor volume.

$$V\frac{dD}{dt} = Q(D_0 - D) + R_D - r_D V \tag{5.1}$$

$$V\frac{dA}{dt} = Q(A_0 - A) + R_A - r_A V \tag{5.2}$$

where D = concentration of the primary electron-donor substrate $[M_D L^{-3}]$*

D_0 = input concentration of the primary electron-donor substrate $[M_D L^{-3}]$

R_D = rate of electron-donor addition (or removal, if negative) separate from the normal liquid flow $[M_D T^{-1}]$

r_D = rate of microbial utilization of the primary electron-donor substrate $[M_D L^{-3} T^{-1}]$

V = volume of the completely mixed reactor segment $[L^{-3}]$

Q = volumetric flow rate $[L^3 T^{-1}]$

t = time $[T]$

A = concentration of the primary electron-acceptor substrate $[M_A L^{-3}]$

A_0 = input concentration of primary electron-acceptor substrate $[M_A L^{-3}]$

R_A = rate of electron-acceptor addition (or removal) separate from normal liquid flow $[M_A T^{-1}]$

X_a = concentration of active biomass $[M_x L^{-3}]$

r_A = rate of utilization of the primary electron-donor substrate $[M_A L^{-3} T^{-1}]$

---

*Shown in square brackets are the dimensions of the units for each parameter. The main dimensions are M = mass, L = distance, and T = time. Subscripts differentiate types of mass, such as D = donor, A = acceptor, and X = biomass. For consistency, all mass must be expressed in units proportional to electron equivalents. Chemical oxygen demand (COD) is the most common mass unit that meets this requirement. If other unit systems are employed (e.g., VSS for biomass), conversion factors are required.

Equations (5.1) and (5.2) have three features that are characteristic for all mass balances.

- The left-hand side gives the change in mass per unit time. For the fairly common case of steady state, the differential becomes zero.

- The right-hand side contains three types of terms. The first is the *advection* term, which defines how much mass comes in with the influent flow and leaves with the effluent flow. The second is a term for inputs (or outputs when R is negative) that enter (leave) separately from the liquid flow. The third includes all the reactions that occur in the reaction volume. In these equations, primary donor is consumed for cell synthesis ($Y_D r_D V$), while primary acceptor is consumed for cell synthesis [$(1 - Y_D)r_D V$] and for oxidation to support cell maintenance ($bX_a V$).

The second essential mass balance is on active biomass.

$$V\frac{dX_a}{dt} = Q(X_{a0} - X_a) + R_x + r_x V \qquad (5.3)$$

where X_{a0} = input concentration of active biomass [$M_x L^{-3}$]
  R_x = rate of biomass input (or withdrawal) separate from the normal liquid flow [$M_x T^{-1}$]
  r_x = rate of net active-biomass growth [$M_x L^{-3} T^{-1}$]

Many target compounds are secondary substrates. The mass balance on a secondary substrate is

$$V\frac{dS}{dt} = Q(S_0 - S) + R_s - r_s V \qquad (5.4)$$

where S = concentration of the secondary substrate [$M_s L^{-3}$]
  S_0 = input concentration of the secondary substrate [$M_s L^{-3}$]
  R_s = rate of input (or removal) of secondary substrate separate from the normal liquid flow [$M_s T^{-1}$]
  r_s = rate of microbial transformation of the secondary substrate [$M_s L^{-3} T^{-1}$]

Expressing the r terms is the primary focus of the next section. Proper mathematical representations are then inserted into Eqs. (5.1) through (5.4). The terms R_D, R_A, R_X, and R_s represent abiotic addition and removal mechanisms. Although they can be very important, especially for primary acceptors and secondary substrates, they are not biological processes and are not discussed further. More information can be found in Namkung and Rittmann (1987), Rittmann et al. (1988), Rittmann (1990), and Smets et al. (1990).

Solving the mass balance equations

Although a detailed discussion of how to solve the mass balance equations is beyond the scope of this chapter, several general principles can be elaborated.

- All the equations are solved simultaneously.

- For simple systems, a closed, analytical solution is possible. Simplifications that make analytical solutions possible are that the system is at steady state, the active biomass is not differentiated, and abiotic (i.e., R) and biotic (i.e., r) rate terms are linear functions of the substrate concentration. A good example of an analytical solution is given by Namkung and Rittmann (1987), who describe the fate of synthetic organic chemicals in activated-sludge treatment.

- When systems are not simple, the solution must be obtained by computer-based numerical methods. Computer solutions are most essential when the system is not at steady state and when the R and r terms are nonlinear.

Computer solutions can range from spreadsheets to complex computer codes that must run on high-speed mainframe computers. Greater computational power is required when the number of species increases, the model must compute results over time and for different locations, and the R and r terms are highly nonlinear. Information on techniques for computer solution can be found in Finlayson (1980), Wang and Anderson (1982), and Smith (1978), for example.

Kinetic Expressions

This section presents and discusses mathematical expressions used for the biotic (r) rate terms. The goal is to represent quantitatively the key biological phenomena that affect primary and secondary substrates. Of special importance are the linkages among the different substrates and the active biomass. Expressions for biomass differentiation are beyond the scope of this chapter.

Primary substrates and active biomass

The net growth rate of active bacteria requires two terms. The first term represents biomass synthesis, which is proportional to primary-substrate utilization, as illustrated in Fig. 5.1. The second term represents the rate at which the biomass is oxidized to provide energy for maintenance, a phenomenon called *endogenous metabolism* (Roels, 1983). When the biomass oxidization rate is represented by a first-order model, the net growth rate of active bacteria can be expressed mathematically as

$$r_x = Y_{\text{eff}}(r_D) - b_{\text{eff}}X_a \tag{5.5}$$

where Y_{eff} = effective true yield coefficient $[M_x M_D^{-1}]$
b_{eff} = effective first-order decay coefficient $[T^{-1}]$

Critical to Eq. (5.5) is the introduction of effective yield and decay coefficients Y_{eff} and b_{eff}. These coefficients explicitly include the interactions among the substrates and are explained below. The development of Eq. (5.1) assumes that the transformation of secondary substrates does not yield increases in biomass.

The utilization rate of the primary electron donor, assuming that this compound limits the growth kinetics, is represented by a pseudo-Monod relationship,

$$r_D = \frac{\bar{q}_{D-\text{eff}}X_a D}{K_{D-\text{eff}} + D} \tag{5.6}$$

where $\bar{q}_{D-\text{eff}}$ = effective maximum specific rate of primary electron-donor utilization $[M_D Mx^{-1} T^{-1}]$
$K_{D-\text{eff}}$ = effective half-maximum-rate concentration for the primary electron donor $[M_D L^{-3}]$

The utilization rate of the primary electron acceptor is linked to the utilization rate of the donor and the rate of biomass growth.

$$r_A = r_D - r_X + f_m b_{\text{eff}}X_a \tag{5.7}$$

where f_m = fraction of electron-donor electrons originally converted to biomass and subsequently transferred to the electron acceptor to support biomass maintenance

The value of f_m can be computed from

$$f_m = \frac{f_d b_{\text{eff}}\Theta_x}{1 + b_{\text{eff}}\Theta_x} \, Y_{\text{eff}} \tag{5.8}$$

where f_d = fraction of newly formed biomass that is biodegradable
Θ_x = biomass retention time $[T]$ at steady state
$\quad = X_a V / r_x$

Effective coefficients for primary substrates

The "effective" coefficients—$\overline{q}_{D\text{-eff}}$, $K_{D\text{-eff}}$, Y_{eff}, and b_{eff}—depend on the kind of primary substrate and the type of interactions between the primary and secondary substrates. Possible relationships for the effective coefficients are summarized in Table 5.1 and discussed here.

The first situation represents a system in which the primary donor is non-self-inhibitory, and no interactions occur between the primary and secondary substrates. Here, the Monod (1949) function is used for the primary-substrate utilization rate; in this case, all the effective coefficients are constant and do not depend on the secondary-substrate concentration. The Monod relationship has an homology to an enzyme-catalyzed reaction in which the enzyme E and primary substrate D combine to form a complex ED, which then dissociates into product R and free (uncombined) enzyme E:

$$D + E = ED$$

$$ED \rightarrow R + E$$

The second system involves a self-inhibitory primary substrate and no interactions between the primary and secondary substrates. In this case, when a large amount of primary substrate is present, the enzyme-catalyzed reaction is diminished by the excess primary substrate. This phenomenon is called *self-inhibition* or *primary-substrate inhibition.* The normal way to express degradation kinetics of a self-inhibitory primary substrate is with the Haldane expression (Andrews, 1968), in which $\overline{q}_{D-\text{eff}}$ and $K_{D-\text{eff}}$ decrease with increasing primary-substrate concentration. The primary substrate appears to be a competitive inhibitor when present at high concentrations, combining with the complex ED to give the nonreactive intermediate ED_2.

$$D + E = ED$$

$$D + ED = ED_2$$

$$ED \rightarrow R + E$$

In reality, self-inhibition probably is more complicated than formation of a nonreactive complex. However, the Haldane expression usually represents the inhibitory effects satisfactorily.

The third system consists of a non-self-inhibitory primary substrate and a secondary substrate that is a competitive inhibitor of the primary substrate. Competitive inhibitors increase the $K_{D-\text{eff}}$ coefficient and have no effect on the $\overline{q}_D - _{\text{eff}}$ coefficient. The competitive-inhibitor secondary substrate S binds to the enzyme E, yielding the unreactive enzyme-inhibitor complex ES:

TABLE 5.1 Effective Coefficients for Primary-Substrate Utilization and Cell Growth

System		Reference	$\bar{q}_{D-\text{eff}}^{\dagger}$	$K_{D-\text{eff}}^{\dagger}$	$Y_{\text{eff}}^{\dagger}$	$b_{\text{eff}}^{\dagger}$
Primary substrate	PS and SS interactions*					
1. Non-self-inhibitory	No interactions	Monod (1949)	$\bar{q}_D$	K_D	Y	b
2. Self-inhibitory	No interactions	Andrews (1968)	$\dfrac{\bar{q}_D}{1+(D/K_1)}$	$\dfrac{K_D}{1+(D/K_1)}$	Y	b
3. Non-self-inhibitory	SS is a competitive inhibitor of PS	Bailey and Ollis (1986)	$\bar{q}_D$	$K_D\left(1+\dfrac{S}{K_2}\right)$	Y	b
4. Non-self-inhibitory	SS is a noncompetitive inhibitor of PS	Bailey and Ollis (1986)	$\dfrac{\bar{q}_D}{1+(S/K_3)}$	K_D	Y	b
5. Non-self-inhibitory	SS is an uncoupler	Gottschalk (1986)	$\bar{q}_D$	K_D	$\dfrac{Y}{[(1+(S/K_4)]}$	$b\left(1+\dfrac{S}{K_5}\right)$

*PS = primary substrate; SS = secondary substrate.

$\dagger$ $\bar{q}_D$ = maximum specific rate of primary-substrate utilization in the absence of inhibition $[M_D M_X^{-1} T^{-1}]$; K_D = half-maximum-rate concentration for primary substrate in the absence of inhibition $[M_D L^{-3}]$; Y = yield coefficient in the absence of uncoupling $[M_X M_D^{-1}]$; b = first-order decay coefficient in the absence of uncoupling $[T^{-1}]$; K_1 = self-inhibition constant for the primary substrate $[M_D L^{-3}]$; K_2 = inhibition constant due to the presence of the secondary substrate associated to the half-maximum-rate primary-substrate concentration $[M_S L^{-3}]$; K_3 = inhibition constant due to the presence of secondary substrate associated to the maximum specific rate of primary-substrate utilization $[M_S L^{-3}]$; K_4 = inhibition constant due to the presence of the secondary substrate associated to the yield coefficient $[M_S L^{-3}]$; K_5 = inhibition constant due to the presence of the secondary substrate associated to the decay coefficient $[M_S L^{-3}]$; D = concentration of primary substrate $[M_D L^{-3}]$; S = concentration of secondary substrate $[M_S L^{-3}]$.

$$D + E = ED$$

$$S + E = ES$$

$$ED \rightarrow R + E$$

In this case, binding of primary substrate and inhibitor to the enzyme are mutually exclusive. Because some enzyme is bound to the ES complex, not all the enzyme is available for catalyzing primary substrate conversion, and the reaction rate is lowered by the secondary substrate. The rate reduction caused by a competitive-inhibitor secondary substrate can be completely offset by increasing the primary-substrate concentration sufficiently; the maximum possible reaction velocity is not affected by the competitive inhibitor. Competitive inhibitors usually are primary-substrate analogs, because they bear close structural relationships to the normal substrate.

The fourth system involves a non-self-inhibitory primary substrate and a secondary substrate that is a noncompetitive inhibitor of the primary substrate. Noncompetitive inhibitors decrease the $\bar{q}_{D-\text{eff}}$ coefficient, while not affecting the KD-eff coefficient. A secondary substrate that is a noncompetitive inhibitor reacts with the metabolic enzyme at a site different from the reaction site. Inhibitor and primary substrate can simultaneously bind to the enzyme, forming the nonreactive or poorly reactive ternary complex EDS.

$$D + E = ED$$

$$S + E = ES$$

$$ED + S = EDS$$

$$ES + D = EDS$$

$$ED \rightarrow R + E$$

In the presence of a noncompetitive inhibitor, no amount of primary substrate addition to the reaction mixture can provide the maximum reaction rate that is possible without the inhibitor. Noncompetitive inhibition is also called *allosteric inhibition*. The name *allosteric* ("other shape") was originally coined for this mechanism because these kinds of inhibitors of enzymic activity react at a place other than the active site. They also may be structurally dissimilar to the primary substrate.

The last system considered consists of a non-self-inhibitory primary substrate and a secondary substrate that uncouples primary-substrate oxidation and biomass synthesis. This case can be mathematically modeled by a decrease in the Y_{eff} coefficient and/or an increase in the b_{eff} coefficient, while keeping constant the $\bar{q}_{D-eff}$ and K_{D-eff} coefficients. Uncouplers normally make the cytoplasmic membrane permeable for protons. As a consequence, a proton motive force within the membrane cannot be established, and ATP cannot be synthesized by electron transport phosphorylation (Gottschalk, 1986). Because of their mode of action, uncouplers are also called *protonophores*.

The mechanisms affecting electron and energy flows due to the presence of competitive and noncompetitive inhibitors and uncouplers are illustrated in Fig. 5.2. Mathematically, competitive inhibitors increase K_{D-eff}, noncompetitive inhibitors decrease $\bar{q}_{D-eff}$, and uncouplers decrease Y_{eff} and/or increase b_{eff}.

Secondary substrates

Secondary substrates can be of the low-concentration (i.e., sub-S_{min}) type, or they can be cometabolic. In some instances, cosubstrates are involved. This section describes the kinetics for these three situations.

The functional relationship for the secondary-substrate transformation rate depends on interactions between the primary and secondary substrates. In the low-concentration case, when the substrates do not interact, the Monod model is used for the secondary-substrate transformation rate (Stratton et al., 1983; Bouwer and McCarty, 1985).

$$r_s = \frac{\bar{q}_s X_a S}{K_s + S} \tag{5.9}$$

where $\bar{q}_s$ = maximum specific rate of secondary-substrate transformation $[M_s M_X^{-1} T^{-1}]$

K_s = half-maximum-rate concentration for secondary substrate $[M_s L^{-3}]$

The interaction between primary and secondary substrates occurs through X_a, which is increased when more primary substrate is available.

The second situation, classical cometabolism, is one in which the primary substrate can directly affect the secondary-substrate transformation. The kinetics are mathematically modeled using a pseudo-Monod expression,

$$r_s = \frac{\bar{q}_{s-eff} X_a S}{K_{s-eff} + S} \tag{5.10}$$

where $\overline{q}_{s-\text{eff}}$ = effective maximum specific rate of secondary-substrate transformation $[M_s M_x^{-1} T^{-1}]$

$K_{s-\text{eff}}$ = effective half-maximum-rate concentration for secondary substrate $[M_s L^{-3}]$

The effective coefficients, $\overline{q}_{s-\text{eff}}$ and $K_{s-\text{eff}}$, can have similar expressions to those shown for systems 3 and 4 in Table 5.1, but D is the compound that is affecting the coefficients for removal of S. When the primary substrate is a competitive inhibitor of the secondary substrate, $\overline{q}_{s-\text{eff}} = \overline{q}_s$ and $K_{s-\text{eff}} = K_s (1 + D/K_2')$; on the other hand, $\overline{q}_{s-\text{eff}} = \overline{q}_s/(1 + D/K_3')$ and $K_{s-\text{eff}} = K_s$ when the primary substrate is a noncompetitive inhibitor of the secondary substrate. The coefficients K_2' and K_3' are the inhibition constants $[M_D L^{-3}]$ due to the presence of the primary substrate associated to the half-maximum-rate secondary-substrate concentration and to the maximum specific rate of secondary-substrate transformation, respectively.

The final situation addresses how to describe secondary-substrate kinetics when a cosubstrate is required. The transformation of the secondary substrate requires inputs not directly related to its transformation. Hence, the rate of secondary-substrate transformation must be linked to the cosubstrate's concentration or utilization rate. In some cases, such as the direct utilization of O_2 in a monooxygenase or dioxygenase reaction (Little et al., 1988), the cosubstrate (O_2 here) can have its own Monod function that effectively controls $\overline{q}_{s-\text{eff}}$:

$$\overline{q}_{s-\text{eff}} = \overline{q}_s \frac{C}{K_c + C} \tag{5.11}$$

where C = concentration of the cosubstrate $[M_c L^{-3}]$

K_c = half-maximum-rate concentration for the cosubstrate $[M_c L^{-3}]$

In other cases, such as the use of ICH_2 in monooxygenase and dechlorination reactions, the cosubstrate is the primary donor, but it has a more indirect role: reducing IC to ICH_2. Sáez and Rittmann (1991) proposed that the secondary-substrate transformation rate depends on the primary cosubstrate utilization rate and biomass oxidation,

$$r_s = \alpha(r_D) + \text{ß}(b_{\text{eff}} X_a) \tag{5.12}$$

where α = amount of secondary substrate transformed per unit of primary cosubstrate consumed $[M_s M_D^{-1}]$

ß = amount of secondary substrate transformed per unit of biomass oxidized $[M_s M_x^{-1}]$

Equation (5.12) represents situations in which the oxidations of the primary substrate and/or biomass provide the electrons required for the secondary-substrate transformation. In the presence of a primary substrate, the generation of electrons via primary-substrate oxidation is usually much more important than the generation by way of biomass oxidation; these cases can be simplified using ß = 0 in Eq. (5.12).

An Example

The coupling of the mass-balance equations and the kinetic expressions allows prediction of the behavior for a given situation. In this section, as an example, we investigate the behavior of a steady-state activated-sludge reactor in the presence of a non-self-inhibitory primary substrate and a secondary substrate that is a competitive inhibitor of the primary donor and requires the primary donor as a cosubstrate.

When the influent active biomass concentration is negligible, the mass-balance equations for a steady-state activated-sludge reactor are as follows.

For active biomass:

$$r_x = \frac{X_a}{\Theta_x} \tag{5.13}$$

For primary substrate:

$$r_s = \frac{D_o - D}{\Theta} \tag{5.14}$$

For secondary substrate:

$$r_s = \frac{S_o - S}{\Theta} \tag{5.15}$$

in which Θ = hydraulic retention time $T = V/Q$.

The kinetic expressions for the active biomass, primary substrate, and secondary substrate rates become:
For active biomass:

$$r_x = Y(r_D) - b X_a \tag{5.16}$$

For primary substrate:

$$r_D = \frac{\overline{q}_D X_a D}{K_D(1 + S/K_2) + D}$$ (5.17)

For secondary substrate:

$$r_s = \alpha(r_D)$$ (5.18)

For this example, the secondary substrate is not an uncoupler, because it does not affect the Y_{eff} and b_{eff} coefficients; the secondary substrate is a competitive inhibitor of the primary substrate, because it increases $K_{D\text{-eff}}$, while keeping $\overline{q}_{D-eff}$ constant; and $\beta = 0$, which means that the endogenous electron supply is negligible.

The effluent secondary-substrate concentration is obtained by combining Eqs. (5.14), (5.15), and (5.18).

$$S = S_0 - \alpha(D_0 - D)$$ (5.19)

The reactor active biomass concentration is obtained by combining Eqs. (5.13), (5.14), and (5.16).

$$X_a = \frac{Y(D_0 - D)}{1 + b\Theta_x} \cdot \frac{\Theta_x}{\Theta}$$ (5.20)

The effluent primary-substrate concentration is obtained by combining Eqs. (5.14), (5.17), (5.19), and (5.20).

$$D = \frac{K_D[1 + S_0/K_2 - \alpha D_0/K_2]}{\overline{q}_D Y\Theta_x/(1 + b\Theta_x) - \alpha K_D/K_2 + 1}$$ (5.21)

Key to the model solution is that the effluent concentrations of D and S depend on, in addition to the kinetic coefficients, the operating parameters D_0, S_0, and Θ_x. To better understand the effect of the operating parameters, the kinetic coefficients shown in Table 5.2 are employed in the quantitative example that follows.

Figure 5.3 shows the effect of the influent primary-substrate concentration on the effluent concentrations D and S for $S_0 = 500$ mg $BOD_{D/L}$ and $\Theta_x = 1$ day. The removal efficiency for both substrates increases as D_0 increases, because the system becomes less inhibited. For $D_0 \leq 9.2$ mg/L, the system is completely inhibited by the secondary substrate, and neither substrate is removed. For 9.2 mg/L$<D_0<$5000 mg/L, the

TABLE 5.2   Kinetic Coefficients Used in the Example

Coefficient	Value*
Biomass growth rate	
Y	0.5 mg SS/mg BOD_D
b	0.1 day^{-1}
Primary-substrate utilization rate	
$\bar{q}_D$	14.4 mg BOD_D/mg SS • day
K_D	1 mg BOD_D/L
K_2	10 mg BOD_s/L
Secondary-substrate transformation rate	
α	0.1 mg BOD_s/mg BOD_D

*SS = suspended solids, a measure of biomass; BOD_D = biochemical oxygen demand for the primary donor substrate; BOD_s = biochemical oxygen demand for the secondary substrate.

degree of inhibition decreases as D_0 increases, and S and D decline gradually and then rapidly as D approaches 5000 mg/L. Finally, for $D \geq 5000$ mg/L, the system is completely noninhibited, because the secondary-substrate concentration within the reactor is driven to zero. The results in Fig. 5.3 illustrate how an increase in the availability of primary substrates allows greater biomass accumulation and provides more electrons for secondary-substrate removal. Both effects cause faster removal of secondary substrate and relieve inhibition.

Figure 5.4 illustrates the effect of the influent secondary-substrate concentration S_0 on the effluent concentrations D and S for $D_0 = 500$ mg BOD_D/L and $\Theta_x = 1$ day. The removal efficiency decreases for both substrates as S_0 increases, because the system becomes more inhibited. For $S_0 \leq 50$ mg/L, the system is completely noninhibited, because the secondary substrate is completely removed. The degree of inhibition increases as S_0 increases for 50 mg/L$<S_0<$27,700 mg/L. Finally, the system is completely inhibited without any substrate removal for $S_0 \geq 27,715$ mg/L. These results show how an inhibitory secondary substrate reduces primary-substrate removal, which reduces the active biomass and decreases the electron flow for the cosubstrate function of the primary substrate.

Figure 5.5 shows the effects of the biomass retention time Θ_x on the effluent concentrations D and S for $D_0 = 500$ mg $BOD_{D/L}$ and $S_0 = 500$ mg BOD_s/L. For $\Theta_x < 0.17$ day, the system is completely inhibited, and neither substrate is removed. The degree of inhibition decreases as Θ_x increases above 0.17 day. Due to the values of the kinetic constants and influent concentrations considered, the secondary-substrate removal is low in this case. However, as Θ_x increases enough, D is driven close to zero, even though S still is large, because of the buildup of biomass by cell retention.

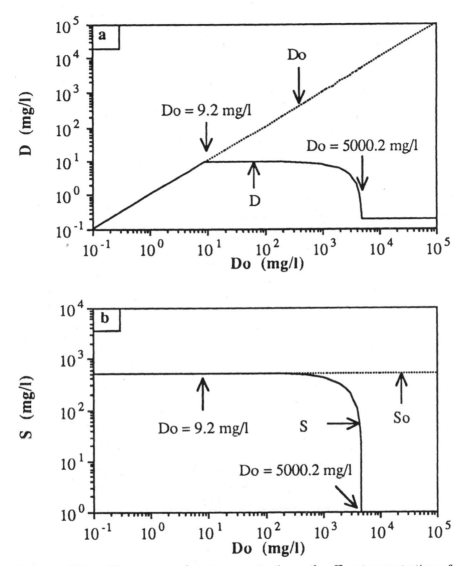

Figure 5.3 Effect of the primary-substrate concentration on the effluent concentrations of (*a*) primary donor substrate and (*b*) secondary substrate. Constant conditions are $S_0 =$ 500 mg BOD_s/L and $\Theta_x = 1$ day.

In summary, modeling shows that the system becomes less inhibited and gives higher removal efficiencies as the influent primary-substrate concentration increases, the influent secondary-substrate concentration decreases, and the biomass retention time increases. These characteristics are direct consequences of the substrate interactions: increased primary-substrate removal gives more biomass accumula-

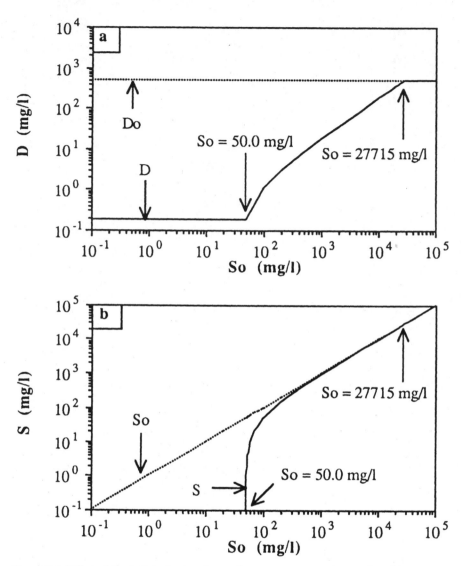

Figure 5.4 Effect of the influent secondary-substrate concentration on the effluent concentrations of (a) primary substrate and (b) secondary substrate. Constant conditions are $D_0 = 500$ mg BOD_D/L and $\theta_x = 1$ day.

tion and a greater electron flow for secondary-substrate removal, reduced secondary-substrate concentrations give lessened inhibition, and greater biomass retention increases biomass accumulation. Once recognized through modeling, these interactions can be taken into account in process design and operation.

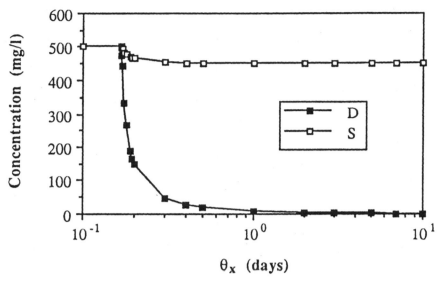

Figure 5.5 Effect of biomass retention time on the effluent concentrations of the primary and secondary substrates. Constant conditions are $D_0 = S_0 = 500$ mg BOD/L.

Conclusion

This chapter has provided the conceptual and quantitative foundation for biological process modeling. The most basic features of a model are:

- It must account for the growth and maintenance of active biomass, which requires mass balances on the primary electron-donor and electron-acceptor substrates, as well as on active biomass.

- It must differentiate secondary substrates, which do not support the biomass, from primary substrates.

- It must correctly account for different types of substrate interactions, including several types of inhibition and cosubstrate effects.

References

Alexander, M. 1981. Biodegradation of chemicals of environmental concern. *Science*, 211:132–138.

Andrews, J. F. 1968. A mathematical model for the continuous culture of microorganisms utilizing inhibitory substrates. *Biotechnol. Bioeng.*, 10:707–723.

Bailey, J. E., and D. F. Ollis. 1986. Biochemical Engineering Fundamentals, McGraw-Hill, New York.

Bouwer, E. J., and P. L. McCarty. 1985. Utilization rates of trace halogenated organic compounds in acetate-grown biofilms. *Biotechnol. Bioeng.*, 27:1564–1571.

Dalton, H., and D. I. Stirling. 1982. Co-metabolism. *Philosoph. Trans. R. Soc. London Ser. B., Biolog. Sci.*, 297:481–496.

Finlayson, B. A. 1980. Nonlinear Analysis in Chemical Engineering, McGraw-Hill, New York.

Godrej, A. N., and J. H. Sherrard. 1988. Kinetics and stoichiometry of activated sludge treatment of a toxic organic wastewater. *J. Water Pollut. Contr. Fed.*, 60:221–226.

Gottschalk, G. 1986. Bacterial Metabolism, 2nd ed., Springer-Verlag, New York.

Kobayashi, H., and B. E. Rittmann. 1982. Microbial removal of hazardous organic chemicals. *Environ. Sci. Technol.*, 16:170A–181A.

Little, C. D., A. V. Palumbo, S. E. Herbes, M. E. Lidstrom, R. L. Tyndall, and P. J. Gilmer. 1988. Trichloroethylene biodegradation by a methane-oxidizing bacterium. *Appl. Environ. Microbiol.*, 54:951–956.

MacDonald, J. A., B. Smets, and B. E. Rittmann. 1992. The effects of energy availability on the conjugative-transfer kinetics of plasmid RP4. *Water Research*, 26:461–468.

Monod, J. 1949. The growth of bacterial cultures. *Annu. Rev. Microbiol.*, 3:371–394.

Namkung, E. and B. E. Rittman. 1987. Estimating volatile organic compound emissions from publicity-owned treatment works, *JWPCF*, 59:670–678.

Rittmann, B. E. 1990. Biotechnological control of hazardous organic contaminants in sewage works, pp. 21–36, in Proc. 4th WPCF/JSWA Joint Technical Seminar, Tokyo, Japan.

Rittmann, B. E., D. Jackson, and S. L. Storck. 1988. Potential for treatment of hazardous organic chemicals with biological processes, pp. 15–64, in D. L. Wise (ed.), Biotreatment Systems, vol. III, CRC Press, Boca Raton, FL.

Rittmann, B. E., and P. L. McCarty. 1980. Model of steady-state-biofilm kinetics. *Biotechnol. Bioeng.*, 22:2359–2357.

Rittmann, B. E., B. Smets, and D. A. Stahl. 1990. Genetic capabilities of biological processes. Part 1. *Environ. Sci. Technol.*, 24:23–29.

Roels, J. A. 1983. Energetics and Kinetics in Biotechnology, Elsevier, Amsterdam.

Sáez, P. B., and B. E. Rittmann. 1989. Discussion of kinetics and stoichiometry of activated sludge treatment of a toxic organic wastewater. *J. Water Pollut. Contr. Fed.*, 61:357–358.

Sáez, P. B., and B. E. Rittmann. 1991. Biodegradation kinetics of 4-chlorophenol, an inhibitory co-metabolite. *J. Water Pollut. Contr. Fed.*, 63:838–847.

Smets, B., B. E. Rittmann, and D. A. Stahl. 1990. Genetic capabilities of biological processes. Part II. *Environ. Sci. Technol.*, 24:162–169.

Smith, G. D. 1978. Numerical Solution of Partial Differential Equations: Finite Difference Methods. Clarendon Press, Oxford, United Kingdom.

Stratton, R. G., E. Namkung, and B. E. Rittmann. 1983. Secondary utilization of trace organics by biofilms on porous media. *J. Am. Water Works Assoc.*, 75:463–469.

Tsien, H. C., G. A. Brusseau, R. S. Hanson, and L. P. Wackett. 1989. Biodegradation of TCE by *Methylosinus trichosporium* OB3b. *Appl. Environ. Microbiol.*, 55:3155–3161.

Vogel, T. M., C. S. Criddle, and P. L. McCarty. 1987. Transformations of halogenated aliphatic compounds. *Environ. Sci. Technol.*, 21:722–736.

Wang, H. E., and M. P. Anderson. 1982. Introduction to Groundwater Modeling, W. H. Freeman, San Francisco.

6

Law and Policy on the Frontier: The Regulation of Bioremediation of Hazardous Waste

David E. Giamporcaro

Section Chief, Biotechnology Program
Office of Pollution Prevention and Toxics
U.S. Environmental Protection Agency[1]
Washington, DC

Let's begin with a prediction: the decade of the 1990s will witness a substantial expansion in the use of bioremediation to degrade hazardous wastes, including the first commercial applications of genetically engineered microorganisms for this purpose.[2] This optimistic prognostication is not based solely on the state of the science of bioremediation. Progress on this front continues at an impressive rate; important achievements are announced almost weekly in the trade

---

[1]The views expressed are those of the author, and not of the U.S. Environmental Protection Agency.

[2]*Biodegradation* is "the natural process whereby bacteria or other microorganisms alter and break down organic molecules" into constituent components. Bioremediation is a subset of biodegradation.(11) There are two principal methods of bioremediation: biostimulation, which is the predominant remedial method, and bioaugmentation.(6) Biostimulation involves the addition of materials such as oxygen, nitrogen, phosphorus, and trace materials to a contaminated site to accelerate the natural biodegradative processes.(20) Bioaugmentation involves introducing microorganisms to a site to augment indigenous populations or increase the rate of degradation.(6)

journals. The future success of bioremediation depends not only on scientific achievements, however, but on regulatory policies and requirements as well (Day, 1990). This optimistic prediction is based, therefore, on the progress that has been made by the U.S. Environmental Protection Agency (EPA) in the past several years in addressing regulatory obstacles and data deficiencies which have impeded the application of this technology. These developments will encourage the commercialization of this technology, and will enable bioremediation techniques to be applied in an environmentally sound manner.

This chapter will discuss the federal statutes and regulations, both existing and under development, which apply to the use of bioremediation in the treatment of hazardous waste. The primary statutes of concern are the Resource Conservation and Recovery Act (RCRA)[3] and the Comprehensive Environmental Response, Compensation, and Liability Act (CERCLA).[4] In addition, policies and regulations implemented under the Toxic Substances Control Act (TSCA)[5] and regulations under the Federal Plant Pest Act (FPPA)[6] may affect the application of the technology. It is not the purpose of this chapter to describe the provisions of these statutes and regulatory programs in full detail; there have been countless treatises and articles which have already done so, and to which the reader may refer. The purpose of this chapter will be to discuss the impact of key provisions of the statutes and regulatory programs discussed herein on the use of bioremediation.

The Promise of Bioremediation

More than a decade after the passage of federal laws intended to redress the problem of treatment and disposal of hazardous waste, the nation still faces a problem of enormous magnitude. About 1200 hazardous waste sites are listed on the National Priorities List (NPL) established under Superfund. An additional 30,000 sites have yet to be evaluated under the Superfund program (Bakst, 1991). The EPA also estimates that an additional 4000 sites may need corrective action under RCRA. Altogether, according to Thayer (1991), "Tens of thousands more sites, including underground storage tanks and chemical or oil spills, are expected to require treatment.

Faced with a task of these proportions, bioremediation offers some clear advantages over conventional hazardous waste treatment tech-

---

[3]42 U.S.C. §6901, et seq. RCRA was enacted in 1976.

[4]42 U.S.C. §9601, et seq. CERCLA was enacted in 1980, and is also commonly referred to as Superfund.

[5]15 U.S.C. §2601, et seq. TSCA was enacted in 1976.

[6]7 U.S.C. §150aa, et seq. The FPPA was enacted in 1957, and is administered by the United States Department of Agriculture (USDA).

nologies. One leading advantage is cost. Bioremediation offers significant savings over the leading hazardous waste treatment technologies. The costs of incineration, for example, range from $300 to $1000 per ton of soil, while land disposal costs between $200 and $300 a ton. By contrast, soil bioremediation costs between $50 and $150 a ton (Glass, 1991). A second major advantage is that bioremediation can be accomplished with less environmental insult to the affected site, and quite possibly to the environment in general. In situ bioremediation, for example, would ordinarily involve minimal disruption of surface soils. In addition, bioremediation avoids some of the significant environmental drawbacks associated with the leading treatment technologies.[7]

Perhaps the greatest advantage of bioremediation, however, is as yet untapped. That is the use of bioremediation technologies as an instrument in pollution prevention. Several experts in the field have cited the application of bioremediation in waste minimization as one of the most promising future applications of the technology (Bakst, 1991; Glass, 1991). The EPA is well aware of this potential. EPA Administrator William Reilly, speaking in June 1991 at a meeting between representatives of the bioremediation industry and the Agency, stated that the various applications of biotechnology "need to become an integral part of our future discussions and actions in pollution prevention" (U.S. EPA, 1991a).

It is little wonder, then, that EPA has taken such great strides over the past two and a half years to assess the scientific, institutional, and regulatory obstacles to the application of this technology. Reilly has clearly stated the Agency's commitment to taking advantage of the promise of bioremediation:

> I think we should continue to develop aggressively the full potential of bioremediation to treat our hazardous waste, and clean up our abandoned sites....I continue to believe that bioremediation has the potential to be a dominant treatment technology for site cleanup in the future. (U.S. EPA, 1991a)

The Regulation of Hazardous Waste

The two statutes which have the greatest bearing on bioremediation, both in the short term and the long term, are RCRA and CERCLA.

RCRA has been described with little exaggeration as the single most complex environmental regulatory program. RCRA establishes a comprehensive system to manage the storage, treatment, and disposal of solid and hazardous waste, including identification of solid and haz-

---

[7]For example, incinerators present air emissions concerns and ash disposal problems (Glass, 1991; Thayer, 1991).

ardous waste, and establishment of treatment, storage, and disposal standards applicable to generators and transporters of hazardous waste, and owners and operators of treatment, storage, and disposal facilities.

CERCLA was enacted to enable and empower EPA to clean up contaminated sites around the country. Since CERCLA is to some extent dependent upon determinations made and requirements imposed under RCRA, the latter statute and its accompanying regulations will be discussed first.

The Resource Conservation and Recovery Act

RCRA is divided into several principal sections, or Subtitles. This discussion will focus on selected provisions of Subtitles C (Hazardous Waste), and I (Underground Storage Tanks).

Subtitle C: Identification, Treatment, Storage and Disposal of Hazardous Waste. Subtitle C requires EPA to identify hazardous waste; to issue permits for facilities that treat, store, and/or dispose of hazardous waste; and to set treatment standards for wastes prior to land disposal (Bakst, 1991). RCRA is structured in such a way that the determination must first be made whether the waste involved meets the statutory definition of solid waste. *Solid waste* is defined, generally, as any solid or liquid form of material which is intended to be discarded.[8]

There are several important exclusions to the definition of solid waste. Domestic sewage, or mixtures of domestic sewage and industrial waste discharged to a Publicly Owned Treatment Works, and point-source industrial wastewater discharges subject to a National Permit Discharge Elimination System (NPDES) permit issued under the Clean Water Act,[9] are not considered solid waste, and therefore are not subject to regulation under RCRA.[10]

EPA has further established two broad categories of hazardous waste: (1) characteristic waste and (2) listed waste.

Characteristic waste is waste which displays one or more of four different characteristics: ignitability, reactivity, corrosivity, and the toxicity characteristic. The first three are largely self-explanatory. EPA regulations establish specific criteria for determining whether wastes

---

[8]42 U.S.C. §6903(27). See also 40 C.F.R. §261.2.

[9]NPDES permits are issued by EPA and establish permissible limits for discharge of pollutants into the navigable waters of the United States. See 33 U.S.C. §1342.

[10]40 C.F.R. §261.4(a).

are ignitable, reactive, or corrosive.[11] The characteristic of toxicity is related to "the potential for certain toxic constituents to leach into groundwater (Bakst, 1991). The toxicity characteristic applies to 40 inorganic and organic compounds which, if present in leachate from a waste in an amount exceeding a regulatory threshold, render the source waste material hazardous.[12]

Listed waste, as the term implies, refers to waste materials or industrial process waste streams which have been specifically designated as hazardous by EPA. Listed waste is divided into four categories, designated as F-, K-, P-, and U-listed wastes. F-listed wastes comprise designated wastes from nonspecific sources. These generally constitute spent chemicals, wastes, and by-products generated in a variety of industrial sectors (Bakst, 1991; Hill, 1991).[13] K-listed wastes are those generated from specific sources. These constitute primarily sludges and by-products generated from specific industrial sources; for example, sludge from the production of various pesticides (Hill, 1991).[14] P-listed wastes constitute specific discarded acutely hazardous commercial chemical products and their off-specification production runs, as well as residue and debris from spills of such products, and residue in containers which held such products (Bakst, 1991).[15] U-listed wastes are specific discarded commercial chemical products or their off-specification variants, and residue and debris from spills of these products (Bakst, 1991).[16] All told, these lists contain about 400 different hazardous chemicals or industrial waste streams (Thayer, 1991).

There are several important rules and policies which have further expanded the scope of the hazardous waste classification.

EPA's "mixture rule" provides that any mixture of a solid waste with a hazardous waste is considered to be hazardous waste (Bakst, 1991). The mixture rule is intended to prevent disposal of RCRA hazardous waste by merely diluting it with solid waste. Most characteristic hazardous waste may be rendered nonhazardous by treatment to remove the condition(s) which triggered the characteristic classification.[17] For example, if the waste contained one of the compounds subject to the toxicity characteristic at a level above the regulatory threshold, it could be rendered nonhazardous by treatment to reduce

---

[11] See 40 C.F.R. §261.21 (ignitability); §261.22 (corrosivity); §261.23 (reactivity).

[12] 40 C.F.R. §261.24.

[13] 40 C.F.R. §261.31.

[14] 40 C.F.R. §261.32.

[15] 40 C.F.R. §261.33(e).

[16] 40 C.F.R. §261.33(f).

[17] 40 C.F.R. §261.3(a)(2)(iii), (d)(1).

the concentration of the particular contaminant below the regulatory threshold.

The situation is different, however, if solid waste is mixed with a listed hazardous waste.[18] In this circumstance, treatment does not change the status of the mixed waste as a hazardous waste. A listed hazardous waste cannot change regulatory status by virtue of being treated.[19]

EPA's "contained-in" policy provides that any environmental media (i.e., soil or groundwater) which contains a listed hazardous waste is considered to be that hazardous waste, and must be treated, stored, or disposed of accordingly (Bakst, 1991).[20] Moreover, EPA's "derived-from" rule provides that any residue generated from the treatment, storage, or disposal of a listed hazardous waste is considered to be that listed waste (Bakst, 1991; Hill, 1991).[21]

The practical effect of the contained-in policy and the derived-from and mixture rules is that soil or groundwater contaminated with listed hazardous waste, or the residue of any treatment process used to degrade a listed hazardous waste—for example, ash from an incinerator or biomass from a bioreactor—must itself be treated as hazardous waste.

On December 6, 1991, the United States Court of Appeals for the District of Columbia Circuit vacated and remanded the derived from and mixture rules to EPA on the grounds that the agency had not provided sufficient notice and comment prior to promulgating the rules in 1980.[22] In order to avoid "a discontinuity in the regulation of hazardous wastes," the court recommended that EPA reenact the regulations on an interim basis.[23]

The EPA reinstated the rules in March 1992 by publishing an interim final rule.[24] The purpose of the interim final rule is to ensure that there is no gap or loophole in the administration of the hazardous waste program. The interim final rule is set to expire on April 28, 1993. During the intervening period, EPA will repropose the mixture and de-

---

[18]40 C.F.R. §261.3(a)(2)(iv).

[19]40 C.F.R. §261.3(c)(1). In order for a listed hazardous waste, or materials such as soil or groundwater which contain a listed hazardous waste, not to be considered hazardous, they must be delisted. 40 C.F.R. §261.3(d)(2).

[20]40 C.F.R. §261.3(b)(2).

[21]40 C.F.R. §§261.3(c)(2)(i).

[22]*Shell Oil Company v. EPA,* 950 F. 2d 741 (D.C. Cir. 1991).

[23]Ibid., Slip Op. at 20–21.

[24]"Hazardous Waste Management System; Definition of Hazardous Waste; 'Mixture' and 'Derived-From' Rules; Interim Final Rule," 57 *Fed. Reg.* 7628 (March 3, 1992).

rived from rules to solicit public comment on options for management of mixed wastes and treatment residues.[25]

The determination that a waste is hazardous triggers a whole series of regulatory requirements regarding the storage, treatment,[26] and disposal of the waste. Any facility that treats, stores, or disposes of RCRA hazardous waste must have an EPA permit, or be operating under what is termed *interim status* pending the issuance of a final permit. It is through the RCRA permits, or regulations applicable to interim-status facilities, that EPA imposes corrective action requirements which govern the cleanup of hazardous wastes. The corrective action requirements are discussed below.

There are some important exceptions to the requirement for a permit or compliance with the interim-status regulations before hazardous waste may be treated or disposed. Generators who generate less than 100 kg per month of hazardous waste (small-quantity generators) are exempt from RCRA.[27] Generators who generate more than 100 kg per month but less than 1000 kg per month may accumulate and treat hazardous waste on site for up to 180 days in tanks or containers which meet minimum regulatory requirements without obtaining an RCRA permit or complying with the interim-status regulations.[28] Similarly, generators of more than 1000 kg per month of hazardous waste may accumulate and treat the waste on site for up to 90 days without triggering the requirement for a permit.[29]

In addition, owners or operators of totally enclosed treatment facilities are not subject to the RCRA permit or interim-status requirements.[30] A *totally enclosed treatment facility* is defined as a facility "which is directly connected to an industrial production process and which is constructed and operated in a manner which prevents the release of any hazardous waste or any constituent thereof into the environment during treatment."[31]

These exceptions are important in that they would allow the use of bioremediation or any other form of treatment of hazardous waste without the need for prior approval of the treatment method under the

---

[25]See "Hazardous Waste Management System; Definition of Hazardous Waste; 'Mixture' and 'Derived From' Rules; Proposed Rule," 57 *Fed. Reg.* 7636 (March 3, 1992).

[26]*Treatment* is defined as "any method, technique, or process,...designed to change the physical, chemical, or biological character or composition of any hazardous waste so as to...render such waste nonhazardous...." 42 U.S.C. §6903(34); 40 C.F.R. §260.10.

[27]40 C.F.R. §261.5(a).

[28]40 C.F.R. §262.34(d).

[29]40 C.F.R. §262.34(a).

[30]40 C.F.R. §264.1(g)(5).

[31]40 C.F.R. §260.10.

RCRA program. EPA has expressly adopted the position that hazardous waste may be treated while it is being stored in tanks or containers during the 90- or 180-day accumulation periods.[32]

The exception for totally enclosed treatment facilities may also have an important bearing on the use of bioremediation as a means of treating process waste streams. Bioreactors which are installed as end-of-pipe treatment systems for industrial effluents may qualify as totally enclosed treatment facilities under RCRA. In addition to offering the advantage of reduced regulatory concerns because of their contained nature, bioreactors can also be designed to address specific industrial waste streams. Integration of bioreactors into existing industrial processes for use in waste minimization should face few technical challenges, as the technology is similar to that which has been used in industrial wastewater treatment for decades (Glass, 1991). The flow-through waste treatment technology characteristic of bioreactor operations involves the same technology as in wastewater treatment (Glass, 1991). In fact, several domestic bioremediation companies are researching the use of contained systems for waste degradation (Garg and Garg, 1990; Glass, 1991).[33]

In circumstances where the above exceptions do not apply, the corrective action provisions of the RCRA program would govern the use of treatment methods to degrade hazardous waste at a treatment, storage, or disposal facility.

The 1984 amendments to RCRA[34] dramatically changed the corrective action program under the statute. Prior to 1984, the RCRA corrective action regulations were principally directed to monitoring the leaching of hazardous constituents into groundwater from surface impoundments, waste piles, land treatment units, or landfills (Hill, 1991).[35] The 1984 amendments required corrective action by any permitted or interim status treatment, storage, or disposal facility to clean up spills or releases of hazardous constituents from any solid waste management unit located at the facility. A *solid waste management*

---

[32]51 *Fed. Reg.* 10146, 10168 (March 24, 1986).

[33] The contained-in policy, and derived from and mixture rules discussed above may not apply to bioreactors which are integrated into industrial process streams for purposes of waste minimization. The RCRA requirements for treatment, storage, and disposal of hazardous waste apply only after a hazardous waste is generated (Day, 1990). Bioreactors which are used to treat industrial process streams to prevent the generation of a hazardous waste therefore would not trigger the RCRA requirements. However, other EPA regulations (e.g., the NPDES permit requirements under the Clean Water Act) may apply to effluents from such bioreactors, depending upon the media to which the effluent is eventually discharged.

[34] The Hazardous and Solid Waste Amendments of 1984, Pub. L. No. 98-616. 42 U.S.C. §6901 et. seq.

[35] See 40 C.F.R. §264.92-264.100.

unit encompasses any structure used to collect, store, treat, or dispose of solid waste.[36]

The corrective action regulations promulgated by EPA in 1985 are very general. EPA required that owners or operators of treatment, storage, or disposal facilities "must institute corrective action as necessary to protect human health and the environment for all releases of hazardous waste or constituents from any solid waste management unit at the facility," and that "[c]orrective action will be specified in the [facility's] permit."[37] In practice, this has meant that EPA can control the cleanup process at an RCRA facility by specifying the treatment methods which may be used to address a release of a hazardous waste at the facility. EPA does so by modifying the permit of an RCRA-permitted facility to specify the treatment method(s) which may be used for corrective action, or by issuing or modifying an administrative order for an interim status facility for the same purpose (Hill, 1991).

Approval of treatment methods which may be used to clean up releases from solid waste management units at treatment, storage, or disposal facilities involves a multistage assessment process. EPA first undertakes a facility assessment, which is then followed by a facility investigation. Finally a corrective measure study is conducted to identify the treatment method(s) which will be applied (Bakst, 1991). The process is lengthy and costly, and has tended to delay cleanups under RCRA in much the same way that cleanups of NPL sites under the Superfund program have been delayed.

Companies that provide bioremediation services on site at an RCRA-permitted facility or an interim-status facility generally are not required to obtain a permit themselves (Bakst, 1991). Such companies are not considered to be either generators of the waste, or owners or operators of the facility. However, EPA must expressly approve the bioremediation project by modification of the facility's RCRA permit, or by issuance or modification of an administrative order for an interim-status facility. And note that a bioremediation company which receives and treats hazardous waste at its own facility *is* required to obtain a permit or comply with the interim-status regulations (Bakst, 1991). In this case, the company would be considered an owner or operator of a treatment, storage, and disposal facility.

---

[36]The definition of *solid waste management unit* has brought an enormous number of structures within the scope of the corrective action requirements. EPA estimates that there are 5700 treatment, storage, and disposal facilities in the United States, which encompass about 80,000 solid waste management units. This does not include the several hundred additional federal facilities which are also subject to these requirements (Bakst, 1991). The RCRA corrective action program "is expected to far outweigh the Superfund program in numbers" (Bakst, 1991).

[37]40 C.F.R. §264.101.

The 1984 amendments to RCRA also prohibited the land disposal of certain hazardous wastes unless the wastes are first pretreated to specified standards.[38] EPA has implemented this requirement by promulgating a series of regulations which prohibit land disposal of different groups of hazardous wastes: solvent[39] and dioxin-containing wastes,[40] the so-called "California list" wastes,[41] and specific listed and characteristic hazardous wastes which are grouped in regulations commonly referred to as the "First Third,"[42] "Second Third,"[43] and "Third Third"[44] rules. The land disposal regulations (LDRs) have become one of the most controversial elements of the RCRA program, and have a direct bearing on the use of bioremediation for corrective action at RCRA facilities.

The prohibition against land disposal does not apply where it can be demonstrated that there will be no migration of hazardous constituents from a disposal unit for as long as the wastes remain hazardous.[45] Persons may therefore be exempted from the LDRs by filing a "no migration" petition with EPA.[46]

In the absence of such a petition, however, the LDRs require that, prior to land disposal, hazardous waste must be pretreated by a specified treatment method (e.g., incineration), or that the waste concentration level meet an established level, referred to as a *treatment standard*. Under the Third Third rule, for example, bioremediation has been identified as a permissible treatment method for 16 different wastes. Even where other treatment methods have been designated, there is a regulatory mechanism which permits the use of alternative treatment methods which are demonstrated to be equivalent to the specified treatment methods.[47] Where a treatment standard has been established, any treatment method capable of achieving the specified treatment standard—including bioremediation—may be used to treat the waste prior to land disposal.

The difficulty lies in the concentration levels which are set in the treatment standards. In implementing the land disposal restrictions

---

[38]42 U.S.C. §6924(d).

[39]40 C.F.R. §268.30.

[40]40 C.F.R. §268.31.

[41]40 C.F.R. §268.32.

[42]40 C.F.R. §268.33.

[43]40 C.F.R. §268.34.

[44]Portions of the third rule were recently remanded to the agency. See "Chemical Waste Management v. EPA," 976 F. 2d 2 (D.C. Cir. 1992) 40 C.F.R. §268.35.

[45]42 U.S.C. §6924(d)(1).

[46]40 C.F.R. §268.6.

[47]40 C.F.R. §268.42(b).

of the 1984 RCRA amendments, EPA required that treatment standards be set based on the waste reduction which could be achieved using the *best demonstrated available technology* (BDAT). Critics have charged that as a result of this requirement, treatment standards have tended to be based on decontamination levels achievable through conventional technologies, such as incineration, and have set artificially high decontamination targets which are difficult to accomplish biologically. (Bakst, 1991; Day, 1990; Glass, 1991). Critics have also asserted that the BDAT requirement has resulted in technology-based, rather than health-based, treatment standards (Day, 1990; Thayer, 1991).

A second area of concern is that the LDRs may restrict the circumstances in which bioremediation technology may be used. The LDRs provide that designated hazardous waste may not be placed "in or on land"[48] unless the wastes are first treated by a specified technology, or the applicable treatment standards are met. Any displacement of material—such as soil or groundwater—which is contaminated with hazardous waste listed in the LDRs may therefore trigger the pretreatment requirement. Critics have argued that this may restrict the use of bioremediation as a primary or secondary treatment method (Day, 1990).

EPA has recently taken measures to address these problem areas. The 1984 RCRA amendments established a four-year exemption from applicability of the LDRs to contaminated soil and debris resulting from a Superfund response action or an RCRA corrective action.[49] This statutory exemption expired on November 8, 1988. EPA has supplemented the statutory exemption through a series of two-year national capacity variances from the application of the LDR requirements to soil and debris contaminated with specific wastes.[50] For example, in the recently promulgated Third Third rule, EPA established a national capacity variance for soil and debris contaminated with various categories of wastes, including those wastes with

---

[48]*Land disposal* is defined as placement of a hazardous waste "in or on...land." 40 C.F.R. §268.2(c).

[49]42 U.S.C. §6924(d)(3).

[50]The national capacity variances for soil and debris contaminated with specified wastes subject to the LDRs are codified at 40 C.F.R. §268.30(c) (variance for soil and debris contaminated with F001–F005 wastes); 40 C.F.R. §268.31(a)(1) (variance for soil and debris contaminated with F020–F023 or F026–F028 wastes); 40 C.F.R. §268.32(d)(1) and (d)(2) (variance for soil and debris contaminated with California list wastes); 40 C.F.R. §268.33(c) (variance for soil and debris contaminated with First Third wastes with a treatment standard based on incineration); 40 C.F.R. §268.43(d) (variance for soil and debris contaminated with Second Third wastes with a treatment standard based on incineration); 40 C.F.R. §268.35(c), (d), and (e) (variance for several different categories of Third Third wastes).

treatment standards based on incineration, vitrification, and mercury retorting.[51]

EPA has acknowledged that the BDAT requirements of the LDRs may not apply to RCRA corrective actions or CERCLA remedial actions until BDAT standards for contaminated soil and debris are established.[52] "BDAT standards were established for process waste streams or for pure wastes themselves, rather than for materials such as soil and debris contaminated with hazardous waste" (Thayer, 1991).

As the national capacity variances expire, however, soil and debris contaminated with wastes subject to the LDRs cannot be land-disposed without pretreatment. In the interim, treatment standards for contaminated soil and debris are established through the issuance of generic variances or site-specific variances.[53] Variances may be granted if "a waste cannot be treated to the specified level, or where the treatment technology is not appropriate to the waste." The petitioner "must demonstrate that because the physical or chemical properties of the waste differs significantly from wastes analyzed in developing the treatment standard, the waste cannot be treated to specified levels or by the specified methods."[54]

The EPA has also proposed several other potential solutions to the problems raised by the LDRs. In July 1990, EPA proposed regulations pertaining to the conduct of corrective action for solid waste management units at RCRA treatment, storage, and disposal facilities.[55] The proposed regulations incorporate several important concepts. For example, the corrective action regulations envision a more flexible site evaluation process. EPA hopes that flexibility in the corrective action evaluative process will enable RCRA sites to move to cleanup status more quickly than has been the case for Superfund sites (Thayer, 1991).

The proposed rule would also adopt a health-based standard for selecting the appropriate level of cleanup at RCRA facilities. For car-

---

[51] 55 *Fed. Reg.* 22520, 22634–22635 (June 1, 1990). The national capacity variance for soil and debris contaminated with hazardous wastes listed in the Third Third rule has been extended to May 8, 1993. The general land disposal restrictions for Third Third wastes went into effect on August 8, 1990.

[52] EPA is developing separate BDAT standards for soil and debris contaminated with wastes listed in the LDRs. The BDAT standards for debris were promulgated in August 1992: "Land Disposal Restrictions for Newly Listed Wastes and Hazardous Debris," 57 *Fed. Reg.* 37194 (August 18, 1992). The rule lists 17 different BDAT treatment technologies—including biodegradation—that can be used to treat hazardous debris.

[53] 40 C.F.R. §268.44.

[54] 40 C.F.R. §268.44(a).

[55] "Corrective Action for Solid Waste Management Units at Hazardous Waste Management Facilities," 55 *Fed. Reg.* 30798 (July 27, 1990). These proposed regulations would replace the general corrective action regulations now found at 40 C.F.R. §264.101.

cinogens, the cleanup standards for hazardous constituents in ground-water, surface water, soils, and air would be based on existing media-specific cleanup standards, such as the maximum contaminant levels established under the Safe Drinking Water Act. Where such standards did not exist, cleanup standards would "be established within the protective risk range of 1×10^{-4} to 1×10^{-6}".[56] Site-specific factors would be used to determine where in that range the cleanup standard would be set.[57]

The proposed rule would also create *corrective action management units,* or CAMUs, which would permit treatment of hazardous waste within a designated area without triggering the LDR treatment standards (Thayer, 1991). If implemented, the concept of CAMUs would constitute an important exception to the applicability of the LDRs. This concept would allow several different solid waste management units at an RCRA facility to be considered a single unit for purposes of the LDRs. Waste could be picked up and placed back on the ground within the confines of a CAMU without triggering the treatment standards of the LDRs, thereby enabling greater flexibility in the selection of treatment methods (Bakst, 1991).[58]

In addition, EPA recently proposed the establishment of a new storage and treatment unit known as a *containment building.*[59] EPA has proposed that certain hazardous wastes, including contaminated debris, could be treated and stored within such a unit without triggering the LDR requirements.[60] Importantly, EPA has recognized that on-site bioremediation projects often take place inside temporary structures, and has solicited comment on what standards should be applied to such *bioremediation treatment buildings.*[61] EPA has also recognized that various types of treatment, including bioremediation, take place in tanks and containers, and has proposed allowing such treatment technologies to be conducted in containment buildings.[62]

---

[56] 55 *Fed. Reg.* 30798, 30826 (July 27, 1990).

[57] Ibid.

[58] Regulations implementing the CAMU concept were promulgated in February 1993. 58 Fed. Reg. 8658 (February 16, 1993).

[59] See "Land Disposal Restrictions for Newly Listed Wastes and Contaminated Debris," 57 *Fed. Reg.* 958, 978 (January 9, 1992).

[60] Under the current LDR program, treatment and storage of hazardous waste inside a structure is considered an indoor waste pile, which is considered prohibited land disposal. 57 *Fed. Reg.* 958, 978 (January 9, 1992).

[61] 57 *Fed. Reg.* 958, 981 (January 9, 1992).

[62] Ibid.

Subtitle I: Underground Storage Tanks.   Subtitle I was added to RCRA as part of the 1984 amendments to the statute.[63] This subtitle imposes requirements on underground storage tanks (USTs) which contain petroleum or hazardous wastes similar to those placed on owners and operators of treatment, storage, and disposal facilities under Subtitle C of RCRA, including requirements for corrective action (Hill, 1991). In the event of a release from a UST, the owner or operator may be required to develop and submit a corrective action plan.[64] In addition, temporary or permanent closure of a UST may trigger corrective action if a release is detected.[65]

Petroleum contamination caused by leaking USTs is regulated primarily by the states. There is considerable variation among the states in establishing treatment standards for cleanup of soil or groundwater contaminated by petroleum or petroleum products. "[T]he numeric criteria that trigger remediation of soils contaminated with petroleum products...will dictate whether bioremediation is a viable treatment technology" (National Governors' Association, 1991).

Research and development under RCRA.   The RCRA regulations also provide for the issuance of research, development, and demonstration (RD&D) permits for hazardous waste treatment facilities which propose "to utilize an innovative and experimental hazardous waste treatment technology or process."[66] Such permits may specify the type and quantity of hazardous waste which may be treated, and may include conditions on the conduct of the research, such as monitoring requirements.

The RD&D permit program has not emerged as an important component of the RCRA regulatory program. EPA has issued few RD&D permits, due in large part to the need to devote resources to development and implementation of other aspects of the hazardous waste program.

A more important avenue for the conduct of research under RCRA is the exemption for treatability studies.[67] *Treatability studies* are defined as studies to determine how a hazardous waste can be effectively treated.[68] Laboratories or testing facilities may test samples of no more than 1000 kg of nonhazardous waste, 1 kg of acute hazardous waste, or 250 kg of soil, water, or debris contaminated with acute hazardous

---

[63] 42 U.S.C. §§6991–6991i.

[64] 40 C.F.R. §280.66(a).

[65] 40 C.F.R. §280.70(a), §280.72.

[66] 40 C.F.R. §270.65(a).

[67] 40 C.F.R. §261.4(e), (f).

[68] See the definition of *treatability study* at 40 C.F.R. §260.10.

waste, without obtaining an RCRA permit or interim status.[69] Such laboratories or testing facilities are subject to notification and record-keeping requirements, and certain limitations on the quantities of hazardous waste which may be stored prior to initiation of treatability studies, and quantities of hazardous waste for which treatability may be initiated in a single day.[70]

The Comprehensive Environmental Response, Compensation, and Liability Act (CERCLA)

The interface between RCRA and CERCLA is twofold. First, the evaluative process used to determine appropriate treatment of contaminated Superfund sites is essentially equivalent to that used to determine corrective action at RCRA facilities. Second, CERCLA provides that, in setting appropriate cleanup standards for a Superfund site, all "applicable or relevant and appropriate requirements" (ARARs) must be met.[71] This requirement ensures that treatment standards established under RCRA must be met in a Superfund cleanup.

The CERCLA assessment process. Before treatment may be undertaken at a Superfund site, EPA, often in conjunction with *potentially responsible parties* (PRPs)[72] must complete an evaluative process to determine the nature and quantity of contaminants at a site, and to determine the most effective treatment method or methods to clean up the site. This process is commonly referred to as the "RI/FS process." The *remedial investigation* (RI) phase involves a detailed characterization of the site. The *feasibility study* (FS) is an evaluation of alternative treatment methods that could be applied at the site.

At the conclusion of this process, which often takes many years, EPA issues a *Record of Decision* (ROD) which reflects the specific treatment methods that must be applied at the site.

Applicable or relevant and appropriate requirements (ARARs). The requirement that legally "applicable or relevant and appropriate

---

[69]Additional quantities may be approved by EPA regional offices, or by state officials in authorized states.

[70]C.F.R. §261.4(f)(1) through (f)(11).

[71]CERCLA Section 121(d)(2)(A), 42 U.S.C. §9621(d)(2)(A).

[72]*Potentially responsible parties* are persons who may be liable for some or all of the cost of cleaning up a Superfund site. PRPs may include persons who owned a waste disposal facility, who arranged for disposal of hazardous substances at such a facility, or transported hazardous substances to such a facility. 42 U.S.C. §9607(a).

requirements" be taken into account before undertaking a cleanup action at a Superfund site conceptually would mean that CERCLA response actions would need to comply with the RCRA LDRs where applicable. Thus, the prohibition against placement of a hazardous waste found at a Superfund site in or on land without pretreatment as specified in the LDRs would apply under Superfund cleanups as well as RCRA corrective actions.

EPA has determined, however, that the BDAT requirements of the LDRs are "inappropriate or unachievable for [contaminated] soil and debris from CERCLA response actions."[73] EPA has established a regulatory presumption that contaminated soil and debris qualifies for a treatability variance,[74] and such a variance will generally be included in a Superfund ROD when treatment of contaminated soil and debris is part of the remedial action.[75] As a result, bioremediation methods used at a Superfund site do not need to meet RCRA BDAT standards. The applicable standards would be determined on a site-specific basis in the ROD.

In addition, EPA applies the concept of an *area of contamination* (AOC) to Superfund site cleanups. An AOC is an area of continuous contamination. Movement of hazardous substances within an AOC does not constitute *placement* within the meaning of the LDRs; hence the pretreatment requirement does not become applicable. This approach has allowed for greater flexibility in the selection of remedial methods at Superfund sites (Bakst, 1991).

Demonstration of alternative or innovative treatment technologies

The regulatory measures discussed above, which EPA has initiated in the past several years to facilitate the use of bioremediation methods to treat hazardous waste are only a small part of the agency's efforts in this regard. EPA has also set in place numerous administrative programs to improve the utilization of this and other innovative technologies.

These efforts have been prompted at least in part by statutory incentives. The 1986 amendments to CERCLA[76] made two important changes to the Act which have a direct bearing upon bioremediation.

---

[73]"National Oil and Hazardous Substances Pollution Contingency Plan; Final Rule," 55 *Fed. Reg.* 8666, 8760 (March 8, 1990).

[74]40 C.F.R. §268.44.

[75]55 *Fed. Reg.* 8666, 8761 (March 8, 1990).

[76]Superfund Amendments and Reauthorization Act of 1986 (SARA), Pub. L. No. 99-499.

Section 121(b) of SARA[77] required EPA to select remedial actions which would result in a permanent and significant reduction of pollutants in a hazardous waste site. Specifically, the section states that EPA "shall conduct an assessment of permanent solutions and alternative treatment technologies or resource recovery technologies that, in whole or in part, will result in a permanent and significant decrease in the toxicity, mobility, or volume of the hazardous substance, pollutant, or contaminant." This language places a premium on those waste treatment methods which result in a complete breakdown of pollutants at a hazardous waste site.

Second, Section 311(b) of SARA[78] mandated EPA to "carry out a program of research, evaluation, testing, development, and demonstration of alternative or innovative treatment technologies...which may be utilized in response actions to achieve more permanent protection of human health and welfare and the environment."

Prompted by these statutory mandates, EPA has initiated several programs to promote innovative technologies, including bioremediation. These programs have been targeted at several problem areas: development of valid and consistent data on the effectiveness of innovative technologies, developing databases to disseminate such information, improving training and professional development in the various types and uses of innovative technologies, and improving financial support for research in the development and application of innovative technologies (Giamporcaro, 1991; Spann, 1991). Several of these programs are summarized below.

Bioremediation Field Initiative.   The Bioremediation Field Initiative was started in 1990 to document performance of full-scale bioremediation field applications (U.S. EPA, 1991b). The purposes of the initiative also include providing technical assistance to remedial project managers and on-scene coordinators on the uses of bioremediation at RCRA and CERCLA sites, and development of a treatability database (U.S. EPA, 1991c).

As of December 1991, six sites had been selected for field evaluation of bioremediation.[79] These sites represent a variety of types of contamination, including soil and groundwater contamination with hydrocar-

---

[77]42 U.S.C. §9621(b).

[78]42 U.S.C. §9660(b).

[79]The six sites are the Libby Superfund site, Libby, Montana; Park City Pipeline Spill, Park City, Kansas; Allied Signal Superfund site, St. Joseph, Michigan; Eielson Air Force Base, Alaska; Hill Air Force Base in Utah; and the Brookhaven Superfund site in Brookhaven, Massachusetts (U.S. EPA, 1991c).

bons (jet fuel) and industrial solvents and waste sludge piles. Site characterization is under way at two sites, and a treatability study is under way at a third site. Bioremediation has begun at the remaining three sites.

The Bioremediation Action Committee. The Bioremediation Action Committee (BAC) was established in 1990 to provide a forum for advancement of the science and practical applications of bioremediation. The BAC operates through six subcommittees whose responsibilities include prioritization of research needs in bioremediation; development and implementation of a national bioremediation oil spill response capability; developing protocols for testing the effectiveness of bioremediation products; identification and collection of bioremediation case study information for public and private sector usage; improving education about bioremediation on the part of scientists, engineers, and the public; and investigating the role of bioremediation technologies in pollution prevention (U.S. EPA, 1991c).

Bioremediation databases. EPA has also established numerous databases which are intended to serve as centralized, widely accessible sources of data on the variety and effectiveness of bioremediation methods. For example, the Alternative Treatment Technology Information Center (ATTIC) contains case study abstracts on various innovative treatment technologies, including 350 bioremediation case studies. The Vendor Information System for Innovative Treatment Technologies (VISITT) includes performance data on innovative treatment technologies, including bioremediation, from 86 vendors.

A related effort is being undertaken by the National Environmental Technologies Applications Corporation (NETAC), a cooperative venture established in 1988 between EPA and the University of Pittsburgh Trust. NETAC is preparing protocols to test the effectiveness of bioremediation products in degradation of oil spills. Once the protocols are established, a Biotechnology Products Evaluation center will be established to develop reliable, comparable data on the effectiveness of available products (Barron, 1991). This system will enable federal and state regulators to rapidly assess alternative products when they must select remedial measures to clean up oil spills.

The SITE program. The Superfund Innovative Technology Evaluation Program (SITE) was established in 1986. Through the SITE program, EPA partially funds projects to assess the effectiveness and cost of innovative and emerging alternative technologies. *Innovative technologies* are those for which there is a lack of consistent and reliable data concerning the applicability and effectiveness of the technology. *Emerging technologies* are to be pilot-scale technologies (Bakst, 1991). To date, 12 sites have been selected under the SITE program for application of bioremediation to treat a variety of contaminated media (Glass, 1991).

Support for bioremediation research.   EPA has also been increasing its bioremediation research programs. In 1987, EPA's Office of Research and Development established the Biosystems Technology Development Program, under which a consortium of EPA research laboratories conduct and sponsor bioremediation research. This program has been recently supplemented by the In-Situ Application Program, which received its initial funding in Fiscal Year 1992 (U.S. EPA, 1991*d*).

These two programs combine with the Bioremediation Field Initiative described above to create a comprehensive research program, which currently encompasses 40 research projects. These projects include studies of aerobic and anaerobic bioremediation of pesticides, bioremediation of PCB-contaminated sediments, and bioreactor treatment of contaminated surface waters. Funds allocated to the program in Fiscal Year 1992 were $5.8 million, which was expected to increase to about $10 million in Fiscal Year 1993 (U.S. EPA, 1991*d*).

The cumulative result of these efforts to date has been impressive. As recently as 1987, bioremediation had been selected as a remedial option at Superfund sites, either alone or in combination with other waste treatment technologies, in only five cases (U.S. EPA, 1988). Similarly, a 1988 case study of remedial selection at 10 Superfund sites conducted by the Office of Technology Assessment (OTA) concluded that bioremediation had been ignored in several remedial assessments (Garg and Garg, 1990).

The picture has changed substantially over the past few years. In 1991, EPA identified 124 RCRA, Superfund, UST, or PCB sites which were "considering, planning, operating, or [had] completed bioremediation." Under Superfund, bioremediation had been selected as a remedial method alone or in combination with other technologies in 31 out of 460 projects, "making it among the leading innovative technologies used" (Thayer, 1991). EPA Administrator William Reilly may have best characterized the accomplishments to date when he said, "I think the progress that has been made is...very much a collaborative progress of industry, government, and the research community—and it indicates what can be accomplished when all of these involved sectors cooperate" (U.S. EPA, 1991*a*).

The Toxic Substances Control Act

The Toxic Substances Control Act (TSCA)[80] was intended to accomplish two principal objectives: to enable EPA to screen new chemical substances prior to their introduction into commerce[81] and to regulate both existing and new chemical substances which present an unreasonable risk of injury to health or the environment.[82] To accomplish these objectives, EPA was also empowered to require testing of chemical substances,[83] and was given broad recordkeeping and data collection authority.[84] TSCA's coverage does not apply to certain categories of chemical substances, such as pesticides, food and food additives, human and animal drugs, and cosmetics; these and other specified chemical substances are subject to regulation under other statutes or by other federal agencies.[85]

The TSCA biotechnology program and genetically engineered microorganisms

The provisions of TSCA which have the most bearing on the use of bioremediation technologies are in Section 5 of the Act, under which EPA currently operates its TSCA biotechnology program. Section 5 of TSCA requires that new chemical substances undergo review by EPA before being introduced into commerce. Manufacturers of new chemical substances must submit a Pre Manufacturing Notice(PMN) to EPA at least 90 days before introducing the new substance into commerce. During the 90-day review period,[86] EPA conducts a review to determine whether the new chemical substance may present an unreasonable risk or injury to health or the environment, and if so, whether to impose regulatory controls "to prohibit or limit the manufacture, processing, distribution in commerce, use, or disposal of such substance or to prohibit or limit any combination of such activities."[87] EPA may also designate by rule specific uses of chemical substances as "significant new uses" which require similar notification to and review by the agency.[88]

---

[80]15 U.S.C. §2601 et seq. TSCA was enacted in 1976.

[81]TSCA Section 5, 15 U.S.C. §2604.

[82]TSCA Section 5(e), 15 U.S.C. §2604(e); TSCA Section 6, 15 U.S.C. §2605.

[83]TSCA Section 4, 15 U.S.C. §2603.

[84]TSCA Section 8, 15 U.S.C. §2607.

[85]TSCA Section 3(2), 15 U.S.C. §2602(2).

[86]The review period may be extended for an additional 90 days. TSCA Section 5(c), 15 U.S.C. §2604(c).

[87]TSCA Section 5(e), 15 U.S.C. §2604(e).

[88]TSCA Section 5(a)(2), 15 U.S.C. §2604(a)(2).

The notification requirements of Section 5 of TSCA apply only to the manufacture or processing of chemical substances for commercial purposes.[89] There has been little question in the implementation of TSCA that for-profit organizations are engaged in commercial purposes, and that research and development (R&D) undertaken by such organizations is similarly for commercial purposes.[90] However, certain research activities of nonprofit and academic institutions which involve the use of new chemical substances may not be for commercial purposes, and hence may not require notification to EPA.

Whether a chemical substance is "new" is determined by whether the substance is explicitly or implicitly included on the TSCA Inventory of Chemical Substances, which EPA is required to maintain under TSCA Section 8(b).[91] Chemical substances are explicitly listed on the TSCA inventory upon receipt by EPA of a Notice of Commencement (NOC) submitted by a manufacturer. The NOC reflects the manufacturer's intent to begin commercial distribution of a chemical substance. In addition, certain chemical substances, such as minerals and ores, are considered to be implicitly included on the TSCA inventory; i.e., they are considered to already be in commerce in the United States, and hence no prior notification or review by EPA is required.[92] Once a chemical substance is included on the TSCA inventory, either explicitly or implicitly, it may be used for any purpose (unless otherwise subject to regulation under TSCA or other statutes), and subsequent notification to EPA by persons other than the original submitter of the PMN is not required.

In 1984, EPA published a proposed Policy Statement[93] in which the Agency clarified that the statutory term *chemical substance* included living organisms, including microorganisms.[94] EPA also proposed that microorganisms produced through specific techniques such as recom-

---

[89] TSCA Section 5(i), 15 U.S.C. §2604(i).

[90] Under Section 5(h)(3) of TSCA, however, chemical substances which are manufactured or processed only in small quantities (as determined by EPA by rule), and solely for R&D, are exempt from the notification requirements of Section 5. 15 U.S.C. §2604(h)(3).

[91] 15 U.S.C. §2607(b).

[92] Asbestos would be an example of a chemical substance which is implicitly listed on the TSCA inventory.

[93] "Proposed Policy Regarding Certain Microbial Products," 49 *Fed. Reg.* 50886 (December 31, 1984). EPA's proposed Policy Statement was published as part of a larger document entitled "Proposal for a Coordinated Framework for Regulation of Biotechnology," 49 *Fed. Reg.* 50856 (December 31, 1984).

[94] *Chemical substance* is defined under TSCA as "any organic or inorganic substance of a particular molecular identity, including (i) any combination of such substances occurring in whole or in part as a result of a chemical reaction or occurring in nature..." TSCA Section 3(2), 15 U.S.C. §2602(2).

binant DNA technology or cell fusion should be considered "new chemical substances" subject to the PMN reporting requirements of TSCA Section 5.

In response to public comments which argued that EPA's proposed approach to oversight of new microorganisms was based on the processes by which such microorganisms were created rather than the potential risk which they might pose, EPA changed its approach to defining which microorganisms would be considered new under Section 5 of TSCA. In 1986, as part of the Coordinated Framework for Regulation of Biotechnology, EPA issued a Policy Statement announcing that intergeneric microorganisms (those which involve the combination of genetic material from source organisms in different genera) would be considered new chemical substances under TSCA Section 5.[95] Persons intending to manufacture such microorganisms for commercial purposes were required to submit a PMN to EPA. The agency also requested voluntary submission of PMNs by persons who intended to introduce intergeneric microorganisms into the environment for research and development purposes, and by persons whose microorganisms were pathogenic or contained genetic material from pathogens.[96]

Under the current TSCA biotechnology program, therefore, persons intending to use intergeneric microorganisms for commercial purposes to biodegrade hazardous wastes would be required to submit a PMN to EPA. Persons intending to introduce intergeneric microorganisms into the environment for bioremediation research would be encouraged to file voluntary PMNs with EPA.

The EPA has not received any PMNs to date for intergeneric microorganisms used for bioremediation purposes. Bioremediation projects to date have tended to focus on the biostimulation or bioaugmentation of naturally occurring microorganisms already present at a site which are capable of degrading hazardous constituents in soil or groundwater (Glass, 1991). However, the bioremediation industry is showing increasing interest in the use of genetically engineered microorganisms for bioremediation. Many experts expect to see genetically engineered microorganisms used for bioremediation in bioreactors, and perhaps in field demonstrations, over the next five to ten years (Garg and Garg, 1990; Glass, 1991; Roy, 1991). As Sayler and Day (1991) note, "The next generation of bioremediation products and

---

[95] "Coordinated Framework for Regulation of Biotechnology," 51 *Fed. Reg.* 23302 (June 26, 1986). The EPA policy statement appears at 51 *Fed. Reg.* 23313.

[96] EPA believes that researchers intending to introduce intergeneric microorganisms into the environment have generally submitted voluntary PMNs. The agency has not received any PMN submissions under the pathogen provisions of the 1986 policy statement.

processes routinely will incorporate genetic engineering to improve strains, gain better process control, guarantee high performance, and monitor environmental activity."

Genetically engineered microorganisms are expected to offer several advantages in bioremediation. Many experts have stated that genetically engineered microorganisms would allow enzymatic degradative pathways to be more precisely controlled, including directing metabolic flow to avoid any potential toxic intermediates (Thayer, 1991). Genetic engineering may also allow greater survivability of microorganisms used in bioremediation by imparting resistance to factors which impede the activity and viability of the microorganisms (Sayler and Day, 1991; Thayer, 1991). Genetically engineered microorganisms may also allow acceleration of degradative pathways, thereby increasing the commercial viability of the technology as a whole (Glass, 1991; Thayer, 1991).

Naturally occurring microorganisms

Naturally occurring microorganisms and intrageneric microorganisms (those whose creation involves the exchange of genetic material between microorganisms in the same genera) are not subject to the notification requirements of TSCA Section 5 under the provisions of the 1986 Policy Statement. These categories of microorganisms are considered to be already existing in nature, and hence implicitly listed on the TSCA inventory.

This does not mean, however, that EPA could not exercise regulatory control over naturally occurring microorganisms, including those used for bioremediation purposes, if the agency were to determine that the use of such a microorganism presented an unreasonable risk to health or the environment. For example, EPA could use its authority under Section 5 to designate specified uses of a naturally occurring microorganism as a significant new use. The agency could also take action against microorganisms of concern under Sections 6 or 7 of the Act, or could require recordkeeping under Section 8 of the Act. In addition, all manufacturers or processors of chemical substances have an affirmative duty under Section 8(e) of TSCA to notify EPA of any "information which reasonably supports the conclusion that [a chemical substance] presents a substantial risk of injury to health or the environment." [97]

The bioremediation industry has been particularly concerned about the possible coverage of naturally occurring microorganisms under the

---

[97]15 U.S.C. §2607(e).

TSCA biotechnology program. This concern stems from a document which was made available by EPA for public comment in 1988.[98] The agency stated in that document that it was considering requiring reporting of all commercial uses involving environmental release of naturally occurring microorganisms ongoing as of or since December 1, 1985. The agency's intention was to establish a list of such uses which would not be subject to reporting. Persons intending to use any microorganisms for a commercial use which was not listed would have been required to submit a significant new use notice to EPA before doing so. Under this scheme, all uses of naturally occurring microorganisms for bioremediation purposes would have been reported to the agency. Any use of a microorganism not included on the list, or use of a listed microorganism for an unlisted use, would have required submission of a significant new use notice.

In July 1991, EPA held a meeting of a subcommittee of its Biotechnology Science Advisory Committee (BSAC) to address scientific issues raised in draft proposed TSCA biotechnology regulations currently under development by the agency.[99] The draft proposed regulations do not propose retaining the approach outlined in the 1988 document—i.e., establishing a list of all commercial uses of naturally occurring microorganisms, and requiring submission of significant new use notices for uses or microorganisms not included on the list.

In the July 1991 draft proposed regulations, EPA stated that it was proposing to change the definition of "new" microorganisms from that adopted in the 1986 Policy Statement. The proposed rule would in all likelihood present several options for defining which microorganisms would be subject to notification under Section 5 of TSCA. EPA's preferred approach is to define *new microorganisms* as those with deliberately modified hereditary traits, with the exception of microorganisms which fall within certain exclusion categories. As under the current program, naturally occurring microorganisms would not be considered new. In addition, under EPA's preferred approach, microorganisms used for bioremediation which fell within one of the exclusion categories would similarly not be subject to screening under Section 5 of TSCA.

In summary, "[b]ioremediation likely will not be significantly impacted by TSCA unless genetically modified, or engineered, microorganisms are employed" (Bakst, 1991).

---

[98] "Biotechnology; Request for Comment on Regulatory Approach," 54 *Fed. Reg.* 7027 (February 15, 1989).

[99] The BSAC is a science advisory committee on biotechnology matters; it is attached to EPA's Office of Prevention, Pesticides, and Toxic Substances (OPPTS).

The bioremediation of PCBs

Polychlorinated biphenyls (PCBs) occupy a unique regulatory position. When Congress enacted TSCA in 1976, it expressly singled out PCBs for regulation. Under Section 6(e) of TSCA,[100] Congress prohibited the manufacture, processing, or distribution of PCBs other than in a totally enclosed manner after January 1, 1978, and required EPA to promulgate regulations governing the disposal of PCBs.[101]

The EPA's PCB disposal regulations apply to PCBs and "PCB Items" which contain PCB concentrations of 50 ppm or greater.[102] *PCBs* as defined by the regulations include mineral oil dielectric fluid and other liquids containing PCB concentrations of 50 ppm or greater but less than 500 ppm, soils and other solid materials contaminated with PCBs, and dredged materials and municipal sewage treatment sludge.[103] *PCB Items* include transformers, capacitors, and manufactured items containing PCBs.[104]

Commercial operating permits.   The PCB regulations provide for the issuance of two types of disposal permits: commercial operating permits and R&D permits (Giamporcaro, 1991). Commercial operating permits may be issued by EPA headquarters or by individual regional offices. The PCB reglations generally provide that PCBs and PCB Items can be disposed of in an incinerator, a chemical waste landfill, or a high-efficiency boiler.[105]

The regulations also contain an important provision which allows for the use of alternative methods of destroying these materials.[106] The alternative methods are required to achieve a level of performance equivalent to incineration or high-efficiency boilers. In order to be considered "equivalent," alternative methods such as bioremediation must reduce PCB concentrations to less than 2 parts per million per remaining PCB congener peak, as measured with gas chromatography.[107] The practical effect of this standard is that the actual PCB concentration in a contaminated sample which is considered to have been disposed will vary according to the number of PCB congeners present in the sample. For example, if the contaminated sample contained only one PCB con-

---

[100]15 U.S.C. §2605(e).

[101]Ibid.

[102]40 C.F.R. §761.60.

[103]40 C.F.R. §761.3.

[104]Ibid.

[105]40 C.F.R. §761.60(a), (b).

[106]40 C.F.R. §761.60(e).

[107]The chemical class of PCBs consists of 209 different chemical structures, which are referred to as "congeners."

gener, the final PCB concentration level would need to be 2 ppm or less. If the sample contained six different PCB congeners, however, the final total PCB concentration in the sample could be 12 ppm or less. This flexibility is conducive to permitting the use of alternative disposal methods.

The EPA has issued 38 commercial operating permits for the disposal of PCBs. Twenty-four of these have been for alternative disposal methods, although only one has involved biological degradation of PCBs.

R&D permits.   The PCB regulations also provide for the issuance of R&D permits for PCB disposal methods.[108] Permits involving the use of less than 500 lb of PCB material are reviewed at the regional level, whereas those involving 500 lb or more are reviewed at EPA headquarters.

To date, EPA has approved 15 R&D permits involving methods of biological degradation of PCBs. Six of these permits have been approved at the headquarters level, and nine at the regional level. All of the research has involved the use of naturally occurring microorganisms, and it has involved a variety of biological degradation methods. The rate of applications for R&D permits has been steadily increasing over the past several years. For example, of the six permits issued by EPA headquarters, the first was issued in 1988 for the use of white rot fungus to treat PCB-contaminated soil. Two additional R&D permits were issued in 1990. One of these was for an in situ study of factors affecting anaerobic bioremediation of pond sediments contaminated with PCBs. The second was for the treatment of PCB-contaminated soils and sediments in a pilot-scale bioreactor. Three R&D permits were issued in 1991. Two were for in situ treatment of contaminated soil and river sediments, and the third project involved a study of the effectiveness of combining photolysis and bioremediation to treat contaminated soils.

PCB disposal at RCRA and Superfund sites.   Although the disposal of PCBs must be carried out pursuant to a commercial operating permit issued under TSCA Section 6(e), the provisions of the RCRA and Superfund programs must also be taken into account when PCB contamination is present at sites covered under either of these programs. Under the RCRA regulations, PCBs are listed as a hazardous constituent which, if present in solid waste, may be a basis for inclusion of such waste as listed hazardous waste.[109] In fact, some P-listed and U-listed wastes contain PCBs. Such wastes are required to

---

[108]40 C.F.R. §761.60(i)(2).

[109]See 40 C.F.R. §261.11(a)(3) and 40 C.F.R. Part 262 App. VIII.

meet the applicable treatment and disposal requirements under both TSCA and RCRA.[110] That is, the PCB concentration in the RCRA-listed hazardous waste must be reduced to a level commensurate with the PCB disposal requirements under TSCA, as discussed above.

In addition, the selection of the eventual treatment method for disposal of PCB-contaminated materials at an RCRA facility or Superfund site is determined under the site assessment processes carried out under those respective programs. The fact that an owner or operator of an RCRA facility, or a potentially responsible party at a Superfund site, has obtained an R&D permit to use a particular technology to degrade PCB-contaminated materials does not foreclose the selection of some other technology for corrective action or remedial or response actions at that site. For example, several of the sites for which PCB R&D permits have been issued under TSCA are in the RCRA corrective action process. These permits contain a statement that the PCB permit cannot be used to bias or predetermine the selection of corrective measures under RCRA at the facility.

In this regard, it is important to note that although there is considerable research under way to access the effectiveness of bioremediation of PCBs, as evidenced by the number of R&D permits issued under TSCA for PCB bioremediation over the past few years, the research is still in its initial stages, and the effectiveness of these methods to degrade PCBs at large scale has not yet been demonstrated. It is hoped that as this research progresses over the course of the next decade additional commercial operating permits for the bioremediation of PCBs can be issued.

The Federal Plant Pest Act

The Federal Plant Pest Act (FPPA), which is administered by the Animal and Plant Health Inspection Service (APHIS) of the USDA, has had more theoretical than practical applicability to date to the use of bioremediation methods for treatment of hazardous waste. The FPPA requires persons who intend to import or to transport a plant pest in interstate commerce to obtain a permit. A *plant pest* is defined under the FPPA as the living stage of any organism "which can directly or indirectly injure or cause disease or damage in any plants or parts thereof, or any processed, manufactured, or other products of plants."[111]

The APHIS regulations implementing the FPPA apply to both nonengineered and engineered organisms which possess plant pest

---

[110]See generally, "Land Disposal Restrictions for New Listed Wastes and contaminated Debris," 57 *Fed. Reg.* 958, 1006 (January 9, 1992).

[111]7 U.S.C. §150aa(c).

characteristics.[112] In June 1987, APHIS promulgated regulations under the FPPA in which it asserted jurisdiction to require a permit prior to the release into the environment or transport of genetically engineered microorganisms which contain genetic material derived from a microorganism listed as a plant pest by APHIS,[113] and which cause injury, disease, or damage in plants.[114]

The requirements of the FPPA would therefore be applicable only in the narrow circumstance where a microorganism used for bioremediation possessed plant pest characteristics, or, if the microorganism was genetically engineered, was derived from a listed plant pest and possessed plant pest characteristics. Although APHIS has issued hundreds of permits under its regulations, none has involved the use of microorganisms for bioremediation purposes.

Conclusion

Much has been accomplished in recent years to address problem areas in the application of bioremediation; much remains to be done. Perhaps the area of greatest achievement has been in establishing an infrastructure for generation and dissemination of data and information on bioremediation. These efforts need to continue.

The area that may need continued close attention is regulatory obstacles or disincentives to the use of this technology. In particular, the expiration of the national capacity variances under the RCRA LDRS may hinder, at least in the short term, the application of this technology. In addition, as BDAT standards for contaminated soil and debris are developed, the potential impact on bioremediation must remain in clear focus.

Finally, promulgation of final biotechnology regulations under TSCA may help to accelerate the development of the next stage in bioremediation—the use of genetically engineered microorganisms. Such regulations would clarify the nature and extent of TSCA's applicability to the use of microorganisms for bioremediation, and thereby help bioremediation companies better assess the investment costs associated with development of genetically engineered microorganisms for this purpose.

---

[112] See 7 C.F.R. Part 330, 7 C.F.R. Part 340.

[113] See 7 C.F.R. §340.2 for the list of microorganisms which are or contain plant pests.

[114] 7 C.F.R. Part 340.

References

Bakst, J. S. 1991. Impact of present and future regulations on bioremediation. *J. Ind. Microbiol.* 8:13–22.

Barron, T. 1991. EPA closing in on standards for oil spill bioremediation. *Environ. Today* 2:1, 59.

Day, S. M. 1990. Federal regulations and policies hold the "veto power" over commercialization of biotreatment—an industry perspective. Hazardous Materials Management, HAZMACON 90, Anaheim, CA.

Garg, S., and D. P. Garg. 1990. Genetic engineering and pollution control. *Chem. Eng. Progr.* 86:46–51.

Giamporcaro, D. E. 1991. Bioremediation waste control more promising. Reprint. BioWorld. Io Publishing, San Mateo, CA.

Glass, D. J. 1991. The promising hazardous waste bioremediation market in the United States. Reprint. Decision Resources, Burlington, MA.

Hill, R. L. 1991. An overview of RCRA: The "mind-numbing" provisions of the most complicated environmental statute. *Environ. Law Rep.* 21:10254–10276.

National Governors' Association. 1991. States' Use of Bioremediation: Advantages, Constraints and Strategies. National Governors' Association, Washington, D.C.

Nicholas, R. B., and D. E. Giamporcaro. 1989. Nature's prescription. *Hazmat World,* pp. 30–36, June.

OTA. 1991. Bioremediation for Oil Spills. Office of Technology Assessment, OTA-BP-0-70, U.S. Government Printing Office, Washington, D.C., May.

Roy, K. A. 1991. Ecova's hopes for success hitched to bioremediation's rising star. *Hazmat World,* pp. 40–43, May.

Sayler, G. S., and S. M. Day. 1991. Special Report: Bioremediation—Experts explore various biological approaches to cleanup. *Hazmat World,* pp. 51–53 (January).

Spann, J. J. 1991. EPA workshop on innovative technologies. *Water Environ. Technol.,* p. 18 (sidebar).

Thayer, A. M. 1991. Bioremediation: Innovative technology for cleaning up hazardous waste. *Chem. Eng. News* 69:23–25, 28, 32–37, 41–44.

U.S. EPA. 1988. ROD (Record of Decision) Annual Report. U.S. Environmental Protection Agency, Washington, D.C., July.

U.S. EPA. 1990. Summary Report on the EPA—Industry Meeting on Environmental Applications of Biotechnology. U.S. Environmental Protection Agency, Washington, D.C., February 22.

U.S. EPA. 1991a. Summary of the Second EPA/Industry Meeting on Environmental Applications of Biotechnology. U.S. Environmental Protection Agency, Washington, D.C., June 14.

U.S. EPA. 1991b. Furthering the Use of Innovative Technologies in OSWER Programs. U.S. Environmental Protection Agency, Washington, D.C., August, Doc. No. 9380.0–17FS.

U.S. EPA. 1991c. Bioremediation in the Field. U.S. Environmental Protection Agency, Washington, D.C., December.

U.S. EPA, 1991d. Bioremediation Research Program Strategy. U.S. Environmental Protection Agency, Washington, D.C., December.

In Situ Bioremediation: Basis and Practices

Carol D. Litchfield

Chester Environmental
Monroeville, Pennsylvania

Basis for in Situ Bioremediation

In situ bioremediation (ISB) is the treatment in place, without excavation, of soils and groundwater contaminated with organic compounds. This chapter on ISB will be concerned only with contamination of the vadose (unsaturated) and saturated zones of an aquifer. It will not cover the treatment of sludge basins, ponds, sediments, or embayments, as these require an entirely different engineering approach although the biological principles are usually the same; nor will the types of laboratory studies so necessary for a successful ISB be reviewed.

In situ biodegradation is a natural process which has been going on since the first microbes and excess organic matter were both present in the soil. At its most fundamental, *biodegradation* is the recycling or repackaging in soils and water of carbon, nitrogen, and other nutrients. This process is essential for the proper maintenance of the carbon and nitrogen cycles in nature. In recent years it has been recognized that in situ biodegradation can also be applied to hazardous wastes, and techniques for detecting and enhancing natural in situ bioremediation have been developed. These are the focus of this chapter.

Two major engineering approaches to the design of ISB have evolved. One deals with shallow groundwater systems and the saturated zone, as diagrammed in Fig. 7.1. Pumping water from a recovery well creates a cone of depression in the saturated zone. The recovered water is passed through a filter to remove suspended solids from the aquifer,

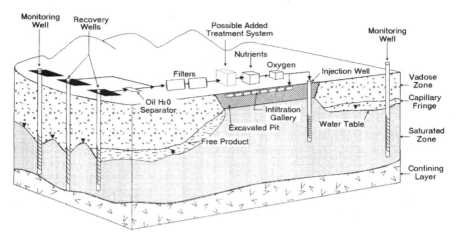

Figure 7.1 Diagram of an idealized in situ bioremediation design.

and it may pass through an aboveground treatment system such as a bioreactor or air stripper. The water is then supplemented with nutrients and an electron acceptor and returned to the aquifer, where it enters close to the source of the contamination—for example, into an excavated tank pit. Here, in the vadose zone, the groundwater mounds, providing nutrients and moisture to the indigenous bacteria in the contaminated unsaturated soil. Gravity and the pumping action of the recovery wells pull the groundwater into the saturated zone, where it passes over the soils containing the contaminants and introduces nutrients needed for microbial degradation. Carbon is seldom a limiting factor in the contaminated subsurface, but phosphorous and/or nitrogen frequently is. It is therefore critical to perform treatability studies to determine which nutrients are limiting natural ISB and what concentrations are most effective at enhancing ISB.

The second approach involves treating the unsaturated soils. While this is a more recent development, several field trials have demonstrated the usefulness of this technique. The basic concept is diagrammed in Fig. 7.2. Here air is forced into the vadose zone at a relatively slow rate to avoid dehydration of the soils and the microorganisms. Moisture is returned to the soils along with the necessary nutrients via a sprinkler or drain field system. Horizontal pipes below the zone of contamination are constructed both to capture the added moisture and to aid in drawing the air into the aquifer. This technique is frequently referred to as *bioventing*.

This chapter will describe ISB, its requirements and limitations, trace its development as an acceptable tool for the remediation of soils and groundwater, and cite case studies of field applications using either traditional ISB or bioventing.

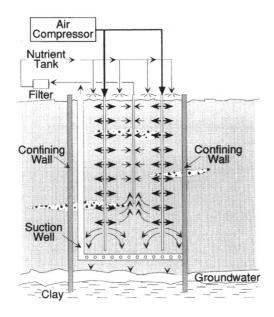

Figure 7.2 Design of a potential bioventing process. Adapted from Lund et al., Ref. 56.

Requirements for ISB

In order for ISB to be enhanced by the addition of limiting nutrients, two major factors must be considered: the site characteristics and the microorganisms.

Site characteristics

For the proper design of an in situ treatment system one needs to know a great deal about the soil and aquifer characteristics of the site. A summary of general geotechnical testing for aquifer soils has been recently published by Sevee,[80] while other pertinent references for hydrogeological tests can be found in Freeze and Cherry[30] and Bouwer.[11] At a minimum these geotechnical tests must determine the exact nature and the horizontal and vertical extent of contamination, whether it impacts the groundwater or just the vadose-zone soils, depth to groundwater, hydraulic conductivity and/or permeability of the soils (i.e., can water and nutrients be moved through the saturated or vadose zone and how rapidly), the specific yield and storage coefficient of the aquifer, the zone of influence of recovery or reinjection wells, groundwater flow direction, the ability of the aquifer to receive the recycled water, the cation exchange capacity of the soils to estimate nutrient sorption onto the soil particles, and the anionic and cationic composition of the soils and the groundwater.

For some of these geotechnical tests standard methods have been developed,[5] (see Table 7.1), and need to be performed whether one is pro-

TABLE 7.1   Some ASTM Standard Methods for Soil Characterization[5]

Geotechnical tests	ASTM number
Particle size	D-422
Moisture content	D-2216
Minimum relative density	D-4254
In situ density	D-2435
Specific gravity	D-854
Atterberg limits	D-4318
Bulk density	D-1587 or D-2937
Description of soils	D-2488
Hydraulic conductivity in vadose zone, comparison of methods, a guide	D-5126

jecting in situ bioremediation or a more traditional remedial strategy. For other tests, Refs. 30 and 80 contain details of accepted practices. In many states these tests must now be performed by a certified hydrogeologist/geologist. The state licensing requirements are changing constantly, so if there are questions regarding a particular state it is best to consult with that state's environmental department.

In addition, an understanding of drilling methods and the installation of monitoring wells is essential to ensure that the proper design is used for optimal ISB monitoring.[19]

For the successful treatment of any contaminated site, as thorough an understanding of the geology, geochemistry, and hydrogeology as possible is essential because one aspect of ISB is the recycling or introduction of water into the near-surface or subsurface soils. It is critical, then, that hydrogeological control be established so that the water does not go into unwanted areas, thereby further spreading the contamination. Also, it is important to prevent excessive use of nutrients, and to prevent the enhancement of microorganisms and the flow of groundwater into areas that are not contaminated.

Knowledge of the geology and geochemistry of the site are also important to prevent unwanted interactions of the soils or groundwater with the added nutrients. This is especially true when adding phosphates to high-calcium soils and groundwater or when adding oxygen to environments containing high amounts of reduced iron or manganese which can be oxidized. Both situations cause precipitation and consequent plugging of the formation. High levels of calcium, iron, or manganese are not insurmountable problems, but they must be known beforehand so that systems can be designed to control or eliminate the problem.

Other site characteristics which much be evaluated for any remedial activity include the below-grade location of buried objects such as electrical cables, water pipes, and sewers, and the status of the aboveground space, including the topography, presence of overhead pipe racks, electrical wires, etc., and location of buildings, parking lots, roads, or other structures and activity areas. All of these can influence where equipment will be placed, what the sources of the contamination are, and even the extent and rate of migration of the contaminant plume.

Microbiological characteristics

The most basic premise for ISB is that microorganisms are present in the subsurface, have adapted to the contaminants, and all of the necessary nutrients are present or can be added for the optimal biodegradation of the contaminants.

This seems like an obvious and fundamental statement, but as recently as the mid-1970s one could find in the literature statements about the extremely low numbers of microorganisms in groundwaters. It was hypothesized that water filtering through soil would lose all microbes and hence groundwater would be essentially sterile. In 1973, McNabb and Dunlap[59] reviewed the existing literature and reported, contrary to the prevailing belief, that microorganisms were indeed widespread in the subsurface. Therefore, they postulated, the potential existed for microbial degradation of organic materials in groundwaters. Ghiorse and Balkwill also demonstrated a diverse population of microorganisms in the subsurface during their study of pristine sites in Oklahoma and Louisiana.[31]

However, it was believed by many scientists that the microorganisms obtained from the subsurface were in fact contaminants derived from surface soils. This issue was not resolved until McNabb and Mallard[60] and Wilson and coworkers[94] described a technique for retrieving uncontaminated subsurface soils. This technique relies on a paring device attached to an extruder used to push the soil from the core barrel.[60,95] A modification of this technique was used during the deep-coring program at Savannah River, South Carolina, where microorganisms were isolated and identified in deep-subsurface cores taken to 265 m.[7] Studies in Germany have also demonstrated the ubiquitous presence of microorganisms in the subsurface.[90] A review of microbial detection methods, distributions, and activities in subsurface environments has been published by Ghiorse and Wilson[32] that clearly shows the advances made in our understanding of microbial ecology of the subsurface during the last ten to fifteen years.

It is not enough, however, for microorganisms to be present; they must also be adapted to the environment and to the contaminants. In 1972, McKee et al.[61] reported that naturally occurring species of the genera *Pseudomonas* and *Arthrobacter* were involved in the disappearance of gasoline adsorbed to soil particles. The gasoline disappeared at the rate of 2 g of gasoline per 1 g of bacterial cells. The next year, Litchfield and Clark[54] reported that significant numbers of hydrocarbonoclastic bacteria were present in groundwater contaminated with petroleum hydrocarbons and that the numbers were related to the concentration of the contaminant. These early studies implied that natural adaptation had occurred to permit the biodegradation of petroleum hydrocarbons.

Similar studies have also been reported for nonpetroleum hydrocarbon contamination. Spain and Van Veld showed that river sediment populations from the Escambia River when placed in ecocores could adapt to repeated exposures to p-nitrophenol.[82] This concept of microbial adaptation to contaminating organics was also examined by Aelion et al., who studied rates of mineralization of selected compounds.[1] Of the five aromatic compounds and ethylene bromide (EB) they tested, only p-nitrophenol demonstrated the expected adaptation response. Phenol, p-cresol, and EB were mineralized so rapidly that the authors concluded adaptation of the aquifer populations had already occurred.[1]

Further work documenting the adaptation or selection of microbes to degrade contaminants was presented by Lee et al. for a creosote-waste site in Conroe, Texas.[49] Microcosms established with soils from the contaminated area at this site had one to two orders of magnitude greater loss of six specific compounds from creosote than was observed in microcosms prepared with soils from a nearby pristine site.[49]

A question arises from the above studies as to whether the contaminants exert selective pressures on the population and thus select for the microorganisms capable of degrading the contaminants and allowing them to survive in greater numbers than all others, or whether there is genetic alteration which then allows the new, contaminant-degrading bacteria to outcompete the other microbes. Circumstantial evidence exists for both theories. Ogunseitan et al. found significantly higher numbers of plasmids in bacteria isolated from a contaminated site than from a pristine site.[69] Jain et al. also demonstrated that naturally occurring catabolic plasmids could be maintained without the introduction of allochthonous strains.[42] More recently, the presence of plasmids in a significant number of bacterial isolates from the Savannah River deep-drilling study was reported.[29] Thus, genetic exchange and natural genetic modifications may occur in response to the introduced organic compounds.

It should also be noted that others have found multiple populations of bacteria coexisting in soils.[47,77] Similar results had been reported by Azam and Hodson for marine microbial assemblages from seawater where nonlinear kinetics were observed during the mineralization of organics over a three order-of-magnitude substrate concentration range. The authors attributed this to different populations with different affinity constants responding to the various concentrations of the substrates.[6] The effect of multiple populations is to cause multiphasic kinetics for biodegradation. The extent to which this might be occurring during an actual ISB has not as yet been investigated. If multiphasic kinetics are operable in contaminated groundwater, such information would have immediate value in responding to the claims that ISB isn't applicable to a certain site because the concentrations are too low or too high.

Finally, for microbes actively to degrade contaminants, they must have sufficient nutrients present. Usually one or more nutrients or micronutrients are deficient in the subsurface, so for optimal degradation it is critical to perform treatability tests to determine which components are necessary and at what concentrations. This involves testing the soils and groundwater for nitrogen sources such as ammonium and nitrate, for phosphates, and for other micronutrients. The importance of these nutrients to microbial degradation was demonstrated by Swindoll et al.[85] and Lewis et al.[50] These latter authors found that the length of the lag periods was longer for microbial populations in nitrogen- or phosphorous-limited environments and the length decreased with the addition of nitrogen or phosphorous.[50] It is important, too, to know whether the system is anoxic or oxidized, as this will affect the oxidation-reduction state of many elements and thus their bioavailability. In 1987, Barker et al. showed that low oxygen levels limited the extent of biodegradation of benzene, toluene, and xylene (BTX).[18] Thus treatability tests typically result in the addition of oxygen or other electron acceptor to the aquifer materials along with nitrogen source and/or phosphate and perhaps trace elements or potassium.

Limitations to ISB

There are four major factors which can limit the application of ISB:

- Time
- Metabolic by-products or recalcitrance
- Geochemistry and hydrogeology
- Environmental factors

The extent to which these factors become inhibitors of the application of ISB is often more a political and economic decision than a scientific one. However, there are situations, especially in the areas of degradation products and control of the hydrogeology, that could limit the application of this technology.

Time

ISB is not an instantaneous, quick fix technology. It requires time for the necessary treatability study, time for the geotechnical work, and time for the microorganisms to grow and degrade the contaminants in situ. If the client is under a consent decree to have a site cleaned within a very limited time frame, it may not be possible to use ISB at that site. Just as often, however, time is given as an excuse for not using ISB because there is a perception that traditional pump-and-treat methods will be quicker and more acceptable to the regulatory community. In most cases, with the exception of excavation, ISB will usually be quicker and more complete than pump-and-treat technologies. If used in conjunction with air stripping or an aboveground bioreactor, the two technologies combined are substantially quicker than either technology alone, while the added cost is usually minimal.

Metabolic by-products or recalcitrance

One of the major reasons for performing treatability studies is to determine the ability of the native population to be enhanced, the extent of biodegradation possible under ideal laboratory conditions, and the potential for unwanted metabolic by-product formation. If a significant population of contaminant-degrading microbes cannot be stimulated, then the site is not a candidate for ISB. Typically, we expect to see at least 20 to 25 percent biodegradation during a 4- to 6-week treatability study. If this does not occur, and in more than 40 cases that has happened only twice, then we would recommend other technologies for the remediation of that site.

An equally serious, but more difficult, case is the production of unwanted metabolic by-products. If biotransformation rather than mineralization occurs, then one must determine the toxicity of these new products. If they are less toxic and/or immobile than the original contaminant, then ISB is still a reasonable choice. If, however, they are more toxic and/or more mobile, risk analyses must be performed to determine whether ISB should still be considered. In most cases, the answer will be no. It may be possible, however, to include an aboveground treatment system (perhaps operated under different redox conditions) to mitigate these transformation products and continue the action of the ISB on the subsurface soils.

One needs to be careful in extrapolating laboratory data to the field, however. We recently had a potentially limiting situation develop in the laboratory for a site containing chlorobenzene. One of the intermediates in chlorobenzene biodegradation is 3-chlorocatechol.[73] When the numbers of chlorobenzene-degrading bacteria were estimated, we discovered that certain colonies appeared purple. This was the result of the accumulation of 3-chlorocatechol. Fortunately, bacteria isolated from nonpurple colonies could utilize 3-chlorocatechol as well as chlorobenzene. Therefore, we could demonstrate that the natural microbial consortium would not accumulate this intermediate during the ISB, and the state regulatory agency approved an in situ bioremediation at this site.[68] Thus, the natural microbial consortium may mineralize complex mixtures better than pure cultures or even laboratory studies would indicate, and so prevent the accumulation of metabolic by-products.

There are compounds, mixtures, and substances which are recalcitrant or for which no practical in situ treatments have as yet been developed. These include some of the polychlorinated biphenyls (PCBs), specifically the PCB congeners greater than 1248; polyaromatic hydrocarbons (PAHs) with six-membered and larger rings; mixed wastes involving radioisotopes; asbestos; and metals such as cadmium, mercury, or chromium. If metals are present at a site, during the treatability studies it must be demonstrated that mobilization of previously immobilized metal complexes does not occur. Subtle changes in pH or organic content can alter the affinity or redox state of a metal and make it more mobile, thus creating a problem where none previously existed. Similarly, microbial metabolism may result in the precipitation of metals as their sulfides under anaerobic conditions. This could be advantageous or result in aquifer plugging. The utility of selective in situ metal precipitation has yet to be investigated in the field.

Geochemistry and hydrogeology

The limiting factor in ISB is control of the groundwater. Of course this is a problem for any remedial technology, but it becomes more critical when one is adding nutrients and/or oxygen to the subsurface. Without control of the distribution of the recycled nutrient-laden water one has no indication as to whether the amendments are reaching the contaminated zone or whether other regions are being impacted by the nutrients. Not only does this prolong the remediation and perhaps lead to incomplete biotreatment, it is also very costly and leads to unnecessary enrichment of the groundwater. Conditions under which this scenario could occur include fractured bedrock or a poorly or improperly defined aquifer.

Although clay is not the best medium for an ISB, as long as water and nutrients can be moved through the aquifer (however slowly), it is possible to implement ISB. A tight formation will take longer for the cleanup levels to be reached, but as long as time is not a critical factor, ISB is possible in silty clay or clayey formations. Under these circumstances any treatment technology will take longer than it will with a more permeable ($>10^{-5}$cm/s) formation.

Environmental factors

The major environmental factors which are often cited as impacting ISB are temperature, pH, and redox state. None of these is necessarily a limitation to ISB. Groundwater temperatures do not change significantly with the seasons. There may be a slight 3 to 4°C increase in shallow, southern groundwaters during the summer and a similar drop for northern groundwaters during the winter, but these are not outside the limits for microbial growth. In fact, the microbial consortium present in the aquifer has adapted to these conditions and is able to metabolize and grow at the natural temperatures.

The pH of the aquifer is normally pH 6–9. There are some cases where the contaminants have decreased the pH to 4, but even here one can demonstrate biodegradation in the laboratory, indicating that the microbial consortium has adapted to the local environmental conditions. The potential danger comes with inducing a drastic pH change due to acid produced during the biodegradation. If the aquifer is not highly buffered, the potential exists, especially with highly chlorinated compounds, for producing large quantities of chloride ion, which can cause a significant drop in the pH. This could have the effect of inhibiting microbial activity, and, as mentioned above, mobilizing compounds which were previously immobilized. However, with carefully designed treatability studies, this potential problem can be discovered and ways devised to avoid it.

The same is true for the redox state and changes which might be necessary in the oxygen availability in the aquifer. Most shallow (less than 200 ft), uncontaminated aquifers contain some oxygen, while deeper aquifers often contain sulfate or nitrate. Where contamination has occurred, though, it is not uncommon to find extremely low dissolved oxygen levels. This may be due to natural ISB. Indeed, Rifai and Bedient, using the computer model Bioplume II, confirmed that oxygen levels increased as the hydrocarbon levels decreased.[74] Similar conclusions were reached by Chang et al.,[16,17] who noted that wells containing <0.9 ppm BTX had significantly higher dissolved oxygen levels than those wells containing >1.0 ppm BTX. Thus, for optimal biodegradation to occur, oxygen may be the most critical limiting factor. It can be replaced by the addition to the subsurface of air, pure oxygen, hydrogen peroxide, or another electron acceptor such as nitrate. The disadvantage of adding oxygen, of course, is that you will also oxidize iron and manganese, which can then precipitate and cause plugging of the aquifer. But if one has performed the necessary studies, described previously, precipitation can be prevented and oxygenation of the aquifer can generally be safely accomplished.

Advantages of ISB

There are four major advantages to enhancing the natural biodegradation rates. The first is that remediation times are greatly reduced over nonenhanced processes or more traditional pump-and-treat technologies. Both the subsurface soils and groundwater are treated at the same time, thus eliminating long-term leaching of the contaminants from the soil particles. When only groundwater is treated via an air stripper or an aboveground bioreactor, the soils serve as a reservoir of contaminants and will slowly leach the organics into the groundwater. This leaching rate is dependent on the partitioning coefficient of the contaminants between the soil particles, soil organic matrix, and the water and gas phases.

Another major advantage of using ISB is that, under aerobic conditions, the contaminants are usually mineralized to CO_2, water, biomass, and salts if appropriate. This eliminates future liabilities to the client which might result from excavation and landfilling the contaminated soil. A third advantage concerns costs. Although incineration is quicker, it is becoming increasingly difficult to obtain permits for incineration and the transportation costs as well as the excavation costs will generally result in a threefold to tenfold higher remediation cost than for ISB. Finally, because ISB is a natural process utilizing the indigenous microbial population, it is generally viewed as more environmentally acceptable. Only inorganic compounds are added to the aquifer, and these are substances which will be consumed as the microbes degrade the contaminants. Thus ISB provides an efficient, cost-effective, and thorough technology for cleaning contaminated aquifers.

Case Studies—Petroleum Hydrocarbons

Historical background

The first reported ISB was by Jamison et al. in 1975.[43] They described the pipeline spill in eastern Pennsylvania that contaminated a dolomite formation with gasoline. After free-product recovery, and the addition of 58 tons of ammonium sulfate, 29 tons of mono- and dibasic phosphates, and sparged air, they estimated that approximately 1080 barrels of gasoline had been biodegraded. This was about one-third of the original 3200-barrel spill.[43] Bacterial numbers were originally 1×10^3 to 1×10^4 colony-forming units per milliliter and reached a peak of 4.2×10^6 colony-forming units per milliliter before the population crashed after the disappearance of the gasoline carbon source.[71]

Since Raymond's initial studies,[71] he and his coworkers have conducted other in situ bioremediations in sandy aquifers in Millville, New Jersey;[72] Long Island, New York;[48] and Watsonville, California.[53] In each case the site geology/hydrogeology, treatability testing, and nutrient and oxygen addition needs were described along with the level of remediation achieved.

Published case studies

Because the pioneering work of Raymond concerned the bioremediation of petroleum hydrocarbons, the majority of applications of ISB to date have involved this group of contaminants. Most of the reported case studies include descriptions of the geology, hydrogeology, contaminant concentrations and distributions between aqueous and solid phases, and hydraulic conductivity, as well as the results of the biotreatability tests. A selection of some of the more thoroughly described petroleum hydrocarbon case studies is listed in Table 7.2.

One of the most completely documented field trials of ISB has been the work at the Traverse Coast Guard Air Station, Traverse City, Michigan.[4,88,89] This site became contaminated with approximately 10,000 gal of jet fuel (JP-4) due to a broken flange on an underground storage tank. The site, which is about 1 mi from East Bay, Lake Michigan, consists of a shallow sand and gravel aquifer. Depth to groundwater is about 14 ft, and the groundwater flow is in a northeasterly direction towards East Bay. A pilot test was designed to evaluate nutrient and hydrogen peroxide additions for the enhancement of ISB. The test area was a plot 30 by 100 ft in which were installed five injection and chemical feed wells, nine 4-in monitoring wells, and twelve small-diameter cluster wells nested at different depths with specially designed probes screened over intervals from about 13 ft to 25 ft below land surface.

The design and operation of the system relied on the computer model Bioplume II. After calculating the oxygen demand of the system, it was determined that a flow rate of 40 gpm was required to mound the water table 1 ft to reach the material trapped in the capillary fringe. The water was split and 29 gpm was directly injected while 11 gpm was treated with the required nutrients prior to injection. Along with nutrient addition, pure liquid oxygen was supplied to the system for approximately 3 months, then hydrogen peroxide was added for the duration of the pilot test. After the addition of the liquid oxygen, several tracer tests were performed to determine the transport of dissolved oxygen, chloride, ammonia, and phosphate.

Throughout the study, in addition to the nutrients and oxygen, benzene, toluene, ethylbenzene, and xylenes (BTEX); pH; conductivity; temperature; water level; and microbial numbers were monitored using standard published procedures. Core materials were collected aseptically before the initiation of the project, after approximately 3 months of operation, and approximately 5 months later.

The results after approximately 9 months of operation showed that as oxygen and nutrients were consumed the amount of BTEX decreased, indicating microbial consumption of the organics. Furthermore, there was up to an order of magnitude increase in the numbers

TABLE 7.2   Selected Published Case Studies of the in Situ Bioremediation of Petroleum Hydrocarbon Spills

Location	Type and amount of contaminant	Full field (F) or pilot(P) test	Geology*	Area	Nutrient addition	Electron acceptor	Duration of treatment	Level of treatment	Problems	Company†	Reference
Watsonville, California	Gasoline, 1000 gal	F	CL, ML, SC, occ. SW	NS‡	NH_4^+, phosphates	H_2O_2	13 mo	>90%	NS	DERS	53
Long Island, New York	Gasoline, 10,000 gal	F	GP, CL lenses	150 × 300 × 7.5 ft	NH_4^+, phosphates	H_2O_2	64 mo	>99%	NS	DERS	48
Northern Indiana	Gasoline	P	GM	200–400 × 600–1000 ft	NH_4^+, phosphates	H_2O_2	6 mo	63–80%	Ca, Mg, Fe	IT	3
Michigan refinery	Petroleum hydrocarbons, NS	P	NS	NS	NS	Aerated water in basin	106 days	78–90%	Rain and flooding	OHM	78
Southern California	Gasoline, NS	F	GM, CL and SM lenses	30,000 ft²	RESTORE™ 375	H_2O_2	6 mo	84 to >99%	Another spill	IT	2, 14
Oakland, California	Gasoline, 5000 ppm in soil	F	GM, ML	8–10,000 yd³	NS	H_2O_2	9 mo	80%	NS	HL	63
Canada	Gasoline, NS	F	Fill GP, fractured bedrock	A + B = 80 × 100 ft C = 70 × 240 ft	NH_4^+, phosphates	H_2O_2	6 mo	A = 95% B = 40–50% C = 85%	Sorbed material	GTI	13
Upper Rhine Valley, Germany	Oil Spill, est. 17 tons	F	SW, ML	900 × 300 ft	NH_4^+, phosphates	Nitrate 300 PPM	24 mo	>95%	Iron and methane	NS	90
Camp Grayling Army Airbase, Grayling, Michigan	Diesel, 16,000–25,000 ppm in soil	F	SW, CL lenses	150 × 150 × 14 ft	NS	Above ground bioreactor aerated water recycled	11 mo	>95%	NS	Hunter	51
Eastern Missouri	Gasoline, 30,000 gal	F	Impermeable till with fractured bedrock	360 × 720 ft	NS	H_2O_2	32 mo	>99%	NS	J. Mathes	9

TABLE 7.2 Selected Published Case Studies of the in Situ Bioremediation of Petroleum Hydrocarbon Spills (Continued)

Location	Type and amount of contaminant	Full field (F) or pilot(P) test	Geology*	Area	Nutrient addition	Electron acceptor	Duration of treatment	Level of treatment	Problems	Company†	Reference
Southern California	Gasoline, 200 ppm total petroleum hydrocarbons	F	SC	Groundwater at 60 ft and approximately 40 × 50 ft on the surface	ACT™ (NH_4^+, phosphates)	H_2O_2	10 mo	>99%	Low permeability	CAA	28
Amsterdam, The Netherlands	BTEX/ mineral oil, 200 and 6000 mg/kg soil, respectively	F	NS	NS	NS	Oxygenated water	3 mo	79% of the oil; 98% BTEX	NS	DRM	83
Arnhem, The Netherlands	Mineral oil, 10,000 mg/kg soil	F	NS	21 ft of unsaturated soil to groundwater	NS	KNO_3	Ongoing	After 2 mo, 5–56%	NS	DRM	83
Eastern Pennsylvania	Gasoline, 900 gal	F	SP, MH	540 × 400	NH_4^+, phosphates	H_2O_2	24 mo	≈99%	Drought lowered water table	GTI	52

*Abbreviations are based on the Unified Soil Classification System.

†Company names have been abbreviated as follows: DERS = DuPont Environmental Remediation; OHM = OH Materials, Inc.; IT = International Technology Corp.; GTI = Groundwater Technology Inc.; Hunter = Hunter Bioscience, Inc.; J. Mathes = John Mathes & Associates; CAA = Cambridge Analytical Associates Bioremediation Systems; DRM = De Ruiter Milieutechnologie B. V.

‡NS = not stated.

of hydrocarbon-degrading bacteria, especially at the shallower depths. The amount of increase depended on the proximity to the oxygen and nutrient sources and the presence of BTEX. Oxygen appeared to be consumed first in the shallower depths and did not break through to the next level until the BTEX level was nondetectible or <10 ppb. As expected for a sandy aquifer the permeability was not affected by the addition of hydrogen peroxide.[88,89]

Problems with oxygen delivery and the substitution of nitrate

There have been several cases where hydrogen peroxide was added to ISB systems, and problems of rapid decomposition[21,38,81] or precipitation[91] were encountered. This has prompted the testing of nitrate as an alternate electron acceptor. One of the first sites where nitrate was used for the anaerobic degradation of petroleum hydrocarbons was in the upper Rhine Valley, in Saarbrücken, Germany.[90] The basic information on the geology and contamination at this site is listed in Table 7.2. ISB was selected for this site because of its size and the presence of buildings over the contaminant plume. Iron and methane were problems that had to be addressed. Ammonium chloride, phosphates, and nitrate were all added to the infiltration water, the nitrate at the level of approximately 4 milligrams per milligram of hydrocarbon. The flushing water was also saturated with oxygen via an aeration system. Both CO_2 and nitrogen gas increased over time. The aliphatic hydrocarbons decreased from an initial level of 2 mg/L to 0.1 mg/L after two years. Originally present at about 5.5 mg/L, the xylenes decreased to 0.08 mg/L over the 24 months of operation. The author reported that the amount of oxygen calculated to be present could not account for the extensive biodegradation, and he concluded that the major electron acceptor in this system was nitrate.[90]

Another major study on ISB under denitrifying conditions was conducted at the U.S. Coast Guard Station at Traverse City, Michigan. The general features of the site were described previously.[4,88]

For the pilot test of denitrification, a plot 30 ft by 30 ft, up-gradient of the aerobic test plot, was treated with sodium nitrate and nutrients (phosphates and ammonium chloride) added through an infiltration gallery using recirculated water. The disappearance of benzene and toluene coincided with the amounts of oxygen supplied to the aquifer through the aerated recirculated water. However, the abrupt increase in nitrate in the recovered groundwater coincided with the disappearance of the m- and p-xylenes. This indicates that nitrate was only being used during the biodegradation of the xylenes. Average soil concentrations of the various contaminants measured decreased: benzene from

0.84 to 0.032 mg/kg, toluene from 33 to 0.013 mg/kg, ethylbenzene from 18 to 0.36 mg/kg, m- and p-xylene from 58 to 7.4 mg/kg, and o-xylene from 26 to 3.2 mg/kg.[41] Alkylbenzene concentrations were also reduced by more than 90 percent, but were still present at levels ranging from 17 to 258 mg/kg dry weight. The authors concluded that in this study at least both aerobic and anaerobic process were involved in the microbial decomposition of BTX and the other components of the JP-4 fuel.[20,40,41]

An example of naturally occurring anaerobic bioremediation of jet fuel apparently occurred at the Miami International Airport International Concourse.[35] When free-product recovery efforts encountered pockets of methane gas in the porous sandy soils, a safe gas recovery system had to be developed which included explosion-proof materials and a means to dilute the 75 percent methane in the atmosphere to 5 percent, which is the lower explosive limit. Several gas recovery units have now been installed,[35] but the basic issue of microbial anaerobic degradation causing the gas production has not been addressed. Here, again, as in the case in Germany,[90] the presence of methane implies naturally occurring anaerobic ISB.

Vadose-zone biotreatment

Often it is not possible to mound the water from the saturated zone into the vadose zone to effect an ISB. This is especially true when there is poor permeability or water-holding capacity or the depth to groundwater is over 10–20 ft. Recent advances in engineering design have resulted in the development of in situ vadose-zone treatments by forcing air into the vadose zone and supplementing the soils with moisture and nutrients (Fig. 7.2). One of the earliest descriptions of this technique was reported by Staps for a pilot-scale operation in The Netherlands.[83] As described by Staps, treatment of the vadose zone by the addition of air, moisture, and nutrients is a combination of physical and biological processes. For the pilot test, 125 kg of gasoline was added to clean, sandy soil which was vented by vacuum applied to drains. Nutrients were applied to maintain a ratio of 100:10:2 C:N:P. Additional drains were constructed to recover the added water. In 250 days of testing, concentrations of gasoline in the sand decreased from 17,000 ppm to <100 ppm. The investigators calculated that 58 percent of the gasoline was removed by evaporation due to the venting, 27 percent was due to microbial degradation, 7.5 percent disappeared in the retrieved water, and 6.8 percent was lost due to uncontrolled evaporation.[83]

Since that experiment, several pilot studies have been conducted on petroleum hydrocarbons. Ely and Heffner obtained a patent for pulling a vacuum on the vadose zone to enhance the natural biodegradation,[25] while Hinchee et al. presented a conceptual approach to enhancing biodegradation via soil venting.[21,39] All of these authors also recognized the need to differentiate between evaporative losses and biodegradation and suggested monitoring the off-gases for CO_2 and O_2. The advantage of these systems is that very large amounts of air can be drawn or pushed through the vadose zone, resulting in the introduction of proportionately larger quantities of oxygen—for example, 1 scfm will introduce 23 lb/day of oxygen into the soils.[58] Another advantage is that gases can penetrate into soil structures more readily than liquids, thus increasing the mass transfer of oxygen and distributing the needed oxygen to more organisms for biodegradation.

This soil-venting technology is well adapted to the situation described by Brown and Crosbie.[12] The majority of the residual gasoline contamination at this site was present at a depth 9–11 ft below land surface in a region of seasonal groundwater fluctuations. Six soil vapor extraction wells were constructed. The collected soil vapors were passed over charcoal columns for polishing before release into the atmosphere. The effluent gas was monitored and the amount of CO_2, initially 11 percent, decreased to 1.4 percent as the average hydrocarbon concentration decreased.[12]

At the Hill Air Force Base, Utah, a JP-4 spill occurred in January 1985 resulting in the release of 27,000 gal of fuel.[22] Approximately 2000 gal were recovered as free product, and the remaining fuel was contained in the unsaturated soils over an area of approximately 1 acre and 50 ft deep. The water table was at 600 ft below land surface. High-volume soil venting was used for approximately 9 months, during which time respiration was monitored. The authors calculated that 15 to 25 percent of the lost hydrocarbons had been biodegraded, and the remainder of the volatiles had been collected from the high-volume soil-venting procedure. They then installed a bioventing system which operated at one-third to one-half the high-volume rate and followed the effects of supplementing with moisture and moisture plus nutrients. Based on extensive field and vent gas monitoring, the authors concluded that 80 to 90 percent of the resulting decrease in petroleum hydrocarbons could be attributed to biodegradation with moisture addition the most important parameter. The combined high-volume and low-volume bioventing reduced the contaminants in the soils from an average of 410 mg/kg prior to venting to an average of 3.8 mg/kg after 21 months total treatment time.[22]

The above results showing the beneficial effects of nutrient addition have not been confirmed by Miller et al. for a jet fuel spill at Tyndall Air Force Base, Florida.[62] There, a 7-month pilot test showed that only 55 percent of the hydrocarbons were biodegraded but that with adjustment of air flow alone, based on 2 to 4 percent oxygen consumption, this could have been increased to 85 percent. However, neither moisture nor nutrient amendment had any impact on the extent of biodegradation.[62]

An enhancement of the soil-bioventing technique is that of adding the air and nutrients just into the water table and pulling a vacuum on the soils to force the moisture-laden air up through the capillary fringe and vadose zone. A modification of this concept has been patented by Corey et al., who refer to the process as *in situ bioventing*.[18] A field test of this system was performed at the Savannah River site.[55] They reported the removal of approximately 16,000 lb of chlorinated solvent contaminants during the 139 days of operation using their process. A horizontal well was drilled 165 ft below a contaminated plume, and a second horizontal well was placed in the vadose zone at 75-ft depth. Purge air was introduced into the lower well, and a vacuum was placed on the upper well. This resulted in an increased removal of about 21 lb/day over horizontal vacuum extraction alone, while horizontal extraction was 5 times more efficient than a vertical vacuum extraction well.[55] In this test no nutrients were added, but microbial monitoring indicated increased biomass and colony-forming units resulting from the increase in air flow to the aquifer. This system is certainly capable of performing as an in situ bioventing process by simply adding any needed nutrients; it would improve the biodegradation rates and the off-gases could be monitored for increased CO_2 production resulting from increased microbial activity. This is an extremely interesting concept and needs further field application and validation, especially for semivolatile compounds.

Case Studies—Creosote- and PCP-Contaminated Sites

In the case studies described above, most of the emphasis was placed on the reduction of the BTEX fraction and occasionally on a reduction in the TPH levels in the soils. (TPHs are extremely complex mixtures of hundreds of aliphatic and aromatic compounds.) Because microorganisms could degrade a significant portion of these complex mixtures, could they also degrade the polyaromatic hydrocarbons (PAHs) associated with wood-treating sites and coke plants or gas works? To the surprise of many, two-, three-, four-, and even five-membered fused rings were found to be biodegradable. The six-membered and larger rings are much more recalcitrant. This recalcitrance is not too surprising when one considers the complexity of humates and lignins and their survival in soils. As is generally true with organic compounds, the higher the molecular weight, the slower the biodegradation rate. Much of the regulatory emphasis, however, has been on the carcinogenic PAHs (which

TABLE 7.3   EPA List of B-2 Level Potentially
Carcinogenic PAHs and Their Structures

Common Name	Structure	MCL* (μg/L)
Benzo(a)anthracene		0.1
Chrysene		0.2
Benzo(k)fluoranthrene		0.2
Benzo(a)pyrene		0.2
Dibenzo(a,h)anthracene		0.3
Benzo(b)fluoranthrene		0.2
Indeno-(1,2,3-c,d)pyrene		0.4

*MCL stands for maximum contaminant level, as
established by the EPA.

contain five or fewer rings) and secondarily on the reduction of total
PAHs in groundwater and soils. The seven potentially carcinogenic
PAHs are listed in Table 7.3.[15]

To date, there have been few reported case studies on the in situ
bioremediation of PAHs. This is no doubt partly because of the perception that these compounds are not biodegradable, partly because
the types of industries that use or produce these substances (coke
plants, gas works, wood-treating sites) are generally localized in industrial zones and not residential areas, and partly because these
compounds are not highly volatile or soluble in water and hence are
not as mobile in groundwater as the BTEXs. Also, degradation rates,
based on laboratory studies, are slower—hence in situ bioremediation
will be slower.

There are numerous laboratory treatability studies that have been
completed on groundwater and aquifer materials contaminated with cre-

osote and/or pentachlorophenol (Refs. 49, 57, 64, 86, 98, 99). All of these studies show that, with nutrient addition and oxygen enhancement, most of the carcinogenic PAHs can be degraded. There are several pilot tests either under way or awaiting final regulatory approvals to begin field demonstrations of the extent and rate of biodegradation.

One such pilot test has been reported by Lund and coworkers, who described their laboratory and field work at a coke oven plant in Karlsruhe, Germany.[56] In the laboratory, they treated the soils with ozone to introduce oxygen into the rings to make them more biodegradable, with the result that by the end of the test period (104 days), approximately 40 percent of the PAHs had been degraded. A field pilot test is under way in which they have constructed walls around the test plots to a depth of 17 m. The approximately 2- to 3-m deep fill layer was excavated, and the concrete and debris were removed, pulverized mixed with surface soil, and reintroduced to the test plot. The plot has been dewatered and the area is being treated by bioventing along with nutrient addition through a sprinkling system.[56] As of this date, no results have been reported.

Chester Environmental has recently completed a laboratory feasibility study on the soils from a wood-treating site in the southeastern United States. Using respirometry, plate counts, and chemical analyses, we demonstrated that, with nutrient enhancement, the microorganisms were able to degrade between 40 and 90 percent of the individual PAHs, including the potentially carcinogenic PAHs and pentachlorophenol. The flask study lasted 6 weeks and did become nutrient-limiting under our test conditions.[98] A field pilot demonstration is planned for the summer of 1993.

Mueller et al. have also demonstrated the natural microbial degradation of the PAHs in the soils at the American Creosote Works Superfund site in Pensacola, Florida.[64] Under their test conditions removal rates ranged from 53 percent of the PAHs containing four or more fused rings to >80 percent of the nitrogen-heterocyclic compounds and >95 percent of the lower-molecular-weight PAHs (three fused rings or less). Their study lasted only 14 days, and during this time there was no biodegradation of pentachlorophenol.[64]

Other workers have shown that under anaerobic conditions many of the PAHs can also be degraded. This recognition started with the observation of the production of methane at a creosote-contaminated site in St. Louis Park, Minnesota.[23,24,34] The authors found that naphthalene and phenolic compounds were disappearing faster than expected, and that methane was being formed but only in the contaminant plume. Hence, natural anaerobic bioremediation of the site was occurring.[24] They then isolated various metabolic groups including obligately anaerobic methanogenic consortia and nitrate-respiring

Pseudomonas stutzeri both from laboratory bioreactors and directly from the aquifer.[23]

Finally in regard to the biodegradation of creosote-contaminated sites, Flyvbjerg et al.[27] demonstrated in laboratory studies that under denitrifying conditions only a small portion of the compounds in creosote-contaminated groundwater was degraded. They suggested that multiple types of anaerobic conditions may be necessary to achieve complete remediation of a site.[27] The establishment of in situ treatment trains utilizing varying redox conditions from methanogenic to aerobic has not as yet been attempted, but could theoretically provide the most complete remediation of a complex, mixed-waste site.

Case Studies—Other Organic Compounds

An early attempt to demonstrate biodegradation of chlorinated solvents was reported by Jhaveri and Mazzacca.[44,45] In their process, both an aboveground bioreactor and in situ treatment were combined. Methylene chloride, *n*-butyl alcohol, acetone, and dimethylaniline had reached the groundwater through a leaking underground process line at the site in Waldwick, New Jersey. The site consists of a surface layer of silt and gravel followed by 8–15 ft of glacial till underlain by approximately 40 ft of semiconsolidated silt and fine sand. This layer is underlain by Brunswick shale for several hundred feet. A suspended-film bioreactor was combined with up-gradient injection of treated water, in situ aeration wells located within the plume, and a down-gradient series of recovery wells. After three years of operation, 90 percent of the contaminated plume had been biodegraded.[44,45]

One of the more pervasive classes of contaminants in groundwater is the halogenated solvents: tetrachloroethylene (PCE), trichloroethylene (TCE), dichloroethylene (DCE), and vinyl chloride (VC), along with their saturated ethane counterparts and chloroform and carbon tetrachloride. The fate of these compounds in the subsurface has recently been reviewed by Vogel et al.[87] It is generally recognized that under anaerobic conditions these compounds can be reductively dehalogenated to the vinyl chloride stage,[70,97] although Rowland and Eisenberg noted complete anaerobic degradation in an anaerobic shallow aquifer in the southeastern United States.[76] Under aerobic conditions, however, methane, other gases, or an aromatic compound can induce degradation and/or cometabolism of TCE.[36,65,92,93] Methane, or other gases such as propane, induces monooxygenases,[92,93] while the aromatic compounds induce dioxygenases;[65] none of these enzymes is extremely specific and therefore will dechlorinate aliphatic solvents. Both of these processes have been field-tested.

A dioxygenase induction system was given a field pilot test[66] with the addition of tryptophan[67] and the pseudomonad *P. cepacia* G4 to groundwater contaminated with TCE. This resulted in a 97 percent reduction in TCE during the 20-day field test.[66] This study was conducted prior to the implication of tryptophan in medical sequelae and has now been replaced by other dioxygenase inducers. A full-scale remediation using a new inducer (patent pending) is scheduled for 1992 following a field demonstration of the reduction of TCE by *P. cepacia* G4 from 700 ppb to 90 ppb in 25 days.[100]

A more extensive test of a monooxygenase induction system was conducted at the Moffit Field U.S. Naval Air Station in California. This work was performed by McCarty and coworkers.[75,79] The background for these field studies had been the earlier findings by Higgins[37] in the United Kingdom and Wilson and Wilson[95] that methane monooxygenase is nonspecific and will dechlorinate the aliphatic solvents. The design of the field test was described by Roberts et al.[75] Beneath an approximately 12-ft-thick surface clay layer, the 4.5-ft shallow sand and gravel aquifer was contaminated with a chlorinated aliphatic solvent, 1,1,1-trichloroethane (TCA). Tracer tests and laboratory studies showed that the aerobic methanotrophs could be induced in the system by supplying oxygen and methane. These gases were alternated to prevent excess microbial growth at the injection wells.[75] TCE, *cis-* and *trans-*DCE, and VC were added to the injection water and their progress through the aquifer was followed via sampling wells placed on approximately 3-ft centers. Retardation factors were calculated and the system was pulsed with methane and oxygen. The study was continued for three years and approximately 30 percent of the *cis*-DCE biodegraded, and over 80 percent of the *trans*-DCE and over 95 percent of the VC degraded during the third 200-day test. However, only between 10 and 20 percent of the TCE was transformed during this third season of testing.[79] The percent TCE and *cis*-DCE biotransformed had been higher during the second season of testing, 10–30 percent and 30–58 percent, respectively.[75] The flow rates used for this field study were fairly rapid (2 m/day), so the degradation proceeded rapidly over the short interval of the study. They also demonstrated that methane monooxygenase was required for the biotransformations in this system and that induction of the enzyme could only be achieved by supplying methane to the methanotrophs on a continual basis.[79]

For anaerobic ISB, Suflita and his coworkers demonstrated that benzoate and phenol in landfill leachates could be degraded under methanogenic or sulfate-reducing conditions, but reductive dehalogenation was limited to the methanogenic system.[84] They later showed that sulfate inhibition of dehalogenation of 2,4,5-trichlorophenoxyacetic acid could sometimes be relieved by the addition of molybdate to their test microcosms.[33] To date, there have been no reported attempts to modify naturally occurring anaerobic dehalogenation by nutrient amendments that included molybdate.

Other workers have noted the degradation by *Phanerochaete chrysosporium* of 2,4,6-trinitrotoluene (TNT) and have suggested applying this fungus to soils contaminated with TNT.[26] This same fungus has also been found to degrade alkyl halide insecticides which also contaminate aquifer soils and groundwater,[46] but again applications to pilot or full-scale ISB treatment systems have not been realized.

There has been one reported in situ bioremediation at an herbicide formulation plant[10] where the aquifer was 35 ft of glacial outwash, with 25 ft of silty sand and clay overlying 10 ft of coarse sand and gravel and underlain by shale bedrock. The contaminant was 4-chloro-2-methylphenol, which was found to be highly biodegradable when aeration was provided to the groundwater. Modifications to the existing pump-and-treat system were made by increasing the number of injection and recovery wells and adding air-lift pumps into the recovery wells. After 6 months of operation, there was a 50 percent reduction in the size of the contaminant plume.[10]

Conclusions

In situ bioremediation or biotreatment of subsurface soils and groundwater is a technology which has come of age. It has been successfully applied to petroleum hydrocarbon contamination for almost 20 years, and more recently to creosote and other organic compounds. It may not be the remedial action of choice in all cases, but it must certainly be evaluated as a potential remedial technology for those classes of compounds for which it is applicable. In some cases, even though the hydrogeology is not optimal, it may be the only method of choice based on risk analysis, costs, and timeliness. The adaptability and versatility of microorganisms should never surprise us, and we should let their abilities be the final determining factor when deciding whether biodegradation is practicable.

Acknowledgments

To provide a summary of the actual field applications, besides the published literature, numerous practitioners of ISB have been contacted and have generously shared their experiences and reports with the author. The author also thanks G. Gromicko and S. F. Nishino for their helpful review of the manuscript, and S. Sanchez for her assistance with the bibliography.

References

1. Aelion, C. M., C. M. Swindoll, and F. K. Pfaender. 1987. Adaptation to and Biodegradation of Xenobiotic Compounds by Microbial Communities from a Pristine Aquifer. *Appl. Environ. Microbiol.* 53:2212–2217.
2. Anonymous. N.D. In Situ Bioreclamation Case Study A. Gasoline Contamination in Southern California. International Technology Corporation. Torrance, CA, 4 pp.
3. Anonymous. 1987. In Situ Bioreclamation: A Case History. Gasoline Contamination in Northern Indiana. International Technology Corporation. Torrance, CA, 5 pp.
4. Armstrong, J. M., W. Korreck, L. E. Leach, R. M. Powell, S. V. Vandegrift, and J. T. Wilson. 1988. Bioremediation of a Fuel Spill: Evaluation of Techniques for Preliminary Site Characterization. pp. 931–943. NWWA/API. Proceedings of Petroleum Hydrocarbons and Organic Chemicals in Ground Water: Prevention, Detection and Restoration, Nov. 9–11, 1988. National Water Well Association, Dublin, OH.
5. American Society for Testing and Materials. 1984. Annual Book of ASTM Standards. Section 04.08 Soil and Rock; Building Stones. ASTM, Philadelphia, PA.
6. Azam, F., and R. E. Hodson. 1981. Multiphasic Kinetics for D-Glucose Uptake by Assemblages of Natural Marine Bacteria. *Marine Ecol. Prog. Ser.* 6:213–222.
7. Balkwill, D. L., J. K. Fredrickson, and J. M. Thomas. 1989. Vertical and Horizontal Variations in the Physiological Diversity of the Aerobic Chemoheterotrophic Bacterial Microflora in Deep Southeast Coastal Plain Subsurface Sediments. *Appl. Environ. Microbiol.* 55:1058–1065.
8. Barker, J. F., G. C. Patrick, and D. Major. 1987. Natural Attenuation of Aromatic Hydrocarbons in a Shallow Sand Aquifer. *Groundwater Monit. Rev.* 7:64–71.
9. Bell, R. A., and A. H. Hoffman. 1991. Gasoline Spill in Fractured Bedrock Addressed with *in Situ* Bioremediation. pp. 437–443. *In* R. E. Hinchee and R. F. Olfenbuttel (eds.), *In Situ* Bioreclamation: Applications and Investigations for Hydrocarbon and Contaminated Site Remediation. Butterworth-Heinemann, Boston, MA.
10. Borow, H. S., and J. V. Kinsella. 1989. Bioremediation of Pesticides and Chlorinated Phenolic Herbicides—Above Ground and In Situ—Case Studies. pp. 325–331. Superfund '89. Proceedings of the 10th National Conference, Nov. 27–29, 1989. Hazardous Materials Control Research Institute. Greenbelt, MD.
11. Bouwer, H. 1984. Elements of Soil Science and Groundwater Hydrology. pp. 9–39. *In* G. Bitton and C. P. Gerba (eds.), Groundwater Pollution Microbiology. Wiley, New York, NY.
12. Brown, R. A., and J. R. Crosbie. 1989. Oxygen Sources for In Situ Bioremediation. pp. 338–344. Superfund '89. Proceedings of the 10th National Conference, Nov. 27–29, 1989. Hazardous Materials Control Research Institute. Greenbelt, MD.
13. Brown, R. A., R. Tribe, and A. Duquette. 1989. Bioreclamation: the three R's of product losses—response, regulation, remediation. *Water Contr.* February:10–11.
14. Brubaker, G. R., and J. H. Exner. 1988. Bioremediation of Chemical Spills. *Basic Life Sci.* 45:163–171.
15. Cerniglia, C. E., and M. A. Heitkamp. 1989. Microbial Degradation of Polycyclic Aromatic Hydrocarbons (PAH) in the Aquatic Environment. pp. 41–68. *In* U. Varanasi (ed.), Metabolism of Polycyclic Aromatic Hydrocarbons in the Aquatic Environment. CRC Press. Boca Raton, FL.
16. Chiang, C. Y., E. Y. Chai, J. P. Salanitro, J. D. Colthart, and C. L. Klein. 1987. Effects of Dissolved Oxygen on the Biodegradation of BTX in a Sandy Aquifer. pp. 451–470. NWWA/API. Proceedings of Petroleum Hydrocarbons and Organic Chemicals in Ground Water: Prevention, Detection and Restoration, Nov. 17–19, 1987. National Water Well Association, Dublin, OH.
17. Chiang, C. Y., J. P. Salanitro, E. Y. Chai, J. D. Colthart, and C. L. Klein. 1989. Aerobic Biodegradation of Benzene, Toluene, and Xylene in a Sandy Aquifer—Data Analysis and Computer Modeling. *Ground Water* 27:823–834.
18. Corey, J. C., B. B. Looney, D. S. Kaback. 1989. In-Situ Remediation System and Method for Contaminated Groundwater. U.S. Patent 4,832,122. The United States of America as represented by the United States Department of Energy, Washington, D.C.

19. Davis, H. E., J. Jehn, and S. Smith. 1991. Monitoring Well Drilling, Soil Sampling, Rock Coring, and Borehole Logging. pp. 195–237. *In* D. M. Nielsen (ed.), Practical Handbook of Ground-Water Monitoring. Lewis Publishers, Chelsea, MI.

20. Downs, W. C., S. R. Hutchins, J. T. Wilson, R. H. Douglass, and D. J. Hendrix. 1989. Pilot Project on Biorestoration of Fuel-Contaminated Aquifer Using Nitrate: Part I—Field Design and Ground Water Modeling. pp. 219–233. NWWA/API. Proceedings of the Conference on Petroleum Hydrocarbons and Organic Chemicals in Ground Water: Prevention, Detection and Restoration, Nov. 15–17, 1989. National Water Well Association, Dublin, OH.

21. Downey, D. C., R. E. Hinchee, M. S. Westray, and J. K. Slaughter. 1988. Combined Biological and Physical Treatment of a Jet Fuel-Contaminated Aquifer. pp. 627–645. NWWA/API. Proceedings of Petroleum Hydrocarbons and Organic Chemicals in Ground Water: Prevention, Detection and Restoration, Nov. 9–11, 1988. National Water Well Association, Dublin, OH.

22. Dupont, R. R., W. J. Doucette, and R. E. Hinchee. 1991. Assessment of *In Situ* Bioremediation Potential and the Application of Bioventing at a Fuel-Contaminated Site. pp. 262–282. *In* R. E. Hinchee and R. F. Olfenbuttel (eds.), *In Situ* Bioreclamation. Applications and Investigations for Hydrocarbon and Contaminated Site Remediation. Butterworth-Heinemann, Boston, MA.

23. Ehrlich, G. G., E. M. Godsy, D. F. Goerlitz, and M. F. Hult. 1982. Microbial Ecology of a Creosote-Contaminated Aquifer at St. Louis Park, Minnesota. *Dev. Ind. Microbiol.* 24:235–245.

24. Ehrlich, G. G., D. F. Goerlitz, E. M. Godsy, and M. F. Hult. 1982. Degradation of Phenolic Contaminants in Ground Water by Anaerobic Bacteria: St. Louis Park, Minnesota. *Ground Water* 20:703–710.

25. Ely, D. L., and D. A. Heffner. 1988. Process for In Situ Biodegradation of Hydrocarbon Contaminated Soil. U.S. Patent 4,765,902. Chevron Research Company, San Francisco, CA.

26. Fernando, T., J. A. Bumpus, and S. D. Aust. 1990. Biodegradation of TNT (2,4,6-Trinitrotoluene) by *Phanerochaete chrysosporium. Appl. Environ. Microbiol.* 56:1666–1671.

27. Flyvbjerg, J., E. Arvin, B. K. Jensen, and S. K. Olsen. 1991. Biodegradation of Oil- and Creosote-Related Aromatic Compounds under Nitrate-Reducing Conditions. pp. 471–479. *In* R. E. Hinchee and R. F. Olfenbuttel (eds.), *In Situ* Bioreclamation: Applications and Investigations for Hydrocarbon and Contaminated Site Remediation. Butterworth-Heinemann, Boston, MA.

28. Fogel, S., M. Findlay, and A. Moore. 1991. Enhanced Bioremediation Techniques for In Situ and Onsite Treatment of Petroleum Contaminated Soils and Groundwater. pp. 201–209. *In* E. J. Calabrese and Paul T. Kostecki (eds.), Petroleum Contaminated Soils, Vol. 2. Lewis Publishers, Chelsea, MI.

29. Fredrickson, J. K., R. J. Hicks, S. W. Li, and F. J. Brockman. 1988. Plasmid Incidence in Bacteria from Deep Subsurface Sediments. *Appl. Environ. Microbiol.* 54:2916–2923.

30. Freeze, R. A., and J. A. Cherry. 1979. Groundwater. Prentice-Hall, Englewood Cliffs, NJ.

31. Ghiorse, W. C., and D. L. Balkwill. 1983. Enumeration and Morphological Characterization of Bacteria Indigenous to Subsurface Environments. *Dev. Ind. Microbiol.* 24:213–224.

32. Ghiorse, W. C., and J. T. Wilson. 1988. Microbial Ecology of the Terrestrial Subsurface. *In* A. I. Laskin (ed.), Advances in Applied Microbiology. 33:107–172. Academic Press, New York, NY.

33. Gibson, S. A., and J. M. Suflita. 1990. Anaerobic Biodegradation of 2,4,5-Trichlorophenoxyacetic Acid in Samples from a Methanogenic Aquifer: Stimulation by Short-Chain Organic Acids and Alcohols. *Appl. Environ. Microbiol.* 56:1825–1832.

34. Godsy, E. M., D. F. Goerlitz, and G. G. Ehrlich. 1983. Methanogenesis of Phenolic Compounds by a Bacterial Consortium from a Contaminated Aquifer in St. Louis Park, Minnesota. *Bull. Environ. Contam. Toxicol.* 30:261–268.

35. Hayman, J. W., R. B. Adams, and J. J. McNally. 1988. Anaerobic Biodegradation of Hydrocarbon in Confined Soils beneath Busy Places: A Unique Problem of Methane Control. pp. 383–396. NWWA/API. Proceedings of Petroleum Hydrocarbons and Organic Chemicals in Ground Water: Prevention, Detection and Restoration, Nov. 9–11, 1988. National Water Well Association, Dublin, OH.

36. Hegeman, G. D., and D. G. Nickens. 1991. Aerobic bacterial remediation of aliphatic chlorinated hydrocarbon contamination. U.S. Patent 5,024,949. BioTrol, Inc., Chaska, MN.

37. Higgins, I. J. Apr. 6, 1982. Biotransformations Using Methane-Utilizing Bacteria. U.S. Patent 4,323,649. Imperial Chemical Industries, Ltd. London.

38. Hinchee, R. E., and D. C. Downey. 1988. The Role of Hydrogen Peroxide in Enhanced Bioreclamation. pp. 715–722. NWWA/API. Proceedings of Petroleum Hydrocarbons and Organic Chemicals in Ground Water: Prevention, Detection and Restoration, Nov. 9–11, 1988. National Water Well Association, Dublin, OH.

39. Hinchee, R. E., D. C. Downey, and T. Beard. 1989. Enhancing Biodegradation of Petroleum Hydrocarbon Fuels through Soil Venting. pp. 235–248. NWWA/API. Proceedings of the Conference on Petroleum Hydrocarbons and Organic Chemicals in Ground Water: Prevention, Detection and Restoration, Nov. 15–17, 1989. National Water Well Association, Dublin, OH.

40. Hutchins, S. R., W. C. Downs, D. H. Kampbell, G. B. Smith, J. T. Wilson, D. A. Kovacs, R. H. Douglass, and D. J. Hendrix. 1989. Pilot Project on Biorestoration of Fuel-Contaminated Aquifer Using Nitrate: Part II—Laboratory Microcosm Studies and Field Performance. pp. 589–604. NWWA/API. Proceedings of the Conference on Petroleum Hydrocarbons and Organic Chemicals in Ground Water: Prevention, Detection and Restoration, Nov. 15–17, 1989. National Water Well Association, Dublin, OH.

41. Hutchins, S. R., W. C. Downs, J. T. Wilson, G. B. Smith, D. A. Kovacs, D. D. Fine, R. H. Douglass, and D. J. Hendrix. 1991. Effect of Nitrate Addition on Biorestoration of Fuel-Contaminated Aquifer: Field Demonstration. *Ground Water.* 29:571–579.

42. Jain, R. K., G. S. Sayler, J. T. Wilson, L. Houston, and D. Pacia. 1987. Maintenance and Stability of Introduced Genotypes in Groundwater Aquifer Material. *Appl. Environ. Microbiol.* 53:996–1002.

43. Jamison, V. W., R. L. Raymond, and J. O. Hudson, Jr. 1975. Biodegradation of High-Octane Gasoline in Groundwater. *Dev. Ind. Microbiol.* 16:305–312.

44. Jhaveri, V. W., and A. J. Mazzacca. 1983. Bioreclamation of Ground and Groundwater. Presented at the 4th National Conference on Management of Controlled Hazardous Waste Sites. Groundwater Decontamination Systems, Inc., Paramus, NJ. 19 pp.

45. Jhaveri, V., and A. J. Mazzacca. 1985. Bioreclamation of Ground and Ground Water by In-Situ Biodegradation. Case History. Presented at the 6th National Conference on Management of Uncontrolled Hazardous Waste Sites. Groundwater Decontamination Systems, Inc., Paramus, NJ. 32 pp.

46. Kennedy, D. W., S. D. Aust, and J. A. Bumpus. 1990. Comparative Biodegradation of Alkyl Halide Insecticides by the White Rot Fungus, *Phanerochaete chrysosporium* (BKM-F-1767). *Appl. Environ. Microbiol.* 56:2347–2353.

47. Kolbel-Boelke, J., E.-M. Anders, and A. Nehrkorn. 1988. Microbial Communities in the Saturated Groundwater Environment II: Diversity of Bacterial Communities in a Pleistocene Sand Aquifer and Their In Vitro Activities. *Microb. Ecol.* 16:31–48.

48. Lee, M. D., and R. L. Raymond, Sr. 1991. Case History of the Application of Hydrogen Peroxide as an Oxygen Source for *In Situ* Bioreclamation. pp. 429–436. *In* R. E. Hinchee and R. F. Olfenbuttel (eds.), *In Situ* Bioreclamation: Applications and Investigations for Hydrocarbon and Contaminated Site Remediation. Butterworth-Heinemann, Boston, MA.

49. Lee, M. D., J. T. Wilson, and C. H. Ward. 1983. Microbial Degradation of Selected Aromatics in a Hazardous Waste Site. *Dev. Ind. Microbiol.* 25:557–565.

50. Lewis, D. L., H. P. Kollig, and R. E. Hodson. 1986. Nutrient Limitation and Adaptation of Microbial Populations to Chemical Transformations. *Appl. Environ. Microbiol.* 51:598–603.

51. Lieberman, M. T., E. K. Schmitt, J. A. Chaplan, J. R. Quince, and M. P. McDermott. 1989. Biorestoration of Diesel Fuel Contaminated Soil and Groundwater at Camp Grayling Airfield Using the PetroClean™ Bioremediation System. pp. 641–654. NWWA/API. Proceedings of the Conference on Petroleum Hydrocarbons and Organic Chemicals in Ground Water: Prevention, Detection and Restoration, Nov. 27–29, 1989. National Water Well Association, Dublin, OH.

52. Litchfield, C. D., C. W. Erkenbrecher, Jr., C. E. Matson, L. S. Fish, and A. Levine. 1988. Evaluation of Microbial Detection Methods and Interlaboratory Comparisons during a Peroxide-Nutrient Enhanced *In Situ* Bioreclamation. pp. 52(1–6). *In* B. H. Olson and D. Jenkins (eds.), Proceedings: International Conference on Water and Wastewater Microbiology, Feb. 8–11, 1988, Vol. 2.

53. Litchfield, C. D., M. D. Lee, and R. L. Raymond, Sr. 1989. Present and Future Directions in *In Situ* Bioreclamation. pp. 587–596. *In* A. J. Borner (ed.), Proceedings of the Second Annual Hazardous Materials Conference/Central, March 14–16, 1989. Tower Conference Management Company, Glen Ellyn, IL.

54. Litchfield, J. H., and L. C. Clark. 1973. Bacterial Activity in Ground Waters Containing Petroleum Products. American Petroleum Institute Publication No. 4211. American Petroleum Institute, Washington, D.C. 37 pp. and appendices.

55. Looney, B. B., T. C. Hazen, D. S. Kaback, and C. A. Eddy. 1991. Full Scale Field Test of the In Situ Air Stripping Process at the Savannah River Integrated Demonstration Test Site (U). Westinghouse Savannah River Company, Aiken, S. C. 90 pp.

56. Lund, N.-Ch., J. Swinianski, G. Gudehus, and D. Maier. 1991. Laboratory and Field Tests for a Biological *In Situ* Remediation of a Coke Oven Plant. pp. 396–412. *In* R. E. Hinchee and R. F. Olfenbuttel (eds.), *In Situ* Bioreclamation: Applications and Investigations for Hydrocarbon and Contaminated Site Remediation. Butterworth-Heinemann, Boston, MA.

57. Mahaffey, W. R., and R. A. Sanford. 1991. Bioremediation of PCP-Contaminated Soil: Bench to Full-Scale Implementation. *Remediation* Summer:305–323.

58. Marrin, D. L., J. J. Adriany, and A. J. Bode. 1991. Estimating Small-Scale Differences in Air Permeability and Redox Conditions for the Design of Bioventing Systems. pp. 457–465. NWWA/API. Proceedings of The Petroleum Hydrocarbons and Organic Chemicals in Ground Water Prevention, Detection, and Restoration, Nov. 20–22, 1991. National Water Well Association, Dublin, OH.

59. McNabb, J. F., and W. J. Dunlap. 1973. Subsurface Biological Activity in Relation to Ground Water Pollution. Ground Water 13:33–44.

60. McNabb, J. F., and G. E. Mallard. 1984. Microbiological Sampling in the Assessment of Groundwater Pollution. pp. 235–260. *In* G. Bitton and C. P. Gerba (eds.), Groundwater Pollution Microbiology. Wiley, NY.

61. McKee, J. E., F. B. Laverty, and R. N. Hertel. 1972. Gasoline in Groundwater. *J. Water Pollut. Contr. Fed.* 44:293–302.

62. Miller, R. N., C. C. Vogel, and R. E. Hinchee. 1991. A Field-Scale Investigation of Petroleum Hydrocarbon Biodegradation in the Vadose Zone Enhanced by Soil Venting at Tyndall AFB, Florida. pp. 283–302. *In* R. E. Hinchee and R. F. Olfenbuttel (eds.), *In Situ* Bioreclamation: Applications and Investigations for Hydrocarbon and Contaminated Site Remediation. Butterworth-Heinemann, Boston, MA.

63. Mote, P. A., D. F. Leland, and D. R. Smallbeck. 1990. Accelerated Site Remediation Using Enhanced In Situ Biodegradation. Harding Lawson Associates, Novato, CA, 15 pp.

64. Mueller, J. G., D. P. Middaugh, S. E. Lantz, and P. J. Chapman. 1991. Biodegradation of Creosote and Pentachlorophenol in Contaminated Groundwater: Chemical and Biological Assessment. *Appl. Environ. Microbiol.* 57:1277–1285.

65. Nelson, M. J. K., S. O. Montgomery, W. R. Mahaffey, and P. H. Pritchard. 1987. Biodegradation of Trichloroethylene and Involvement of an Aromatic Biodegradative Pathway. *Appl. Environ. Microbiol.* 53:949–954.

66. Nelson, M. J. K., J. V. Kinsella, and T. Montoya. 1990. *In Situ* Biodegradation of TCE Contaminated Groundwater. *Environ. Progr.* 9:190–196.

67. Nelson, M. J. K., and A. W. Bourquin. 1990. U. S. Patent 4,925,802. Ecova Corporation, Redmond, WA.

68. Nishino, S. F., J. C. Spain, L. A. Belcher, and C. D. Litchfield. 1992. Chlorobenzene Degradation by Bacteria Isolated from Contaminated Groundwater. *Appl. Environ. Microbiol.* 58:1719–1726.

69. Ogunseitan, O. A., E. T. Tedford, D. Pacia, K. M. Sirotkin, and G. S. Sayler. 1987. Distribution of Plasmids in Groundwater Bacteria. *J. Ind. Microbiol.* 1:311–317.

70. Parsons, F., G. B. Lage, and R. Rice. 1985. Biotransformation of Chlorinated Organic Solvents in Static Microcosms. *Environ. Toxicol. Chem.* 4:739–742.

71. Raymond, R. L., V. W. Jamison, and J. O. Hudson. 1976. Beneficial Stimulation of Bacterial Activity in Groundwaters Containing Petroleum Products in I. Physical, Chemical Wastewater Treatment. *AIChE Symp. Ser.* 73:390–404.

72. Raymond, R. L., V. W. Jamison, J. O. Hudson, R. E. Mitchell, and V. E. Farmer. 1978. Field Application of Subsurface Biodegradation of Gasoline in Sand Formation. Final report submitted to American Petroleum Institute, Washington, D.C. 137 pp.

73. Reineke, W., and H.-J. Knackmuss. 1984. Microbial Metabolism of Haloaromatics: Isolation and Properties of a Chlorobenzene-Degrading Bacterium. *Appl. Environ. Microbiol.* 47:395–402.

74. Rifai, H. S., and P. B. Bedient. 1987. Bioplum II—Two Dimensional Modeling for Hydrocarbon Biodegradation and *In Situ* Restoration. pp. 431–450. NWWA/API. Proceedings of Petroleum Hydrocarbons and Organic Chemicals in Groundwater: Prevention, Detection and Restoration, Nov. 17–19, 1987. National Water Well Association, Dublin, OH.

75. Roberts, P., L. Semprini, G. Hopkins, and P. McCarty. 1989. Biostimulation of Methanotrophic Bacteria to Transform Halogenated Alkenes for Aquifer Restoration. pp. 203–217. NWWA/API. Proceedings of the Conference on Petroleum Hydrocarbons and Organic Chemicals in Ground Water: Prevention, Detection and Restoration, Nov. 15–17, 1989. National Water Well Association, Dublin, OH.

76. Rowland, M. A., and T. N. Eisenberg. 1989. Anaerobic Biodegradation of Trichloroethylene in a Shallow Aquifer. pp. 188–191. Proceedings of Third National Outdoor Action Conference on Aquifer Restoration, Ground Water Monitoring, and Geophysical Methods. May 22–25, 1989. National Water Well Association, Dublin, OH.

77. Schmidt, S. K., and M. J. Gier. 1990. Coexisting Bacterial Populations Responsible for Multiphasic Mineralization Kinetics in Soil. *Appl. Environ. Microbiol.* 56:2692–2697.

78. Schmitt, E. K., and J. A. Caplan. N.D. In-Situ Biological Cleanup of Petroleum Hydrocarbons in Soil and Groundwater. O.H. Materials Corp., Findlay, 14 pp., OH.

79. Semprini, L., P. V. Roberts, G. D. Hopkins, and P. L. McCarty. 1990. A Field Evaluation of In-Situ Biodegradation of Chlorinated Ethenes: Part 2, Results of Biostimulation and Biotransformation Experiments. *Ground Water.* 28:715–727.

80. Sevee, J. 1991. Methods and Procedures for Defining Aquifer Parameters. pp. 397–448. *In* D. M. Nielsen (ed.), Practical Handbook for Ground-Water Monitoring. Lewis Publishers, Chelsea, MI.

81. Spain, J. C., J. D. Milligan, D. C. Downey, and J. K. Slaughter. 1989. Excessive Bacterial Decomposition of H_2O_2 During Enhanced Biodegradation. *Ground Water.* 27:163–167.

82. Spain, J. C., and P. A. Van Veld. 1983. Adaptation of Natural Microbial Communities to Degradation of Xenobiotic Compounds: Effects of Concentration, Exposure Time, Inoculum, and Chemical Structure. *Appl. Environ. Microbiol.* 45:428–435.

83. Staps, J. J. M. 1988. Developments in in situ biorestoration of contaminated soil and groundwater in the Netherlands. pp. 379–390. *In* Z. Filip (ed.), Biotechnologische In-situ-Sanierung. Fischer Verlag, Stuttgart/New York.

84. Suflita, J. M., and S. A. Gibson. 1985. Biodegradation of Haloaromatic Substrates in a Shallow Anoxic Ground Water Aquifer. pp. 30–32. *In* N. N. Durham and A. E. Redelfs (eds.), Proceedings of the Second International Conference on Ground Water Quality Research. Oklahoma State University Printing Services, Stillwater, OK.

85. Swindoll, C. M., C. M. Aelion, and F. K. Pfaender. 1988. Influence of Inorganic and Organic Nutrients on Aerobic Biodegradation and on the Adaptation Response of Subsurface Microbial Communities. *Appl. Environ. Microbiol.* 54:212–217.
86. Thomas, J. M., and C. H. Ward. 1989. In Situ Biorestoration of Organic Contaminants in the Subsurface. *Environ. Sci. Technol.* 23:760–766.
87. Vogel, T. M., C. S. Criddle, and P. L. McCarty. 1987. Transformations of Halogenated Aliphatic Compounds. *Environ. Sci. Technol.* 21:722–736.
88. Ward, C. H., J. M. Thomas, S. Fiorenza, H. S. R. Rifai, P. B. Bedient, J. M. Armstrong, J. T. Wilson, and R. L. Raymond. 1988. A Quantitative Demonstration of the Raymond Process for *In Situ* Biorestoration of Contaminated Aquifers. pp. 723–743. NWWA/API. Proceedings of Petroleum Hydrocarbons and Organic Chemicals in Ground Water: Prevention, Detection and Restoration, Nov. 9–11, 1988. National Water Well Association, Dublin, OH.
89. Ward, C. H., J. M. Thomas, S. Fiorenza, H. S. Rifai, P. B. Bedient, J. T. Wilson, and R. L. Raymond. 1989. *In Situ* Bioremediation of Subsurface Material and Ground Water Contaminated with Aviation Fuel: Traverse City, Michigan. pp. 83–96. Proceedings of the 1989 A&WMA/EPA International Symposium on Hazardous Waste Treatment: Biosystems for Pollution Control. Air and Waste Management Association. Pittsburgh, PA.
90. Werner, P. 1985. A New Way for the Decontamination of Polluted Aquifers by Biodegradation. *Water Supply* 3:41–47.
91. Wetzel, R. S., D. H. Davidson, and C. M. Durst. N.D. Effectiveness of In Situ Biological Treatment of Contaminated Groundwater and Soils at Kelly Air Force Base, Texas. Science Applications International Corporation, McLean, VA. 18 pp.
92. Wilson, B. H., and M. V. White. 1986. A Fixed-Film Bioreactor to Treat Trichloroethylene-Laden Waters from Interdiction Wells. pp. 425–436. *In* Proceedings of the Sixth National Symposium and Exposition on Aquifer Restoration and Ground Water Monitoring, May 19–22, 1986. National Water Well Association, Dublin, OH.
93. Wilson, J. T., and B. H. Wilson. 1985. Biotransformation of Trichloroethylene in Soil. *Appl. Environ. Microbiol.* 49:242–243.
94. Wilson, J. T., J. F. McNabb, B. H. Wilson, and M. J. Noonan. 1983. Biotransformation of Selected Organic Pollutants in Ground Water. *Dev. Ind. Microbiol.* 24:225–233.
95. Wilson, J. T., Jr., and B. H. Wilson. 1987. U.S. Patent 4,713,343. The United States of America as represented by the Administrator of the U.S. Environmental Protection Agency, Washington, D.C.
96. Wilson, S. B., and R. A. Brown. 1989. In Situ Bioreclamation: A Cost-Effective Technology to Remediate Subsurface Organic Contamination. Groundwater Monitoring Review 9:173–179.
97. Wood, P. R., R. F. Lang, and I. L. Payan. 1985. Anaerobic Transformation, Transport, and Removal of Volatile Chlorinated Organics in Ground Water. pp. 493–511. *In* C. H. Ward, W. Giger, and P. L. McCarty (eds.), Ground Water Quality. Wiley, New York, NY.
98. Baker et al., personal communication.
99. Litchfield et al., personal communication.
100. Nelson, personal communication.

Use of Altered Microoganisms for Field Biodegradation of Hazardous Materials

Michael A. Gealt

Department of Bioscience and Biotechnology
Drexel University
Philadelphia, Pennsylvania

Morris A. Levin

Maryland Biotechnology Institute
University of Maryland
College Park, Maryland

Malcolm Shields

Center for Environmental Diagnostics
University of West Florida
Pennsacola, Florida

The large amount of hazardous waste generated and disposed of has given rise to environmental conditions requiring remedial treatment. The use of landfills has traditionally been a cost-effective means to dispose of waste. However, increased costs of transportation and decreasing numbers of landfill sites now necessitate the examination of treatment processes that can be carried out on site (land farming, composting), and, preferably, in situ. Thus, economics dictate the exploration of bioremediation techniques as potentially environmentally sound cost reduction methods.

It is difficult to understand why many people consider bioremediation to be a new approach for waste reduction. The fact that we are not buried in organic matter is testimony to the effectiveness of the microbial carbon cycle. We have now chosen to concentrate attention on discrete aspects of this cycle that can reduce environmental pollutant levels. Although the first commercial application of bioremediation took place more than forty years ago, our knowledge of such processes dates back to primitive humankind. The current use of biological treatment processes (biotechnology) is far more widespread than is evident from the response of the public. Sewage treatment plants are easily identifiable large-scale examples, and certainly septic tanks offer an even more widespread example. The major change in our thinking is the emergence of the idea that perhaps the microorganisms involved may be moved to or be recruited at the site of contamination, rather than the waste being sent to the vicinity of the microorganisms.

Although the use of genetically engineered microorganisms has been considered, to date most bioremediation has been accomplished by enhancing the growth of indigenous microorganisms, or by augmenting the microbial population with exogenous organisms isolated from the site in question or from similar sites (Fox, 1992). Such microorganisms may be superior degraders because selection or mutagenesis procedures have enhanced their ability to degrade a particular material. For example, case studies have shown that crude oil contamination in soil can be reduced to levels acceptable to regulatory agencies in less than 20 weeks with total costs of $35 to $120 per cubic yard utilizing only indigenous microbes. This is considerably less than nonbiological alternatives (Hildebrandt and Wilson, 1991; Levin, UNIDO Biotechnology Manual, in press). Unfortunately, as reported by Fox et al. (1990), there are some soils which, when analyzed in feasibility tests, indicate a lack of indigenous organisms able to carry out mineralization. Clearly, the addition of foreign bacteria or fungi is indicated when there is a dearth of indigenous microorganisms. However, the mere addition of known degrading microorganisms to a contaminated site is far from a guarantee of enhanced biodegradation (see Compeau et al., 1991, for an example where additional organisms did not enhance pentachlorophenol (PCP) degradation).

If the capabilities or the environmental viability of the exogenous organisms could be enhanced, their addition to a contaminated environment might improve the degradation process, either kinetically or by broadening the spectrum of compounds degraded. In some cases, organisms with enhanced capabilities have arisen spontaneously. In addition, procedures currently employed to procure modified strains include both conventional methods and genetic engineering. Conventional methods of obtaining altered bacteria, i.e., selection and mutagenesis, are reliable—they have proven effectiveness in producing altered phenotypes—but are slow and tedious. Additionally, one is never certain that the selection protocol will give rise only to the phenotype desired. Selection protocols must be carefully tailored to identify and isolate desired phenotypes.

While to date the greatest success with biotechnology has been in the medical and pharmaceutical arenas, it is easy to envision future successes of environmental biotechnology. With advances in genetic engineering techniques there are now ways to optimize specific bacteria for field use. The first organism designed for use in the field was an oil-degrading bacterium developed by Chakrabarty. However, the technique he used (plasmid-assisted breeding) is not considered genetic engineering by regulatory definitions today, and the organism was never used in the field (Levin, 1983). The oil-degrading *Pseudomonas* strain was created through the introduction of various degradative plasmids to allow the mineralization of xylene, toluene, naphthalene, octane, salicylate, and other simple aromatics and aliphatics. This demonstrates an attractive area for bioremediation: namely, an easily degradable pollutant in a relatively inaccessible location that makes physical treatment (vitrification, incineration, etc.) extremely expensive. Examples include leaking underground storage tanks, oil spills, and industrial waste streams. There is little need for recombinant technology if the pollutants are readily attacked.

The primary application of recombinant DNA (rDNA) technology in bioremediation will be in dealing with *recalcitrant wastes:* wastes that are found to be difficult to degrade because of a paucity of microbial pathways that can recognize or transform them, especially at a rapid rate. Unfortunately this class of pollutants is very broad, including many pesticides, heterocyclic compounds, substituted and halogenated aromatics, and aliphatics (methanes, ethanes, and ethenes). Direct selection is not useful in these cases because either the enzyme or pathway is not available or is not inducible solely by the recalcitrant compound.

There are now several organisms which have been successfully modified to enhance their degradative capabilities (Rojo et al., 1987; Don et al., 1985; Kukor and Olsen, 1990; Zylstra et al., 1989) in the laboratory. But not only have these strains not been field-tested, neither have the several strains which have arisen following spontaneous mutation or traditional mutagenesis and selection (Rojo et al., 1987; Knackmuss, 1976; Taeger et al., 1988). One strain, *Pseudomonas B13,* has been tested in the field, but only for survival, not for the degradation of chloroaromatic compounds for which it was designed (Krumme et al., 1991).

Bioaugmentation, as currently practiced, uses naturally occurring organisms. The added organisms either furnish an associative consortium or, perhaps most importantly, significantly increase the titer of degraders. The delay in utilization of laboratory-bred microorganisms results in part from unclear regulatory procedures promulgated by federal, state, and local agencies as to which organisms are sufficiently modified to warrant regulation as novel (i.e., engineered) organisms. In an advertising circular for *The Bioremediation Report* (published by COGNIS, Inc., Santa Rosa, CA 95407) the author noted that "...bioremediation is very much an enigma. Albeit, a well studied enigma." It is anticipated that the enigma will be clarified within the next few years. This chapter will examine the advantages and disadvantages of using natural and modified organisms from scientific and regulatory perspectives.

Modification of Organisms by Conventional Techniques

Advantages

The advantages of using selected microorganisms for bioremediation in combination with traditional engineering techniques have been demonstrated in many ways over many years. In some cases a traditional engineering practice, e.g., soil venting, has been combined with the use of native organisms to degrade organics, giving rise to bioventing. This has been used successfully in the field to treat petroleum hydrocarbons (Hoeppel et al., 1991). Composting has also been used successfully to degrade munitions (Williams and Myler, 1990).

Contaminated sites are usually excellent sources for isolation of these organisms. The principles of adaptation and enrichment both contribute to this. *Adaptation* is the natural process that allows organisms to survive in the increased presence of toxic substances. It may be the result of natural physiological modifications resulting in increased concentration of cellular components capable of binding the toxicant, alteration of surface binding molecules resulting in increased exclusion of the toxic materials, or enhanced transport of the toxic material back to the environment.

Additionally, materials brought into the cell present the metabolic process with potential for new substrates for energy production (resulting in enrichment), although in some cases enzymes act on the toxic material without energy production in what has been termed *cometabolism*. In cases where no energy is derived from the component being transformed, energy from utilization of other substrates substantially improves mineralization rates, but may adversely affect the length of time over which cells are able to carry out the transformation (Alvarez-Cohen and McCarty, 1991a, 1991b).

Enrichment is a primary mechanism for altering the structure of the native microbial community to increase the proportion of degrading organisms. Increased numbers of resistant organisms result from their ability to utilize toxic organic compounds as alternate energy sources (and thus detoxify them) and the decrease in competitors for nutrients present in the environment (as microorganisms sensitive to the toxicants decrease in number). Increases in cometabolizers occur because the cometabolizers decrease levels of toxic compounds within their microenvironment. Increases in metabolizers occur because they are able to obtain energy from the degradation of the toxic material. The selection and enrichment processes can be utilized in the laboratory, where such changes in community structure enable the study of particular physiologic responses—i.e., frequencies of enrichment and degradation rates. In the field, natural enrichment occurs and is undoubtedly a primary factor in applications such as composting, bioventing, land farming, sewage treatment, and activated-sludge acclimation.

Civil engineers have known for some time that a toxic insult to a bioreactor system, such as a sewage treatment trickling filter, results in the loss of its selected microbial community. The ability of such natural communities to adapt to ranges of environmental insult are well known. However, the natural restoration of the community is a time-consuming process that can be tremendously shortened by the addition of organisms from a similarly enriched community at another location, via direct importation of microorganisms or addition of cells grown from preserved (dried or frozen) samples from similar communities. This result is apparent from the work of Crawford and coworkers on the land farming of dinoseb (see Chap. 10, by Roberts et al.).

In a few cases it has been shown that specific community members may have the capability to adapt, in fact representing the evolution of control mechanisms and combinations of pathways on a time scale few would have predicted possible (Focht, 1987; Kellogg et al., 1981). This is also borne out in the study of the degradation of the recalcitrant anthropogenic compounds for which a natural environmental cycle is doubtful. Specific examples demonstrated in laboratory situations include polychlorinated biphenyls (PCBs) (Unterman, 1991; Pettigrew-Pope, 1990; Bedard et al., 1984; Kellogg et al., 1981; Sayler, 1991; Furukawa et al., 1987), chlorobenzenes (Spain, 1990), 2,4-dichlorophenoxyacetic acid (2,4-D) (Pemberton et al., 1979; Timmis et al., 1987), and 2,4,5-trichlorophenoxyacetic acid (2,4,5-T) (Sangodkar, 1988). The bioremediation of PCBs, at a level of 5–10 mg/kg, with *Alcaligenes A5* (Shields, 1985) has been shown in 60-kg field lysimeters. Several different strains of bacteria in pure culture have been demonstrated to degrade both chlorinated and nonchlorinated aromatic solvents, including benzene, toluene, xylene, chlorobenzene, and di- and trichlorobenzene in soil slurries (Oldenhuis et al., 1989). The cometabolic degradation of chloroaliphatic solvents by environmental isolates suitably induced by appropriate substrates has been amply demonstrated as well (Fox et al., 1990; Harker, 1992). The use of selected strains in the field is only beginning due to regulatory uncertainty, and, perhaps, to public resistance.

There are two major advantages of using selected mutants, whether they arise spontaneously or following mutagenesis. First, from an economic perspective, they are generally regarded as natural occurrences and not novel organisms requiring extensive testing: minimal regulatory compliance (compared to the use of engineered microorganisms) is required before introduction. The regulatory factor can often save the manufacturer considerable time and money in preparing the products for sale. Second, the selected mutants are recent isolates. In most cases, these microorganisms are more similar to field isolates than are genetically engineered bacteria because characterized strains are most often used for genetic engineering, strains which have been out of the natural environment sometimes for decades. Mutants selected from recently isolated strains may be more successful when reintroduced to their native environment. It is more likely that these strains will have retained their environmental hardiness and thus will function successfully when reintroduced to their environment.

Finally, selected strains may act as part of a consortium rather than acting alone. A consortium may possess sufficient genetic variety so that entire pathways of degradation may be present, even though the information is in different members of the population. Thus, although no single organism contains the desired degradative pathway, the pathway is still operative. While it is possible to engineer all the requisite genes into a single organism, this will probably take much longer and be more costly than getting organisms with complementing capabilities to harmonize.

Disadvantages

When compared with modification by genetic engineering, there are some distinct disadvantages to mutation and selection protocols. First of all, the modifications made are not specific. The mutagens likely to be used act fairly nonspecifically. Spontaneous mutations are random. A selection scheme must be used to find the desired phenotype, and, no matter how cleverly designed, may not be sufficiently specific to yield only a single desired phenotype. Many selection processes (breeding) will require development and use of nontraditional genetic selection and counterselection techniques. As a result, the observed capabilities are the result of an unknown alteration or alterations, and when the system fails it is impossible to determine the cause.

Second, results are limited by those abilities, i.e., by that genetic information, possessed by the bacteria or fungi on hand. The pathway adaptations or constructions must occur because of the direct growth advantage of the newly combined or induced genes to the microbe or must depend on the ability of the researcher to monitor very specific biochemical transformations occurring within a mutagenized population.

Although soil microorganisms have the capacity to evolve enzymes to attack newly encountered chemicals, the rate of development is very slow. It is possible to demonstrate adaptation to specific substrates in the laboratory (Engesser et al., 1980). However, proteins evolve at very different rates (Ayalah, 1992; Ornstein, 1991a). The time required for a 1 percent difference to be established between proteins, called a *unit evolutionary period* (UEP), is expressed in millions of years, generally in the range of 2 to 20. This is clearly a problem common to many recalcitrant anthropogenic compounds that have no naturally occurring analogs.

Modification of Organisms by Genetic Engineering Techniques

For the last decade, one of the promises held out to bioremediation practitioners has been the direct intervention of molecular biology in the biochemistry of catabolic pathways to produce more effective microorganisms for use in treating wastes. The process, involving either direct manipulation of nucleotides of existing gene sequences by now common gene-cloning techniques or creation of a specific enzyme as a result of computer simulation followed by insertion into a selected host, have been reported and applied in the pharmaceutical and/or agricultural sectors (Don et al., 1985; Eaton and Timmis, 1986; Kukor et al., 1988; Kukor and Olsen, 1990; Paulsen and Ornstein, 1991; Paulsen et al., 1991). Applications remain limited to the pharmaceutical and agricultural sectors.

As noted above, no microbes produced as a result of these molecular biology techniques have been used in bioremediation. In a 1985 symposium on the subject of intentional or accidental release of recombinant bacteria to the environment, the subjects of nitrogen fixation (*Rhizobium*), viral vaccines (vaccinia; rabies), ice nucleation by *Pseudomonas,* bacterial pesticides, and herbicide-resistant plants were addressed. The concept of biodegradation did not even rate consideration at this time (Halvorson et al., 1985).

Advantages

Although several major problems must be faced, the ability to genetically engineer microorganisms creates new opportunities. For example, the degradation of trichloroethylene (TCE) is a cometabolic phenomenon in all bacteria for which it has been described (Ensley, 1991). As such, it represents an excellent candidate for genetic manipulations. As already noted by Russel et al. (1992) in their review of aerobic TCE degradation systems, the requirements for aromatic or aliphatic inducers of TCE-metabolizing enzymes, such as toluene, phenol, methane, or isoprene oxygenases, present difficulties for both bioreactor and in situ applications. They also noted, however, that this difficulty might be alleviated by manipulation of the proper genetic sequence governing oxygenase production in the respective strain (Shields and Reagin, 1992).

Many problems must be resolved before field implementation of these newly created organisms, including:

Design of strain. Most contaminated sites are quite unique, having distinct combinations of physical parameters that will affect the biological processes required to deal with toxic materials. Strains designed for introduction may have to be uniquely designed for individual sites so that their suite of genetic capabilities complement the environmental hazards they are about to face. Such microorganisms could be constructed so that they were able to degrade wastes at a faster rate and to degrade a wider variety of materials (Ornstein, 1991*b*).

Genetic engineering. Even after many years of experience with its techniques, accomplishing genetic manipulation is not trivial and is frequently quite time-consuming. The time required to manipulate and test means additional cost to the producer. Another important factor is that data requirements for risk assessment have not yet been completely defined. However, as experience with traditionally derived microorganisms expands, risk assessment procedures will improve. This will permit researchers to devise means to incorporate safety features into engineered cell lines and to develop test protocols to assess both the usefulness and the hazards of bioremediation. This would finally allow a true cost-benefit analysis, where cost is not only a dollar amount but also includes the environmental costs of continued inaction and/or possible adverse effects from the pollutant or the treatment.

Safety engineering. Research into development of specific safety features to limit survival and spread of microorganisms engineered for bioremediation purposes is ongoing. Cuskey (1992) reviewed concepts and applications of biological containment involving the addition of conditionally controlled lethal genes ("suicide genes") to engineered microorganisms. Addition of the gene(s) would conceivably limit the ability of the microorganism to disperse beyond a selected zone or deprive the bacterium of survival ability after a particular environmental constituent was depleted.

Reintroduction to field. Manipulated strains frequently have been used in the laboratory for many years. During this time they have not been challenged with the rigors of the field; as a result, subtle nonselected alterations have frequently occurred which render the bacterium less suitable for the field than was its ancestor. As a result, upon reintroduction, insufficient numbers may establish themselves to carry out the remediation.

All of these problems are resolvable. The potential advantages—greater range and rate of degradative ability—conferred on bacteria through molecular biology intervention continue to drive laboratory research. Advances in regulatory structure (see Chap. 6 by Giamporcaro) encourage industrial development. Above all, continued and increased waste production provides a very strong incentive to increase efficacy in the biodegradation of waste material.

Disadvantages

Several factors have prevented the realization of the promise of genetic engineering. One is a persisting strong belief that all possible routes of microbial biodegradation exist, or at least that such a plethora of enzymes and pathways exist that any desired pathway can be developed by simple rearrangement using nonrecombinant techniques (Gibson et al., 1990; Pemberton et al., 1979; Don et al., 1985). The idea that one should exhaust existing resources has acted as a strong deterrent to embarking on the expensive and less certain avenues for development of efficacious microorganisms through genetic engineering. The ease with which microbes have been isolated or genetically manipulated by techniques other than those of molecular biology, techniques which result in products which are effective and do not require regulatory review, has retarded the growth of the recombinant-based industry.

Proponents of the exploitation of existing degradative pathways have acknowledged that some chemicals possess structural elements or substituents that confer upon the molecule a high degree of resistance to enzymatic attack or are present in mixtures incompatible for the effective degradation of the toxic component (Timmis et al., 1987; Gerger and Shields, 1992) and require custom-designed pathways and/or enzymes (Rojo et al., 1987, 1988). However, the success of traditional strategies involving, for example, either mating or using controlling population pressures, transfer of existing genes between organisms, or alteration of regulatory genes (Rojo et al., 1987, 1988) at relatively lower cost than would be incurred using the newer biotechnology techniques has had the effect of inhibiting research into actual genetic engineering of microorganisms for bioremediation. All of these traditional techniques presuppose that the genetic information is available, i.e., can be readily selected for either in the laboratory or under field conditions.

From a purely economic perspective, a major factor is regulatory uncertainty coupled with public resistance. Few industries that might be expected to foster the major research in this area feel compelled to do so in the face of very feeble guarantees of a return on their investment. The cost of running the regulatory gauntlet could easily total in the millions of dollars.

Another major factor is the dearth of basic scientific information concerning relevant degradation pathways and their modes of regulation. Sources of research funding for development of such data are very few. In those cases where funding *is* available there is almost always a direct need for immediate implementation, as opposed to developing basic information. As a result the track record for development of recombinant organisms for bioremediation is not only dismal, it is virtually unpopulated!

Acknowledgment

The authors wish to express their appreciation to James Harvey, Deputy Program Manager, Technical Resources Inc., Gulf Breeze, Florida, for his aid in assembling the material used in this chapter.

References

Alvarez-Cohen, L., and P. L. McCarty. 1991a. Effects of toxicity, aeration, and reductant supply on trichloroethylene transformation by a mixed methanotrophic culture. *Appl. Environ. Microbiol.* 57:228–235.

Alvarez-Cohen, L., and P. L. McCarty. 1991b. Product toxicity and cometabolic competitive inhibition modeling of chloroform and trichloroethylene transformation by methanotrophic resting cells. *Appl. Environ. Microbiol.* 57:1031–1037.

Ayalah, F. J. 1992. Evolution. In: The Genetic Revolution (ed. B. D. Davis), John Hopkins Press, Baltimore, pp. 178–196.

Bedard, T. R., P. Allenza, and T. G. Lessie. 1984. Hyperinduction of enzymes of the phosphorylative pathway of glucose dissimilation in *Pseudomonas cepacia*. *Curr. Microbiol.* 11:143–148.

Compeau, G. C., W. D. Mahaffey, and L. Patras. 1991. Full-scale bioremediation in contaminated soil and water. In: Environmental Biotechnology (ed. G. Omenn). Plenum Press: New York, pp. 91–110.

Cuskey, S. M. 1992. Biological containment of genetically engineered organisms. In: Microbial Ecology, Principles, Methods, and Applications, (ed. M. A. Levin, R. J. Seidler, and M. Rogul). McGraw-Hill: New York, pp. 911–918.

Don, R. H., A. J. Weightman, H. J. Knackmuss, and K. N. Timmis. 1985. Transposon mutagenesis and cloning analysis of the pathways for degradation of 2,4-dichlorophenoxyacetic acid and 3-chlorobenzoate in *Alcaligenes eutrophus* JMP134(pJP4). *J. Bacteriol.* 161:85–90.

Eaton, R. W., and K. N. Timmis. 1986. Characterization of a plasmid-specified pathway for catabolism of isopropylbenzene in *Pseudomonas putida* RE204. *J. Bacteriol.* 168:123–131.

Engesser, K.-H., E. Schmidt, and J.-J. Knackmuss. 1980. Adaptation of *Alcaligenes eutrophus B9* and *Pseudomonas* sp. B13 to 2-fluorobenzoate as growth substrate. *Appl. Environ. Microbiol.* 39:68–73.

Ensley, B. D. 1991. Biochemical diversity of trichloroethylene metabolism. *Annu. Rev. Microbiol.* 45:283–299.

Focht, D. 1987. Performance of biodegradative microorganisms in soil; Xenobiotic chemicals as unexploited metabolic niches. In: Environmental Biotechnology (ed. G. Omenn). Plenum Press: New York, pp. 15–31.

Fox, J. L. 1992. Assessing the scientific foundations of bioremediation. *ASM News* 58:483–485.

Fox, B. G., J. G. Borneman, L. P. Wackett, and J. D. Lipscomb. 1990. Haloalkene oxidation by the soluble methane monooxidase from *Methylosinus trichosporium* 0B3: Mechanistic and environmental implications. *Appl. Environ. Microbiol.* 29:6419–6427.

Furukawa, H., T. D. Gaffney, and T. G. Lessie. 1987. Insertion-sequence-dependent rearrangements of *Pseudomonas cepacia* plasmid pTGL1. *J. Bacteriol.* 169:224–230.

Gerger, R. R., and M. S. Shields. 1992. Expansion of the range of chloroaromatic compounds degraded by the trichloroethylene degrading strain *Pseudomonas cepacia.* Abs. AMS Q-210, (ed. H. Halvorsen, D. Pramer, and M. Rogul. 1985. Engineered organisms in the environment: Scientific issues. AMS, Washington, D.C.

Gibson, D. T., G. J. Zylstra, and S. Chauhan. 1990. Biotransformations catalyzed by toluene dioxygenase from *Pseudomonas putida F1.* In: *Pseudomonas:* Biotransformations, Pathogenisis, and Evolving Biotechnology (ed. S. Silver, A. M. Chakrabarty, B. Iglewski, and S. Kaplan). ASM: Washington, D.C.

Halvorsen, H. O., D. Pramer, and M. Rogul. 1985. Engineered organisms in the environment: Scientific issues. AMS, Washington, D. C.

Harker, Alan R. 1991. Potential use of genetically engineered microorganisms in the remediation of environmental pollution. In: Proceedings of the International Symposium on Ground Water in Practice. Nashville, TN, 7/29–8/2/1991. ASCE: New York, pp. 232–237.

Hildebrandt, W. W., and S. B. Wilson. 1991. On-site bioremediation systems reduce crude oil contamination. *J. Petrol. Technol.* 43:18–22.

Hoeppel, R. E., R. E. Hinchee, and M. F. Arthur. 1991. Bioventing soils contaminated with petroleum hydrocarbons. *J. Ind. Microbiol.* 8:141–146.

Knackmuss, J. H., W. Beckmann, E. Dorn, and W. Reinecke. 1976. Zum mechanismus der biologischen persitenz von halogenierten und sulfonierten aromatischen Kohlenwasserstoffen (On the mechanism of the biological persistence of halogenated and sulfonated aromatic hydrocarbons). *Zentralbl. Bakteriol.* (orig B) 162:127–137.

Kellogg, S. T., D. K. Chatterjee, and A. M. Chakrabarty. 1981. Plasmid-assisted molecular breeding: New technique for enhanced biodegradation of persistent toxic chemical. *Science* 214:1133–1135.

Krumme, M. L., R. L. Smith, S. M. Thiem, and D. F. Dwyer. 1991. Enumeration of *Pseudomonas* species B13 following injection into a sand and gravel aquifer. Abs. ASM Q71.

Kukor, J. J., and R. H. Olsen. 1990. Molecular cloning, characterization and regulation of a *Pseudomonas pickettii PK01* gene encoding phenol hydroxylase and expression of the gene in *Pseudomonas aeruginosa PA01c. J. Bacteriol.* 172:4624–4630.

Kukor, J. J., R. H. Olsen, and D. P. Ballou. 1988. Cloning and expression of the catA and catBC gene clusters from *Pseudomonas aeruginosa PAO. J. Bacteriol.* 170:58–65.

Levin, M. A. 1992. Bioremediation Safety Issues. United Nations Industrial Development Office (UNIDO). In press.

Levin, M. A., G. H. Kidd, R. H. Zaugg, and J. R. Swarz. 1983. Applied Genetic Engineering. Noyes Press: Park Ridge, New Jersey pp. 15–16.

Oldenhuis, R., R. L. J. M. Vink, D. B. Jansenn, and D. B. Witholt. 1989. Degradation of chlorinated aliphatic hydrocarbons by *Methylosinus trichosporium* 0B3b expressing soluble methane monooxygenase. *Appl. Environ. Microbiol.* 55:2819–2826.

Ornstein, R. L. 1991a. Why timely bioremediation of synthetics may require rational enzyme redesign: Preliminary report on redesigning Cytochrome P450cam for TCE dehalogenation. In: On-Site Bioreclamation (ed. R. E. Hinchee and R. F. Olfenbuttel). Butterworth-Heinemann, Boston, pp. 509–514.

Ornstein, R. L. 1991b. Using molecular dynamics simulations on carmbin to evaluate the suitability of different continuum dielectric and hydrogen atom models for protein simulations. *J. Biomol. Struct.* 7:1019–1042.

Paulsen, M. D., and R. L. Ornstein. 1991. A 175-psec molecular dynamics simulation of camphor bound cytochrome p450 sub (cam). *Proteins* 11:184–204.

Paulsen, M. D., M. B. Bass, and R. L. Ornstein. 1991. Analysis of active site motions from a 175-psec molecular dynamics simulation of camphor bound Cytochrome p450 sub(cam). *J. Biomol. Struct. Dyn.* 9:187–203.

Pemberton, J. M., B. Corney, and R. H. Don. 1979. Evolution and spread of pesticide degrading ability among soil microorganisms. In: Plasmids of Medical, Environmental, and Commercial Importance.

Pettigrew-Pope, Tom. 1990. Increasing role for bugs in waste cleanup. *Waste Age* 21:86–88.

Rojo, F., D. H. Pieper, K. H. Engesser, H. J. Knackmuss, and K. N. Timmis. 1987. Assemblage of ortho cleavage route for simultaneous degradation of chloro- and methylaromatics. *Science* 238:1395–1398.

Rojo, F., J. L. Ramos, D. Pieper, K.-H. Engesser, H.-J. Knackmuss, and K. N. Timmis. 1988. Laboratory evolution of novel catabolic pathways. In: Biosensors and Environmental Biotechnology (ed. C. P. Hollenberg, H. Sahm). Fischer Verlag: Stuttgart, pp. 65–74.

Russel, H., J. Matthews, and G. Sewell. 1992. TCE removal from contaminated soil and ground water. U.S. Environmental Protection Agency, EPA/540/S-92/002: Washington, D.C.

Sangodkar, U. M. X., P. J. Chapman and A. N. Chakrabarty. 1988. Cloning, physical mapping and expression of chromosomal genes specifying degradation of the herbicide 2,4,5-trichlorophenoxyacetic acid (2,4,5-T) by *Pseudomonas cepacia* AC1100. *Gene* 71(2):267–277.

Sayler, G. S. 1991. Contribution of molecular biology to bioremediation. *J. Haz. Mater.* 28:13–27.

Shields, M. S., and M. J. Reagin. 1992. Selection of a *Pseudomonas cepacia* strain constitutive for the degradation of trichloroethylene. Appl. Environ. Microbiol. 58:3997–3983.

Shields, M. S., S. W. Hooper, and G. S. Sayler. 1985. Plasmid mediated mineralization of 4-chlorobiphenyl. *J. Bacteriol.* 163:882–889.

Spain, J. C., 1990. Metabolic pathways for biodegradation of chlorobenzene. In: *Pseudomonas:* Biotransformations, Pathogenisis, and Evolving Biotechnology. (ed. S. Silver, A. M. Chakrabarty, B. Iglewski, S. Kaplan). ASM: Washington, D.C., pp. 197–206.

Taeger, K. H., J. Knackmuss, and E. Schmidt. 1988. Biodegradability of mixtures of chloro- and methyl-substituted aromatics: Simultaneous degradation of 3-chlorobenzoate and 3-methylbenzoate. *Appl. Microbiol.* 28:603–608.

Timmis, K. N., F. Rojo, and J. L. Ramos. 1987. Prospects for laboratory engineering of bacteria to degrade pollutants. In: Environmental Biotechnology (ed. G. Omenn). Plenum Press: New York, pp. 61–79.

Unterman, R. 1991. What is the Km of disappearase? In: Environmental Biotechnology for Waste Treatment (ed. G. S. Sayler, R. Fox, and J. W. Blackburn). Plenum Press: New York, pp. 159–162.

Williams, R. T., and C. A. Myler. 1990. Bioremediation using composting. *BioCycle* 31:78–80.

Zylstra, G. J., L. P. Wackett, and D. T. Gibson. 1989. Trichloroethylene degradation by *Escherichia coli* containing the cloned *Pseudomonas putida F1* toluene dioxygenase genes. *Appl. Environ. Microbiol.* 55:3162–3166.

Applied Biotreatment of Phthalate-Impacted Soils: Recognizing and Overcoming Technical Challenges

A. Keith Kaufman

Bioremedial Services Division
RESNA Industries, Incorporated
Los Angeles, California

Cheryl C. Kreuger

Bioremedial Services Division
RESNA Industries, Incorporated
Los Angeles, California

Applied bioremedial technology is being utilized with increasing frequency as a desirable form of environmental cleanup due to the many practical and economic advantages associated with its appropriate utilization. Numerous case histories documenting effective bioremedial treatment of groundwater and soil impacted by a variety of environmental pollutants can be found in the literature (see Refs. 1, 2, 3, 5, 6, 9, 10, 11, 14, 15, 16). Regulatory agencies have also contributed their share to the continued growth of the industry,[12] especially as a result of the highly publicized bioremedial treatment of portions of the *Exxon Valdez* oil spill[4] and other maritime incidents.

The vast majority of bioremedial applications have been associated with the cleanup of environmental pollution due to petroleum hydrocarbons such as crude oil, diesel, and other refined products. Frequently, biotreatment methods reported for these applications are presented as almost routine, with little information presented, if any, associated with the potential (or actual) impacts on treatment performance due to any one of a variety of physicochemical and/or microbiological parameters. Few biotreatment product or service providers will publish descriptions of remedial failures, irrespective of cause, because of the obvious deleterious effects such publicly disseminated information would have on their prospective business.

In an effort to provide a more realistic perspective on applied bioremedial site mitigation, this chapter will discuss our efforts toward the cleanup of approximately 1600 yd^3 of soil impacted with butylbenzyl phthalate (plasticizer resin)[13] resulting from a train derailment in northern Arizona. While successful, operation and maintenance of bioremedial treatment was challenged by numerous physical and biological influences, the interpretations of which were essential in bringing the project to cost-effective and efficient completion. While clearly not your typical run-of-the-mill bioremediation job, the considerations raised in this chapter will undoubtedly have applicability to many biotreatment operations, even those which involve only "simple" petroleum hydrocarbons.

Background

In the late winter of 1987, a train derailment occurred in a sparsely populated area of northern Arizona resulting in the environmental release of butylbenzyl phthalate, a plasticizer resin.[13] Additionally, various quantities of urea crystals, soybean oil, cowhides, and cornmeal were concurrently discharged to the surrounding soils. Upon completion of emergency response activities, a remedial investigation was initiated in order to determine the extent of contamination and evaluate remedial options. This investigation subsequently revealed that approximately 1600 yd^3 of soil had been impacted with the phthalates in conjunction with the other materials, heterogeneously mixed, as well as railroad debris comprised of train parts. Contamination was generally confined to the upper 6 ft of soil and groundwater was not impacted. Although a search of the literature did not reveal precedent for field bioremedial cleanup of phthalate-impacted soil, knowledge of the chemical composition of the material supported the potential applicability of aerobic (above ground) biodegradation of this compound, and a comprehensive biotreatability investigation was subsequently initiated.

The Biotreatability Investigation

A biotreatability investigation is frequently performed for prospective bioremedial treatments in order to characterize the physicochemical and microbiological conditions which must be effected on site in order to optimize bioremedial activity and efficiency.[8,11] At the outset of such an investigation, it is important to address whether a biostimulation or bioaugmentation approach will be implemented. This determination is generally made on the basis of the following basic information:

- *For biostimulation.* Large proportion of specific contaminant biodegraders within a site-specific microbial population *and* contaminant is the only (or dominant) source of carbon available.

- *For bioaugmentation.* Small proportion of specific contaminant biodegraders within a site-specific population *and* contaminant is not the only (or dominant) source of carbon available.

Considering the significant presence of alternative carbon sources as well as the recency of site contamination (which meant there would not have been sufficient time for development of a large population of phthalate degraders), subsequent biotreatability efforts were directed toward bioaugmentative treatment. Upon establishing optimal biodegradative conditions and characterizing the few native microbial species best suited for the process, a bench-scale pilot investigation was additionally performed on site-specific samples of soil. Data obtained from this portion of the evaluation were used to validate bioremedial feasibility as well as formulate a predictive time frame for field bioremedial completion. This information was then used in the design of a work plan for client and agency review. An average concentration of 100 ppm butylbenzyl phthalate in soils was approved as the cleanup objective for the project.

Upon initiation of the biotreatability investigation, we were very much aware of the atypical conditions which prevailed at the project site (although we did not know at the time just *how* atypical these conditions were to be), most notably the presence in soil of competitive organic substrates (soybean oil and cornmeal), and expected temperature differentials associated with northern Arizona climatology (subfreezing in winter to 100°F in the summer). In addition, high concentrations of urea were not viewed favorably due to the potential inefficiencies associated with this compound as a nitrogen source[7] and its potential (and actual) liberation of ammonia gas (this required our utilizing full-face respirators for the first 6 months of field operations). With these factors in mind, we conducted the biotreatability investigation in a manner which we hoped would generate the following:

1. Biodegradative isolates which would be metabolically acclimated for activity on butylbenzyl phthalate, and not competitively inhibited by the presence of the other organics.

2. Phthalate-biodegrading isolates that would have activity over a wide temperature and pH range.

3. Phthalate-biodegrading isolates that utilize urea and ammonia as sources of nitrogen.

After a 6-week investigation, three indigenous isolates were selected for pilot-testing and prospective field utilization. In addition, phosphates and other nutrients required to maintain optimal soil pH and replenish inorganic constituents assimilated by organisms as a function of their biodegradative metabolism were also identified.

Upon completion of these analyses, pilot-scale testing utilizing both test and control systems revealed that the test (bioaugmented) system reduced phthalate levels from 38,000 ppm to 20,000 ppm in just 15 days, equating to a 47 percent substrate decrease. This activity correlated with an observed increase in viable biodegradative cell numbers. No substantive changes were observed in either phthalate levels or microbial concentrations in the controls evaluated.

Although we believed, due to the variety of unique physicochemical features associated with this project, that achievement of such kinetics in the field was unlikely, we also believed that the cumulative results of the above investigations strongly supported the feasibility and efficacy of full-scale bioremedial treatment.

Field Implementation and Results

An aboveground, lined bioremedial treatment cell was constructed in such a manner as to accommodate the 1600 yd^3 of impacted soils in a layer approximately 18 in thick. The impacted soils were overlaid atop 6 in of clean soil to act as a buffer against liner breakage during subsequent soil-mixing operations. A hydraulic delivery system was emplaced along the treatment cell's perimeter to provide uniform application of appropriate aqueous bionutrients and phthalate-degrading microorganisms during treatment. Clear plastic covering was utilized over the treatable soils periodically to enhance soil warming and moisture retention.

Field bioremedial treatment, consisting of bionutrient and phthalate-degrading microorganism application every 2 weeks with mixing weekly, was initiated in late July of 1989, and a 40 percent reduction in phthalate concentration was noted for the first 2 months of treatment. Phthalate reduction over time correlated well with observed increases in both general and selective (phthalate-degrading) microbial enumeration of soil samples. During the ensuing colder months, however (when temperatures dropped to below freezing), continued phthalate

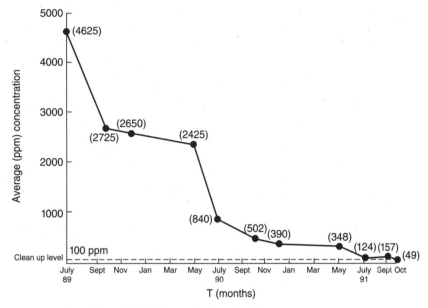

Figure 9.1 Butyl benzyl phthalate reduction in soils.

biodegradation could not be demonstrated, even though continued elevation of microbial population size, as well as nitrogen conversions, were observed for the same period. Upon return of warmer weather, active biodegradation of phthalates was again observed with a rate of reduction very similar to the rate observed for the previous spring and summer (Fig. 9.1). While this period resulted in another substantial drop in phthalate concentration, levels had not yet reached those established as the cleanup objective. As a result, an additional season of treatment was necessary for remedial completion. By October of 1991, levels of phthalates had been reduced to an average of 49 ppm, representing an overall concentration reduction of approximately 99 percent.

Major Operational and Performance Challenges

Urea crystals

While historically used as a soil amendment in measured amounts, urea is not necessarily a desirable choice for nitrogen supplementation,[11] especially when it is present in significant concentrations. The most obvious concern is the prospective dangers associated with urea decomposition and consequent ammonia liberation. In fact, the most immediate noticeable impact at the derailment site for many months after the accident was overwhelming ammonia odor. Levels of ammonia gas significantly exceeded acceptable exposure limits, necessitating the use of full-face respirators for more than 6 months of on-site operations.

Although the fact is not widely known or appreciated, urea (and its ammonium constituents) is effective as a nitrogen source only when the bulk of microorganisms targeted for its use are urease producers (urease is an enzyme, produced only by specific microorganisms) and/or can utilize nitrogen from ammonia (only selected populations can). Such characteristics, while recognized for many years as diagnostically significant for the identification of microorganisms, are rarely considered in the practical world of bioremediation. Unfortunately, failure to recognize the importance of such criteria can affect the efficiency and ultimate success of bioremedial treatment.

As noted in Table 9.1, of the three species of phthalate-degrading isolates selected for bioaugmentation, only one was able to effectively utilize nitrogen from ammonia. Nitrate and nitrite, which are generally more efficient sources of nitrogen for oxidative (aerobic) metabolism (since, as oxidized components, these compounds can provide oxygen as well as nitrogen), were clearly better suited for the phthalate-degrading organisms. Unfortunately, urea and ammonia prevailed at the impacted site.

We believed, and the evidence would appear to validate (Table 9.2), that much of the ammonium could be reoxidized to nitrate and/or nitrite via vigorous and frequent soil mixing. Furthermore, previously performed biotreatability analyses had also indicated the presence of naturally occurring bacterial nitrifiers which, we anticipated, would act upon the ammonium, converting it to the more oxidized form for utilization by the biodegraders. While it would be difficult to determine what portion of the conversions was attributable to a given process, the end result was that, over time, ammonium levels decreased, while oxidized forms of nitrogen increased. Such action served to enhance the metabolic efficiency of the phthalate-degrading microbes, since without their efficient utilization of nitrogen (for incorporation into phthalate-degrading enzymes and other proteins) efficient biodegradation of phthalates could not have occurred.

TABLE 9.1 Characteristics and Features of Phthalate-Degrading Organisms Selected for Soil Bioremediation

Identification	Nitrogen from:			Growth range	
	NO_3	NO_2	NH_4	pH	Temperature
Alcaligenes faecalis	+	+	+	7.75–8.75	6–35°C
Pseudomonas stutzeri	+	+	−	6.75–7.75	18–25°C
Pseudomonas fluorescens	+	+	−	6.75–8.75	6–35°C

TABLE 9.2   Bioprocess Monitoring Results during Field Implementation

Date	Soil/air temperature (°C @ 1 P.M.)	pH	NO$_3$[a]	NO$_2$[b]	NH$_4$[c]	CFU/g soil × 10^6 (selective)[d]	P[e]
7/89	31/27	9.2	ND[f]	ND	7500	0.71	4625
10/89	26/21	9.0	ND	ND	1800	48	2725
1/90	3/−2	8.8	ND	ND	923	336	2650
5/90	22/21	8.9	25	25	800	280	2425
7/90	25/24	8.6	25	25	940	360	840
10/90	22/22	8.7	50	40	1075	221	502
12/90	2/−2	8.7	ND	40	750	311	390
5/91	14/13	8.7	ND	25	875	250	348
7/91	29/29	8.6	ND	25	390	163	124
8/91	28/27	8.7	ND	375	317	151	225
9/91	21/19	8.4	ND	375	225	190	157
10/91	27/22	8.7	ND	370	190	407	49

[a]NO$_3$; nitrate in ppm (detection limit = 1.0 ppm).

[b]NO$_2$; nitrite in ppm (detection limit = 1.0 ppm).

[c]NH$_4$; ammonium in ppm (detection limit = 1.0 ppm).

[d]Selective CFU/g soil; number of specific phthalate microbial degraders per gram of soil. CFU = colony-forming units.

[e]P; average phthalate concentration in ppm (n = 4 composite samples).

[f]ND; not detected.

Alternate carbon sources and temperature

Without question the most influential of factors which directly impacted the efficiency and duration of this project were the dual and apparently linked parameters of alternate carbon (substrate) sources and temperature.

While clearly influential to the kinetics of biodegradative metabolism, temperature alone does not necessarily make or break a bioremediation project, as long as microbial characteristics are matched with anticipated temperatures expected during field implementation at a given site. Knowing that temperatures will be low, for example, would indicate the need for psychrotolerant (cold-tolerant) and/or psychrophilic (cold-loving) microorganisms. Examples of such organisms are those which spoil food in the refrigerator. Conversely, expectations of high temperatures would suggest the utilization of thermotolerant (heat-tolerant) and/or thermophilic (heat-loving) microorganisms. The difference between the "tolerant" and "loving" representatives from the temperature groups is that the former can survive and grow (but not necessarily thrive) at a given temperature range, while the latter grow *preferentially* at that range. Utilizing this information in bioprocess development can enable the use of bioremediation in a wider variety of climates, even though kinetics at colder temperatures will almost always be slower due to molecular thermodynamics.

In northern Arizona, air temperatures fluctuate between subfreezing in the winter, and 100°F in the summer. Clearly, it was necessary to utilize, to the degree possible, organisms which had a broad range of temperature tolerances. During biotreatability screening, phthalate-degrading microorganisms were selected, in part, for their ability to grow in an environment which would be subject to wide temperature fluctuations (Table 9.1). In addition to utilizing clear plastic covering which produced a greenhouse-like effect on soils and prevented their freezing, application of these organisms yielded an increasing bacterial population size even through the winter months (Table 9.2). Unlike during the warmer months however, a concurrent reduction in phthalate concentration was not observed (Fig. 9.1). If both the general and selective phthalate-biodegrader counts continue to increase over the winter months, then why is phthalate concentration not decreasing? The answer most likely relates to the presence of the alternate carbon substrates, cornmeal and soybean oil, concurrently released during the derailment.

Since microorganisms cannot grow unless carbon is utilized, we know that *some* form of carbon is being utilized, since microbial growth *is* observed. Furthermore, since we know that the phthalate is not being utilized, it is highly probable that the alternate carbon sources— cornmeal and soybean oil—are being metabolized. But if alternate substrate degradation was occurring in the winter months, why did it appear *not* to occur in the warmer months?

The answer most likely relates to the combined effects of bacterial genetics and physiology. The inducible enzymes required for phthalate degradation are typically not innate (constitutive) in most microorganisms, since most organisms have never "seen" phthalates and would therefore not have evolved the genetic and metabolic machinery to degrade the compound. As a result, adaptive and metabolic processes necessary for synthesis of the required phthalate-degrading enzymes will be very energetically demanding. Metabolic enzymes required for degradative processes involving cornmeal (starches) and soybean oil (fats) will likely already be possessed by most organisms. As a result, their de novo synthesis, with its concomitant energy demands, would not be required. While it would be inefficient for phthalate degraders to revert back to "easy" carbon substrate (cornmeal/soybean oil) metabolism in the summer, the data from this project suggest that such a reversion *does* occur in colder weather. It is apparent that under such conditions, less energy is required to revert to the easy substrate degradation enzymatic pathway than to maintain active biosynthesis of phthalate-degrading enzymes. While there may be other interpretations of the data and observed field results, such an explanation fits the circumstances and is consistent with recognized microbial physiology, genetics, and biochemistry.

Conclusion

Biotreatability investigations performed at the onset of this project demonstrated that, under appropriate conditions, phthalates can biodegrade efficiently and rapidly. The term "appropriate" clearly did not apply to this particular bioremedial cleanup project, and efforts to anticipate associated physicochemical and microbiological challenges, as well as to attempt to understand the significance of laboratory and field observations, were crucial to the successful outcome of this work. While unable to control certain aspects of this unusual project [such as dealing with twice the normal annual rainfall and a snowy season that, in one year, lasted 8 months (it is a local regulation that NO field equipment may be transported on public roads when snow is observable on the ground)], attention to these variables undoubtedly helped to make a challenging set of circumstances manageable to a degree which enabled fulfillment of our remedial objectives. While the overall cost of this project exceeded initial estimates, the remediation was still performed and completed for less than the projected cost of alternative cleanup methods considered (landfill disposal, fixation, oxidation).

All too frequently, aboveground bioremediation (particularly involving petroleum hydrocarbons) is trivialized as something that is almost perfunctory and routine, requiring little more than addition of water and nutrients, with periodic mixing. While such an approach can work at times, the project discussed above exemplifies some of the complexities which may challenge the bioremedial practitioner. There are many other examples of such projects where "atypical" or "unusual" physicochemical and microbiological conditions warrant more than cursory consideration. As this and many other projects serve to illustrate, bioremediation can be successful even under extraordinary conditions, as long as those conditions, and their potential impact on applied bioremedial treatment, are carefully and thoughtfully evaluated.

References

1. Applied BioTreatment Association. 1989. Compendium of Case Histories. Applied Biotreatment Association, Washington, D.C.
2. Barnhart, M. 1987. Case history: Biological soil treatment in Buffalo, N.Y. *Haz. Mat. Waste Mgmnt.* September–October, pp. 42–44.
3. Basta, N. 1987. Better biological processes boost wastewater treatment. *Chem. Eng.* April 27, pp. 14–15.
4. Berkey, Edgar. 1991. Evaluation process for the selection of bioremediation technologies for Exxon Valdez oil spill. *In* Environmental Biotechnology for Waste Treatment. Plenum Press, New York/London.
5. *Chemical Marketing Reporter.* 1987. Microbes used to fight waste in groundwater. September 28, p. 27.
6. Compeau, Geoffrey C., et al. 1991. Full-scale bioremediation of contaminated soil and water. *In* Environmental Biotechnology for Waste Treatment. Plenum Press, New York/London, pp. 91–109.

7. Frankenberger, W. T., Jr. 1988. Use of urea as a nitrogen fertilizer in bioreclamation of petroleum hydrocarbons in soil. *Bull. Environ. Contam. Toxicol.* 40:66–68.
8. Frankenberger, W. T., Jr. 1992. The need for a laboratory feasibility study in bioremediation of petroleum hydrocarbons. *In* Hydrocarbon Contaminated Soils and Groundwater. Lewis Publishers, Chelsea, MI, pp. 237–293.
9. Heyse, E., et al. 1986. *In situ* aerobic biodegradation of aquifer contaminants at Kelly Air Force Base. *Environ. Progr.* 5:207–211.
10. Hildebrandt, W. W., and S. B. Wilson. 1992. The use of modern on-site bioremediation systems to reduce crude oil contamination on oilfield properties. *In* Hydrocarbon Contaminated Soils and Groundwater. Lewis Publishers, Chelsea, MI, pp. 487–499.
11. Kaufman, A. K., and C. C. Krueger. 1988. Bioremediation of fuel-contaminated soil: A case history. *In* Proceedings, Hazmacon, 1988. Anaheim, CA, pp. 743–752.
12. Kaufman, A. Keith. 1991. The technical, economic, and regulatory future for bioremediation: An industry perspective. *In* Environmental Biotechnology for Waste Treatment. Plenum Press, New York/London, pp. 47–51.
13. Monsanto Company. Material Safety Data. Santicizer 2051 Plasticizer. Monsanto Company, St. Louis, MO.
14. Piotrowski, M. R. 1992. Full-scale, in situ bioremediation at a superfund site: A progress report. *In* Hydrocarbon Contaminated Soils and Groundwater. Lewis Publishers, Chelsea, MI, pp. 371–400.
15. Skiba, Robert S., et al. 1992. Biological treatment: Soil impacted with crude oil. *In* Hydrocarbon Contaminated Soils and Groundwater. Lewis Publishers, Chelsea, MI, pp. 409–415.
16. Suflita, J. M., et al. 1988. Anaerobic biotransformation of pollutant chemicals in aquifers. *J. Ind. Microbiol.* 3:179–194.

Field-Scale Anaerobic Bioremediation of Dinoseb-Contaminated Soils*

D. J. Roberts, R. H. Kaake, S. B. Funk, D. L. Crawford, and R. L. Crawford

Center for Hazardous Waste Remediation Research
Department of Bacteriology and Biochemistry
University of Idaho
Moscow, Idaho

The three general approaches to remediation of soils are physical, chemical, and biological methods. Physical treatments include incineration and vitrification. These processes convert the waste to an inert form which still may have to be disposed of as hazardous waste (e.g., incineration ash). Some processes, such as stabilization, decrease the mobility of a contaminant in soil by the addition of stabilizing agents (cement or plastic resins) which bind or envelop the waste into an impermeable matrix, preventing contaminant migration. Other treatment methods, such as steam stripping or extraction procedures, remove and recover the contaminants from the medium on the basis of their physical properties (volatility or solvent solubility). Biological treatment methods utilize the metabolic diversity of microorganisms to transform toxic, recalcitrant compounds into harmless molecules which may provide energy or metabolic precursors for the microorganisms.

The complexity of the soil environment presents a unique challenge to the bioremediation industry. Although biodegradation is a natural process and a necessary part of nutrient cycling in the soil environ-

---

*Publication No. 91521 of the Idaho Agricultural Experiment Station.

ment, many of the compounds added to soil as a result of human activity are toxic and recalcitrant under the conditions present in the environment. However, the growing list of publications presenting laboratory results indicating successful biodegradation of many anthropogenic, recalcitrant compounds has prompted a new faith in bioremediation as a useful technology for the treatment of contaminated soils.

Many laboratory studies have relied on the disappearance of a compound as a measure of biodegradation. This can be misleading, since in some instances disappearance of a specific molecule may occur concomitantly with its transformation to a more toxic compound. An example is the conversion of the relatively nontoxic herbicide 2,4-dichlorophenoxyacetic acid (2,4-D), to the mutagenic compound 2,4-dichlorophenol by a genetically engineered soil organism.[43] The production of 2,4-dichlorophenol from 2,4-D may have been responsible for the toxic effects on the natural population of the soil seen after the addition of this organism to 2,4-D-contaminated soils.[14] This example demonstrates the need to understand the difference between results obtained from laboratory studies and from environmental applications of bioremediation. Sims et al.[44] presented a discussion of approaches to bioremediation of contaminated soils, pointing out the need for thorough site characterization, treatability studies, and possibly the integration of physical, chemical, and biological remediation methods into a treatment train to achieve complete cleanup of a contaminated site.

Current Soil Bioremediation Practices

The focus of research into bioremediation of contaminated soils must encompass not only the nature of the compound and its transformation intermediates, but also the environment in which it is present. Physical parameters such as temperature, pH, and redox potential, as well as the presence of other contaminants and the binding affinities the contaminating compounds may have for the soil, affect the removal of target chemicals from contaminated soils. Another very important factor in the implementation of soil treatment technologies is the nutrient status of the soil. Many soils are nutrient-limited, and nutrients such as oxygen, nitrogen, and phosphate must be supplied to ensure that the microorganisms are in an active metabolic state. *Bioaugmentation,* the process of nutrient addition to contaminated environments to stimulate biological destruction of contaminants, has been used successfully for some time for the remediation of oil spill contamination in soils.[1,3]

Another important soil characteristic to be considered when planning a remediation program is the biological competency of the soil. Many soils will contain the appropriate microbial population to degrade the contaminants present in that soil, especially if the soil has low concentrations of contaminants that have been present for extended periods of time. Often the native flora can be stimulated to degrade a contaminant by nutrient addition alone. Some soils, however, may be incompetent because of the toxicity of the contaminant to the microorganisms required to carry out the metabolism of the compound, or because of other environmental factors.

To improve the biological competency of a particular soil, microbial inoculants are often added. Microbial inoculants are most successful if they are organisms obtained from the environment in which they will be applied. Bioenhancement of contaminant degradation using laboratory strains of pure cultures and/or genetically engineered organisms under environmental conditions has been reported but has not seen the success expected, probably due to competition for nutrients from the natural microflora.[40] The addition of manure and sewage sludge to contaminated soils has been used successfully to improve their degradative capabilities.[15,35]

Many soil contaminants are amenable to remediation by aerobic methods. These methods, described below, are the most widely used and understood of the soil treatment methods. Each method can be improved by bioaugmentation and bioenhancement procedures as described above. Table 10.1 summarizes the advantages and disadvantages of current soil bioremediation technologies.

Land farming

Land farming is an aerobic treatment method that is applicable to many types of contaminated soils. In land farming, the contaminated soil is treated in above-grade treatment beds. The treatment beds are usually pits lined with a high-density polyurethane liner that is then covered with clean sand to allow drainage. Perforated pipes collect the drainage, which can be treated separately or recycled. The contaminated soil is then spread over the sandy layer. In the United States, regulations require that such pits contain a layer of sand and a leachate collection system under the polyurethane liner in order to recover any leaked material if the liner is breached.

The land-farming process can be optimized by the dilution of contaminated soil with clean soil to reduce initial toxicity, as well as by controlling physical parameters such as aeration, pH, soil moisture content, and temperature. Aeration is often accomplished by tilling the soil, or, in more mechanized systems, by forced aeration. When forced aeration is used, the plots should be covered and the exiting air cleaned through filters. To achieve temperature control, hot air, or the "greenhouse effect," can be employed in a closed system. Land farming has been widely implemented at petroleum refinery sites and at sites contaminated with

TABLE 10.1 Comparison of Biological Remediation Technologies

Technology	Advantages	Disadvantages	Application/contaminant	References
Land farming	Simple procedure Inexpensive Currently accepted method	Slow degradation rates Residual contamination often not removed High exposure risks May require long incubation periods	Surface contamination Aerobic process Low to medium contamination levels Pentachlorophenol Oil and gasoline PAH	9,15,19,40,44,54
Composting	More rapid reaction rates Inexpensive Self-heating	Needs bulking agents Requires aeration Nitrogen addition often necessary High exposure risks Residual contamination Incubation periods are months to years	Surface contamination Aerobic process Can treat high contamination levels Nitroaromatic explosives Aerobic sewage sludges Oil and gasoline	9,23,27,34,38,49, 50,54,59
In situ	Relatively inexpensive Low exposure risks Excavation not required	Low degradation rates Less control over environmental parameters Need good hydrogeological site characterization Incubation periods are months to years	Deep contamination Aerobic or nitrate reducing conditions Low to medium contamination levels Oil and gasoline Chlorinated aromatics Chlorinated hydrocarbons	1–6,12,13,22,30, 32,33,39–42,45,54
Slurry bioreactor	Good control over parameters Good microbe/compound contact Enhances desorption of compound from soil Fast degradation rates Incubation periods are days to weeks	High capital outlay Limited by reactor size High exposure risks	Surface contamination Recalcitrant compounds Soils that bind compound tightly Aerobic or anaerobic process	7,10,11,16,29,31, 36,53,55,59

polynuclear aromatic residues (PNAs), or pentachlorophenol (PCP) (e.g., sites connected with the wood-preserving industry).

Composting

Composting is an aerobic bioremediation technology similar to land farming. In this technique, the contaminated soil is mixed with wood chips, straw, or some other bulking agent to provide porosity for air flow. Composting can be carried out in a bioreactor with a forced air supply to provide aeration, or in open piles (windrows) that are periodically reformed to facilitate oxygen contact. The addition of bulking materials is also used to enhance microbial activity by supplying a readily utilizable carbon source. The aerobic metabolism of large amounts of carbon creates heat, so composting treatments often run at high temperatures. Fast degradation rates can be obtained, even though oxygen contact tends to be poor.

With composting, as with any biological treatment method, environmental parameters such as moisture content and pH need to be monitored and adjusted. The nitrogen content of a compost pile requires particular attention due to the high carbon contents of the bulking agents. Manure is often added to composting operations as a source of nitrogen, as well as a source of organisms. Composting operations are usually run as batch operations, reusing a portion of an old compost pile as the inoculum for new ones. Experimental composting operations have been performed on soils contaminated with explosive nitroaromatic compounds such as 2,4,6-trinitrotoluene (TNT).[17,23,27,59]

Slurry reactors

Contaminated wastewaters have been treated in bioreactor systems, such as municipal sewage treatment plants, for years. The application of bioreactors such as solid-state fermentation reactors and gas-liquid fermentors to multiphase systems is a growing industry and is receiving much attention by engineers. Brauer[10,11] described several such bioreactor systems. Innovative research is now showing that soils can also be treated in specialized bioreactors termed *slurry reactors*. In this process, a contaminated soil is mixed with at least 30 percent aqueous medium in a reactor vessel. The reactor is usually equipped with a mixing system to ensure maximum contact between the microbial population, the target molecules, and nutrients, and to prevent feedback inhibition. The reactor operator can also control pH, temperature, oxygen contact, and moisture content. Slurry reactors can be run either aerobically or anaerobically.

The implementation of slurry reactors to treat contaminated soils offers the opportunity to shorten treatment times for contaminated soils from 6 to 8 months to 1 or 2 weeks or months. The better process control and availability of contaminants to the microorganisms provided by mixed slurry systems are advantages over other biotreatment systems. This technology may prove the most advantageous for the remediation of soils contaminated with highly recalcitrant compounds and for soils that are difficult to treat by other techniques.

In situ treatment

All of the above bioremediation technologies require excavation of the soil from the contaminated site. Excavation of contaminated soil is a very expensive operation requiring specially certified personnel and in some cases specialized equipment. In situ bioremediation, or *bioreclamation,* is a technology that does not require the excavation of large amounts of soil. This method involves the injection of nutrients and an oxygen supply in an aqueous medium directly into the contaminated environment.[1,13,33,39] The aqueous treatment medium flows through the contaminated soil and is then retrieved by extraction wells down-gradient from the injection well and oxygenated again before recirculation into the contaminated site. More aggressive methods for in situ treatment allow the combination of steam stripping, groundwater treatment, and biological remediation. Hydrogen peroxide is often used to provide oxygen in cases where soil permeability to oxygen is limiting.[5,22] An anaerobic in situ treatment has also been performed.[42] In this case, acetate was added as a supplemental carbon source to stimulate nitrate and sulfate-reducing organisms to degrade halogenated aliphatic compounds.

The in situ process is usually applied when excavation of the soil would be difficult, such as when contamination is very deep, or when vast amounts of contaminated soil are involved, making excavation unfeasible. The in situ processes require an extensive understanding of the hydrogeology of the sites to be treated. In situ processes may be precluded when irregular geology prevents the transport of oxygen and nutrients through the contaminated area. Careful monitoring of groundwater is required to protect aquifers. Reaction rates are usually slow for in situ processes due to nutrient, oxygen, and sometimes temperature limitations. Determinations of the efficiency of in situ treatment methods are often difficult because of problems in obtaining truly representative soil samples. The occurrence of pockets of untreated soil is difficult to monitor or avoid.

Degradation of Nitroaromatic Compounds

Nitroaromatic compounds are important in the chemical industry and are used to manufacture thousands of consumer products.[20] These products are represented by at least four classes of chemicals. The largest and most well known of these classes is the polyurethanes, which are manufactured from 2,4- and 2,6-dinitrotoluene. The second largest class of nitroaromatic compounds is the hazardous energetic nitroaromatic compounds such as TNT. Nitroaromatic compounds are also used in large quantities as pesticides and herbicides by the agricultural industry, and as pharmaceuticals.

The current technologies used to treat soils contaminated with nitroaromatic compounds, such as the herbicide 2-*sec*-butyl-4,6-dinitrophenol (dinoseb) entail physical methods such as incineration or the haul-and-store method. These methods are costly and are not environmentally sound. Inefficient incineration can produce hazardous emissions and carbon dioxide. The latter compound is a greenhouse gas, and its buildup in the atmosphere is thought to be harmful to the environment. The haul-and-store method merely stockpiles hazardous waste in one area. Under current regulations, long-term liability for the hazardous wastes remains with the generator of the waste, even after disposal.

Investigations are under way to develop biological methods to remediate soils contaminated with nitroaromatic compounds, but none have been scaled up to a commercial level. In order to evaluate a bioremediation scheme for contaminated soils, the effects of the contaminants on soil microorganisms and the effects of the soil microorganisms on the contaminants must be identified. Although many sites in the U.S. Pacific Northwest have been contaminated with nitroaromatic pesticides and herbicides through spillage and crop-dusting operations, there have been relatively few investigations into the biodegradation of these compounds.

Laboratory studies done with soil enrichments have shown that dinoseb is biodegradable. Doyle et al.[15] reported $^{14}CO_2$ evolution from ^{14}C-dinoseb added to soil, whether the soil was amended with sewage sludge, dairy manure, or left unamended. Stevens et al.[51] tested the abilities of the natural microbiota of several Idaho soils to degrade dinoseb. The results indicated that some soils had the ability to transform dinoseb, but that the presence of nitrate and high levels of dinoseb were inhibitory to dinoseb degradation in most soils. Radiotracer studies were not performed, so mineralization of the dinoseb could not be inferred. The metabolic pathways for dinoseb metabolism were not determined in either of these studies.

The three different types of substituents on the dinoseb molecule (nitro, hydroxyl, and *sec*-butyl) present sites for multiple attack on the molecule. The metabolism of this compound in soil could possibly involve a network of intermediates, rather than a single pathway, making the determination of a complete biodegradation sequence difficult. Studies on the metabolism of dinoseb and an analogous herbicide, 2,4-

dinitro-o-cresol (DNOC), have revealed that the initial attack on the molecules by bacteria is probably at the nitro groups.

Early studies concerning the degradation of dinoseb and DNOC showed that a *Pseudomonas* sp. could initiate an alteration of DNOC to 6-amino-2-methyl-4-nitrophenol and 3-methyl-5-nitrocatechol.[53] Other researchers have enriched for and isolated from various soils several strains of slow-growing aerobic *Arthrobacter*-like organisms and two strains of *Pseudomonas* able to metabolize DNOC. These organisms decolorized DNOC and produced nitrite. No other metabolic intermediates were reported. Although 4-nitrophenol and 2,4-dinitrophenol were attacked by these bacteria, dinoseb and other nitroaromatic compounds were not.[24] Wallnoefer et al.[55] found that the nitro group in the *ortho* position of DNOC or dinoseb was converted to an acetamido group by an *Azotobacter* sp.

In a recent study, Stevens et al. used a chemostat to enrich for and isolate organisms capable of degrading dinoseb from dinoseb-contaminated soil. These organisms were able to transform dinoseb to reduced products under microaerophilic and denitrifying conditions, but were unable to completely mineralize the molecule. The reduced products polymerized, forming a multimeric precipitate. The chemostat enrichment procedure also produced an anaerobic consortium that could completely degrade dinoseb to acetate and CO_2.[52] The actual pathway of dinoseb degradation under anaerobic conditions has yet to be elucidated.

A large portion of the literature concerning the degradation of nitroaromatic compounds has dealt with the explosive nitroaromatic compounds and has been reviewed recently.[26] Most of the results have shown that, whether under aerobic or anaerobic conditions, the first step in the degradation of nitroaromatic compounds is the reduction of the nitro groups to hydroxylamino, then to amino substituents, in both mammalian and prokaryotic systems. Under aerobic conditions, the hydroxylamino moieties can polymerize, forming azo linkages which are more recalcitrant and possibly more toxic than the original parent compound.

Under anaerobic conditions, the amino compounds are formed and are stable.[37] Hallas[18] examined the anaerobic degradation of several nitroaromatic compounds and found reduction of the nitro group to be the major mechanism of compound alteration. Toxicity studies on TNT, 2,6-dinitrotoluene, and 2,4-dinitrotoluene have indicated that the amino or hydroxylamino transformation products are primarily responsible for their toxicity.[21,28,58]

Investigations into the aerobic degradation of mononitrophenolic compounds have indicated that the ring system can be oxidized by mono- and dioxygenases, producing nitrite and catechols that can then undergo oxidative ring cleavage.[46,47] Recent publications have indicated that the dinitrotoluenes and possibly TNT can be mineralized aerobically by a pseudomonad[48] or by the white rot fungus *Phanerochaete chrysosporium.*[17] This type of work is still in its infancy, but it may lead to promising new treatment methods for wastes contaminated with nitroaromatic compounds.

Bench-Scale Studies of the Bioremediation of Dinoseb-Contaminated Soils

The bioremediation scheme developed in our lab arose from the observation of Stevens et al.[52] that anaerobic cultures were capable of metabolizing dinoseb to acetate and CO_2. The majority of the literature indicated that the initial pathway for the degradation of nitroaromatic compounds was reductive. Anaerobic conditions which would favor reductive processes could be highly favorable for the complete degradation of nitroaromatic compounds. Anaerobic conditions would also allow reductions to amino compounds to proceed at a rapid rate so that polymerization of hydroxylamino intermediates would not occur.

The first step in the implementation of the remediation plan was to make the soil anaerobic. It has long been known that saturated soil environments eventually become anaerobic as the available oxygen is utilized by heterotrophic microorganisms.[8] The rate of this process depends on the amount of carbon available to the heterotrophs and on their metabolic state. Once a soil has been rendered anaerobic by heterotrophs utilizing dissolved oxygen as a terminal electron acceptor for the metabolism of the carbon source supplied, low redox potentials can be maintained quite readily by maintaining saturated conditions. The diffusion of O_2 is about 10^{-5} times slower through water than through air. Gaseous diffusion essentially ceases when the fraction of the air-filled pore space in soil is below 0.12;[57] therefore in saturated environments very little oxygen enters either the soil or aqueous phase.

We have used several inexpensive carbon sources to establish anaerobic conditions in soil. We tested soluble carbon sources such as glucose as well as insoluble carbon sources such as starch as co-substrates or as supplemental energy sources for dinoseb-degrading consortia.[50] The application of insoluble starch was found to support rapid oxygen depletion in uncontaminated soils flooded with either water or phosphate buffer. This carbon source also supported long-term maintenance of redox potentials well below -200 mV, which is the redox value below which even the strictest anaerobes (the methanogens) can grow. Waste products of the Idaho potato-processing industry served as a readily available and inexpensive source of insoluble starch. This material has been described elsewhere,[25] and will hereafter be referred to as *starch* for simplicity.

The procedure developed was to flood the soil with phosphate buffer and to add a carbon source to the soil/buffer slurry. The application of this procedure to the bioremediation of dinoseb-contaminated soils was tested in several bench-scale experiments using soils from two different sites.[25] These soils differed in their exposure history and in the concentrations of dinoseb and other contaminants present. Both soils were from rural airstrips that had been used for crop-dusting operations. A sandy loam soil from an airstrip in Ellensburg, Washington, was contaminated with dinoseb as well as other herbicides and fertilizers as a result of the washing of crop-dusting equipment over several decades. This soil was an excellent prototype of chronic low-level contamination over an extended period and represents a large portion of the actual examples of contamination found in the northwestern United States.

A silt loam soil from an airstrip near Hagerman, Idaho, was also obtained. Contamination at this site resulted from leakage from dinoseb storage barrels. This soil was an example of soil that has received acute, high-level contamination over a short time period.

Table 10.2 compares the inorganic constituents of the two soils. The inorganic constituents observed in the two soils indicated that the soils contained ample inorganic nutrients to support microbial metabolic activity. The presence of nitrate, which has been seen to inhibit dinoseb degradation,[52] in both soils had to be taken into consideration when a bioremediation procedure was designed. The presence of sulfate in the acutely contaminated soil might also affect the degradation of dinoseb. Sulfate might be used as an electron acceptor by sulfate-reducing bacteria, which could compete for the starch or fermentation products of the starch. Both soils had obviously been contaminated with fertilizing compounds as well as the organic compounds. This is not unexpected, due to the use of these airstrips for agricultural purposes. The cocontamination of pesticide- or herbicide-contaminated soils with fertilizers will probably occur at most dinoseb-contaminated sites.

Bench-scale experiments have revealed that the remediation of dinoseb from contaminated soils was possible in static flask cultures containing 2% starch in 50% soil/buffer mixtures that were stirred only occasionally. These experiments were performed using up to 300 g of

TABLE 10.2   Inorganic Parameters of Test Soils

Inorganic parameter	Chronically contaminated soil	Acutely contaminated soil
Nitrate (ppm)	134	294
Ammonium (ppm)	144	217
Sulfate (ppm)	Not detected	84
pH	7.58	7.52
P (ppm)	58	45.1
K (ppm)	480	288

SOURCE: Reprinted with permission from Kaake et al. (1992), Ref. 25.

TABLE 10.3 Herbicide Remediation from Contaminated Soil

| Compound* | Concentration | | | Percent removal |
	Initial soil (mean†)	Final aqueous	Final soil (mean†)	
Dinoseb	160,000	1.02	74	99.95
MCPP‡	1,210	17§	440§	63.64
Ioxynil	888	0.17§	10§	98.87
2,4-D	153	0.35§	94	38.70
Dicamba	106	2.60	24	77.04

*Herbicide and pesticide analyses were performed by Manchester Laboratories, Manchester, Washington. EPA methods 8150 and 8080 were used. The compounds listed above were the only ones detected.

†Units are µg/kg; results are the average of analyses of three samples.

‡MCPP = (4-chloro-2-methylphenoxy)propionic acid.

§Compounds were not detected; values represent one-half of the detection limit for the compound.

SOURCE: Reprinted with permission from Kaake et al. (1992), Ref. 25.

contaminated soil. The experiments to determine the biodegradability of the contaminants in the chronically contaminated soil revealed that it was microbiologically competent. Flooding of the soil with an equal volume of 50 mM phosphate buffer at pH 7 and addition of 2% (w/w) of the starch provided the soil with the nutrients necessary to create an anaerobic environment in which dinoseb and most of the other contaminants present could be degraded. Analyses showed that nitrate removal began almost immediately and was complete within 4 to 7 days. Dinoseb degradation began after a 2- to 3-day lag period and was usually complete within 10 to 20 days.[25] The redox potential decreased to below −200 mV within 4 to 5 days. The inoculation of the soil remediation test mixtures with laboratory enrichment cultures did not produce results significantly different from those produced by uninoculated cultures.

Table 10.3 summarizes the results of the analyses of pooled, treated soils from the replicates of one such bench-scale remediation experiment. Of the five initial contaminants, only 2,4-D appeared to be somewhat recalcitrant to anaerobic remediation. The final concentration of dinoseb detected in the soil (74 µg/kg) was well below our target remediation limit of 2.5 mg/kg.

Bench-scale experiments with the acutely contaminated soil revealed that the biological competency of this soil was improved by the addition of a 5% (v/w) laboratory enrichment culture, or by the addition of 10% or 50% (w/w) of dried, previously treated soil.[25] The dried treated soil was soil from the chronically contaminated site at Ellensburg, Washington, that had been treated as described above, and then dried and stored at 4°C for use as inoculum. The addition of 10% treated soil appeared to be the most efficient inoculum.[25]

Pilot-Scale Studies of the Bioremediation of Dinoseb-Contaminated Soils

Bench-scale experiments do not always imply the technology will be successfully scaled up. Therefore, we felt it was important to demonstrate the remediation of dinoseb from contaminated soils in larger-scale experiments. An initial demonstration of the technology was conducted on site with 45 to 50 kg of the chronically contaminated soil in a stirred reactor or in static reactors (plastic tubs).

The soils were excavated by shovel and sieved through a 6.35-mm screen on site. Personnel wore dust filters and protective clothing. The soils were then weighed into the reactor vessels, which contained the appropriate amount of phosphate buffer. The starch supply was then added, and the tanks were mixed thoroughly. The reactor and tubs were incubated in a small travel trailer that had been renovated for this purpose. The static tanks (in triplicate) received 45 kg of chronically contaminated soil; 45 L of pH 7, 50 mM phosphate buffer; and 0.9 kg of starch. Control cultures contained either uncontaminated soil, starch, and buffer, or contaminated soil and buffer without starch. The demonstration was performed twice at this scale. Temperature, pH, and redox potential were monitored every 15 min (first demonstration) or 2 h (second demonstration) through use of remote probes and an A/D data-logging unit controlled by a personal computer. Aqueous and soil samples were taken periodically and assayed for concentrations of dinoseb, volatile organic acids, nitrate, ammonium, hydrolizable starch, and total soluble reducing sugars.

The results of the second demonstration are presented in Figs. 10.1 through 10.3. The dinoseb concentration in the aqueous phase was measured directly by high-pressure liquid chromatography (HPLC); the dinoseb concentration in the soil was measured by HPLC analyses of acetonitrile soil extracts. Both methods are described elsewhere.[25] The results presented in Fig. 10.1 demonstrate that dinoseb degradation rates are rapid under this treatment protocol, and that the mixed tank allowed dinoseb removal from the soil to proceed more rapidly than in the static tanks. The redox potential in the static tanks dropped faster than in the stirred tank (Fig. 10.2). No polymerization products were observed in either the static tanks or the stirred tank. This suggests the initial drop in redox potential was sufficient to prevent polymerization reactions from occurring.

Figure 10.3 presents the accumulation and removal of dinoseb and the three most prominent biotransformation intermediates in the soil of the stirred tank. The intermediates were not identified, so concentrations could not be determined. These results demonstrate the necessity of analyzing intermediate compounds. If the concentration of dinoseb had

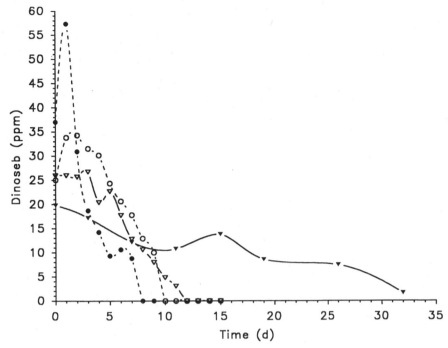

Figure 10.1 Dinoseb removal from soil and aqueous phases of 50-kg treatments of the chronically contaminated soil. Reactors received 45–50 kg of contaminated soil, 1 kg of starch, and 45–50 L of 50 mM phosphate buffer pH 7. ○ = stirred reactor, aqueous. ● = stirred reactor, soil. ▽ = static reactor, aqueous. ▼ = static reactor, soil.

been used as the only parameter to establish whether the treatment was finished, the treatment might have been stopped prematurely, after only 9 to 10 days. Although very little is known about the intermediates, we suspect that the intermediate identified as 1 in Fig. 10.3 may still retain some toxic properties. The ultimate removal of this intermediate seemed to bring about a reduction in toxicity of the soil. When this intermediate was removed, sulfate-reducing organisms present in the soil began to metabolize the fermentation products of the starch degradation, producing a black precipitate of iron sulfide visible in the soil, and fungi began to grow on the surface of the reactors. We are currently developing methods to test the toxicity of the soil at various times during the treatments to determine at which time the toxicity is eliminated. This work will help to prove the environmental safety of the technology and provide valuable information about the toxicity of the transformation intermediates. The intermediates will also be identified.

A larger-scale demonstration of this technology was performed using the acutely contaminated soil from the airstrip site near Hagerman,

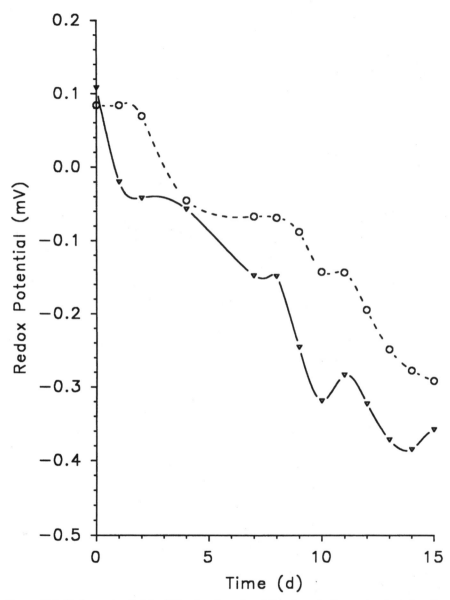

Figure 10.2 Redox potential in 50-kg treatments of the chronically contaminated soil. Reactors are those of Fig. 10.1. ○ = stirred reactor. ▽ = static reactor.

Idaho. The demonstration was carried out in a lined pit on site. All personnel wore Tyvek suits, full-face respirators, gloves, and protective footwear during all procedures done in the pit. The demonstration consisted of three stages. Stage 1 and 2 were performed in 2600 L fiberglass static reactors.

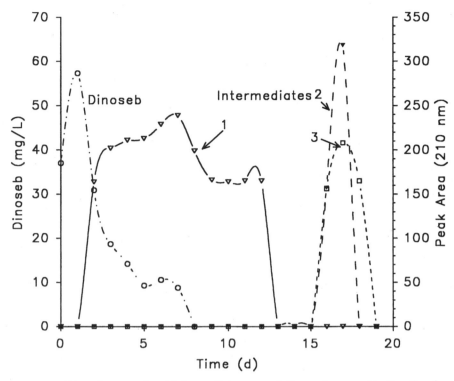

Figure 10.3 Dinoseb removal and intermediate accumulation and removal in the stirred-reactor soil extractions. Unidentified intermediates were quantified using peak area. DNOC was used as an internal standard to assure extraction efficiency and detector performance.

Stage 1 consisted of three of these reactors, which were loaded with approximately 350 L of irrigation water, 2268 g of K_2HPO_4, and 686 g of KH_2PO_4. Ingredients were mixed until the salts dissolved. Approximately 315 kg of contaminated soil was then added to each reactor by means of a backhoe; 6.3 kg of starch was then added, as well as 35 kg of dried, treated soil obtained from the treatment of the chronically contaminated soil. The pH of the contents of each reactor was adjusted to 7 with saturated sodium hydroxide or concentrated phosphoric acid as required, and pH, temperature, and redox electrodes were installed in each reactor. The data was collected and stored in an A/D data-logging unit connected to a 12 V deep-cycle battery. The reactors were covered with 6-mil visqueen, which was secured under the lip of each tank with an elastic cord.

The contents of the reactors were sampled in replicate at time 0 and then every two or three days thereafter. Reactors were mixed once after 15 days, incubation. There was no obvious evaporative loss of the aque-

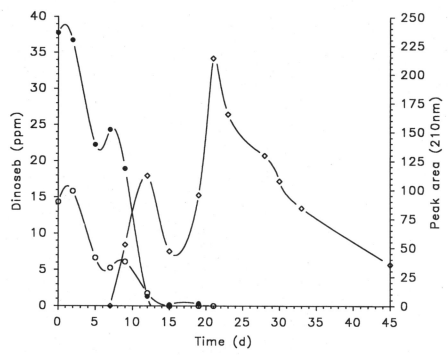

Figure 10.4 Dinoseb removal and intermediate accumulation and removal in 315-kg treatments of acutely contaminated soil. Reactors received 350 L of 50 mM phosphate buffer pH 7, 6.3 kg of starch, and 35 kg of treated soil as inoculum. ○ = aqueous dinoseb. ● = soil dinoseb. ◇ = unidentified intermediate.

ous phases during the incubations. Stage 1 was carried out during the month of August. The temperature in the tanks cycled daily between 25° and 32°C; the average temperature was 28°C. The pH of the aqueous phase remained within 0.2 pH units of 7. The redox potential in the aqueous phase of the reactors dropped rapidly and was below 0 mV by day two. The dinoseb was removed from the aqueous and soil phases by 15 days (Fig. 10.4). The treatments were incubated a total of 45 days to allow the concentration of intermediate 1 (which corresponds to intermediate 1 in Fig. 10.3) to decrease (Fig. 10.4).

Stage 2 consisted of five reactors set up the same as in stage 1, except that the soil contents from stage 1 were split evenly between the five reactors as inocula. The contents of the reactors were allowed to incubate 13 days. This stage was designed to generate inocula for the third stage, so samples were taken only initially and after 13 days. The dinoseb was removed from the soil and aqueous phases by the end of the incubation, but intermediate 1 was still present.

The contents of the five reactors from stage 2 were used as inocula for the stage 3 static reactors. These reactors were capable of holding 6000 L each, and were loaded with approximately 2000 L of irrigation water,

12.4 kg of K_2HPO_4, 3.7 kg of KH_2PO_4, 32 kg of starch substrate, and approximately 2000 kg of contaminated soil. Temperature, redox potential, and pH were monitored as described for stage 1. Soil and aqueous phase samples were taken at time zero and then once a week. The incubation was carried out throughout the months of October to May.

Average temperatures of the aqueous phase were in the mid-twenties initially, but as the climate reflected the change of season, the temperatures in the reactors underwent large fluctuations. The pH was maintained within 0.2 pH units of 7, and the redox dropped to below 0 mV within 2 to 3 days. The biological activity was slowed during the winter months, but resumed when the weather became warmer. By mid-May the soil and aqueous phases no longer contained any detectable dinoseb or biotransformation intermediates. The aqueous phase was removed and applied to an uncontaminated area of land nearby as wastewater. The soil was tilled into the pit on the contaminated site.

The results of HPLC analyses of samples from the stage 3 reactors are summarized in Fig. 10.5. This figure demonstrates again the importance of monitoring the intermediates of the degradation of a com-

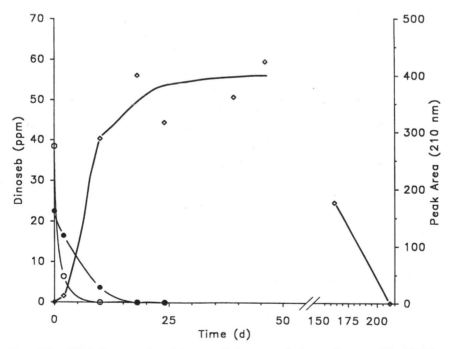

Figure 10.5 Dinoseb removal and intermediate accumulation and removal in 2000-kg treatments of the acutely contaminated soil. Reactors received approximately 2000 kg of contaminated soil, 2000 L of phosphate buffer pH 7, 40 kg of starch and 1 L of treated soil slurry from the 315-kg reactors. ○ = aqueous dinoseb. ● = soil dinoseb. ◇ = unidentified intermediate.

pound rather than only the compound itself. The overall incubation time for dinoseb removal from the soil/buffer mixture at this scale was not significantly different from the incubation times observed in smaller-scale experiments, suggesting that treatments on an even larger scale can be accomplished within similar time frames. This is encouraging, since these were static tanks. Stirred tanks would presumably require even less time for the bioremediation process to occur, as indicated by our experiments with the chronically contaminated soil near Ellensburg, Washington.

Commercialization

The technology described above is presently in the patent review process and will be commercialized in the near future. The commercial methods applied will be site-specific, but we have proposed three application procedures for this technology.

The process will lend itself well to large anaerobic slurry reactors that can be placed on a truck bed and hauled from site to site to carry out soil remediation at sites such as rural airstrips. These sites usually contain relatively small amounts of soil (50–500 m^3) that have low levels of multiple contaminants. The reactor design can be relatively simple, since only pH and temperature must be controlled. The removal and control of oxygen concentration and redox potential are inherent to the system, as long as sufficient amounts of supplemental carbon are supplied. The mixing of the soil ensures more rapid desorption of the compounds from the soil and short treatment times. Other biodegradable contaminants that are not removed by the anaerobic process may be removed by operating the reactor aerobically following the anaerobic stage.

A second application of the technology would be an innovative variation of land farming that we have termed *anaerobic land farming*. This method would use a lined pit similar to that used in land farming, but drainage of the pit would not be necessary. The pit liner would not only serve to isolate the treatment system from uncontaminated areas, but would keep the aqueous phase in the treatment area, allowing soil saturation. Mixing of the contaminated soil with the carbon source, soil inoculum, and buffer by putting it through a hopper before putting it in the pit would be necessary to ensure complete wetting of the soil. Periodic gentle mixing of the contents of the anaerobic pit would be beneficial, ensuring rapid rates of compound solubilization and degradation and preventing the formation of pockets of untreated soil. This type of application would be most suitable to large sites that contain enough soil to make the expense of digging and lining a pit at each site more favorable than the expense, both in time and money, of running several batches of soil through a slurry reactor.

The process could also be run as an in situ procedure by using conventional methods to deliver the supplemental carbon source and other nutrients, but omitting the oxygenation process. If the nutrient transport system were run as a closed loop, inadvertent oxygenation would be avoided. The effects of rendering large tracts of soil carbon-rich and anaerobic must be thoroughly investigated before this type of remediation can be performed.

Cost Comparisons

Economic considerations play a major role in the selection of a soil remediation process. We have estimated the costs of various soil remediation technologies on the basis of personal communications from remediation specialists or from the literature. Figure 10.6 presents a comparison of total estimated costs for different treatment methods. The components of the cost of each treatment are explained below.

Incineration

Incineration is often employed when immediate cleanup of hazardous wastes is necessary. The costs associated with incineration become very high, especially when long-distance hauling is necessary.

A facility in Texas charges $600 per ton to incinerate dinoseb-con-

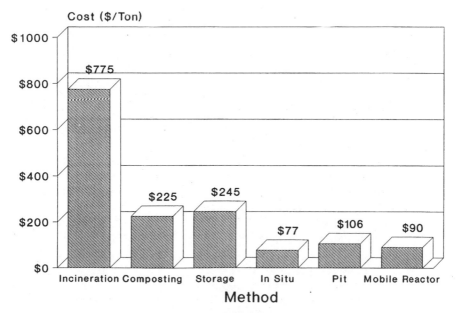

Figure 10.6 Cost comparison of various remediation techniques.

taminated soil. Added to this is the cost of excavation and transportation of the soil to and from the incinerator. The average cost of excavation and transportation of soil is $100 to $150 per ton. The adjusted total cost for incineration would then be $700 to $850 per ton.

Storage

Dinoseb-contaminated soil has also been hauled to landfill sites that will permanently store the soil. These facilities merely store contaminated soils; the waste is not actually eliminated. New federal guidelines will not allow the dumping of dinoseb-contaminated soil above 2.5 ppm after May 8, 1992.

Envirosafe, a hazardous waste storage facility in southern Idaho certified under the Resource Conservation and Recovery Act (RCRA), charges $120 per ton to store contaminated soil. This again excludes the cost of excavation and transportation of the soil, which would bring the total cost to $220 to $270 per ton for site owners in our region.

Anaerobic land farming

We have obtained from the Environmental Services Division of the Morrison Knudsen Company of Boise, Idaho, the design for a pit to treat large quantities of contaminated soil. The majority of the expense for the anaerobic land-farming pit is incurred for the construction of the pit liner. A liner to meet or exceed government specifications would be approximately 6 ft thick and would be composed of several materials. First, a 2-ft layer of clay would be laid down, then a layer of high-density polyethylene (HDPE). This layer represents the secondary containment portion of the liner. The primary section would consist of 12 in of granular material, a geotextile sheet, another 2 ft of clay, a layer of HDPE, and finally 12 in of granular material to protect the polyethylene. Also, a floating plastic sheet or foam would be used to prevent evaporation.

Such a pit can accommodate from 600 to 10,000 tons of soil. Upon completion of the degradation process, the soil would be backfilled to its original location, and the synthetic portion of the liner would be removed and disposed of. A single pit could be used to successively remediate several batches of soil.

On a scale accommodating 5000 tons of soil, the average cost is $40 per ton for a 335-m^2 pit. The approximate cost to excavate the soil and monitor the pit would be $55 to $78 per ton. This brings the total for soil remediation to $95 to $118 per ton (costs calculated from data supplied by Morrison Knudsen).

Composting

Although composting is an aerobic process, we are presenting its costs for comparison to other technologies. The cost range for composting can vary greatly depending on the materials being composted and the size of the operation.

In 1978, the U.S. Army proposed a composting operation that would treat 3 tons of TNT-contaminated soil per day at an average cost of $287 per ton.[38] The initial construction cost was estimated at $376,750 to cover the construction of a permanent large-scale facility that would be used to treat contaminated soil that is continually produced.

Modern facilities for composting have become more efficient and less expensive due to the incorporation of bioreactors. For example, sewage sludge can be composted in a bioreactor for $150 per ton. For hazardous waste, an additional $50 to $100 per ton is added to cover the cost of sampling and monitoring equipment, increased safety standards, and transportation of the contaminated soil. Composting of contaminated soils should therefore cost from $200 to $250 per ton.

Mobile treatment facility

As described above, we have proposed the construction of a mobile soil-slurry reactor that would be used for on-site treatment of contaminated soils. Although no prototype has been constructed, the cost for a truck and trailer, with a bioreactor premounted on the trailer, has been calculated to be about $228,000. This bioreactor mimics the design of a 12-m^3 cement mixer that can be sealed and equipped with components for sampling and rotating the soil/buffer mixture. The average cost of operating this unit would be $90 per ton, which includes loading and operation of the bioreactor.

In situ treatment

To avoid the cost of soil excavation and meeting some of the more complex regulations pertaining to the disposal of hazardous-waste-contaminated soil, in situ remediation has been investigated. Several monitoring wells would have to be drilled throughout the contaminated area, and an extraction well used to recirculate an aqueous solution through the soil to keep it saturated and prevent groundwater contamination. Monitoring wells would be drilled periodically throughout the process to obtain random samples of soil in and around a site.

The cost range for this method of treatment varies greatly depending on the number and depth of wells, knowledge of the hydrogeology at the site, and extent of contamination. We have found estimated costs for the treatment of two hydrocarbon-contaminated sites by this method. The first reported a cost of $130 per cubic meter[12] to treat a gasoline-contaminated site in which 4600 cubic meters of soil were contaminated. A second report was found for a site in which 4600 cubic meters of soil were contaminated with hydrocarbon solvents. The estimated cost of this treatment was $59 per cubic meter.[41] These costs represent approximately $118 per ton and $54 per ton, respectively.

Summary

We have demonstrated a soil bioremediation technology for the removal of the recalcitrant nitroaromatic herbicide dinoseb from large quantities of soil. This technology is simple and inexpensive compared to current physical methods employed for soil remediation. The commercialization of this process could make use of either a slurry reactor design or an anaerobic land-farming pit design, depending on the site characteristics.

The technology consists of stimulating the natural organisms present in the soil by flooding the soil with a pH 7 buffer and adding a rich carbon source to provide an energy source for heterotrophic organisms. The metabolic activity of the heterotrophic organisms removes oxygen and nitrate from the soil slurry (both may inhibit dinoseb degradation), lowering the redox potential and allowing dinoseb degradation to occur.

We have found that at least one soil contaminated with chronic levels of dinoseb over long periods of time contained the appropriate organisms to facilitate the treatment procedure, needing only biostimulation with nutrients and carbon. The long exposure period and low contamination levels apparently provided the conditions necessary to allow the proliferation of an indigenous population of dinoseb-resistant heterotrophic organisms and anaerobic fermentative organisms capable of dinoseb metabolism. Other contaminants present in this soil were not inhibitory to the dinoseb degradation process and were largely removed from the soil during the anaerobic treatment.

In another soil that had been acutely contaminated with high levels of dinoseb, the appropriate organisms were either not present, or were inhibited by the high levels of dinoseb. The augmentation of this soil with previously treated soil from the chronically contaminated site provided the necessary organisms for the bioremediation to occur.

The remediation of contaminants other than dinoseb from multiply contaminated soils suggests that the anaerobic treatment procedure is very versatile. Investigations into the remediation of soils contaminated with other aerobically recalcitrant compounds may reveal additional applications of this technology.

Acknowledgment

We would like to acknowledge the excellent advice and technical assistance of Keith Stormo and Karen Crossley. Dr. Doug Sell of the J. R. Simplot Company, Boise, Idaho, has been instrumental in arranging the Hagerman demonstration and supplying the starch. Mr. Marvin Thorne of the Morrison Knudsen Company in Boise, Idaho, has supplied information regarding land-farming pit design.

This research was supported by the U.S. Environmental Protection Agency under the Emerging Technologies Program (U.S. EPA Award CR816818010) and by the J. R. Simplot Company, Boise, Idaho. We thank Wendy Davis-Hoover of the U.S. EPA, Cincinnati, Ohio, for her suggestions and assistance and the Washington State Department of Ecology's Mike Cochran for arranging quality assurance analytical chemistry services for the Ellensburg, Washington, demonstration.

References

1. Aggarwal, P. K., J. L. Means, and R. E. Hinchee. 1991. Formulation of nutrient solutions for *in situ* biodegradation, pp. 51–66. *In* R. E. Hinchee and R. F. Olfenbuttel (eds.), *In Situ* Bioreclamation: Applications and Investigations for Hydrocarbon and Contaminated Site Remediation, Butterworth-Heinemann, Stoneham, Mass.
2. Alfoldi, L. 1991. Hydrogeologic considerations for *in situ* bioremediation, pp. 33–50. *In* R. E. Hinchee and R. F. Olfenbuttel (eds.), *In Situ* Bioreclamation: Applications and Investigations for Hydrocarbon and Contaminated Site Remediation, Butterworth-Heinemann, Stoneham, Mass.
3. Atlas, R. M. 1991. Bioremediation of fossil fuel contaminated soils, pp. 14–32. *In* R. E. Hinchee and R. F. Olfenbuttel (eds.), *In Situ* Bioreclamation: Applications and Investigations for Hydrocarbon and Contaminated Site Remediation, Butterworth-Heinemann, Stoneham, Mass.
4. Balba, M. T., A. C. Ying, and T. G. McNeice. 1991. Bioremediation of contaminated land: Bench scale to field applications, pp. 464–476. *In* R. E. Hinchee and R. F. Olfenbuttel (eds.), On-Site Bioreclamation: Processes for Xenobiotic and Hydrocarbon Treatment, Butterworth-Heinemann, Stoneham, Mass.
5. Barenschee, E. R., P. Bochem, O. Helmling, and P. Weppen. 1991. Effectiveness and kinetics of hydrogen peroxide and nitrate-enhanced biodegradation of hydrocarbons, pp. 103–124. *In* R. E. Hinchee and R. F. Olfenbuttel (eds.), *In Situ* Bioreclamation: Applications and Investigations for Hydrocarbon and Contaminated Site Remediation, Butterworth-Heinemann, Stoneham, Mass.
6. Bell, R. A., and A. H. Hoffman. 1991. Gasoline spill in fractured bedrock addressed with *in situ* bioremediation, pp. 437–443. *In* R. E. Hinchee and R. F. Olfenbuttel (eds.), *In Situ* Bioreclamation: Applications and Investigations for Hydrocarbon and Contaminated Site Remediation, Butterworth-Heinemann, Stoneham, Mass.
7. Black, W. V., R. C. Ahlert, D. S. Kosson, and J. E. Brugger. 1991. Slurry-based biotreatment of contaminants sorbed onto soil constituents, pp. 408–422. *In* R. E. Hinchee, and R. F. Olfenbuttel (eds.), On-Site Bioreclamation: Processes for Xenobiotic and Hydrocarbon Treatment, Butterworth-Heinemann, Stoneham, Mass.
8. Bouwer, E. J., and G. D. Cobb. 1987. Modeling of biological processes in the subsurface. *Water Sci. Technol.* 19:769–779.
9. Bradford, M. L., and R. Krishnamoorthy. 1991. Consider bioremediation for waste site cleanup. *Chem. Eng. Prog.* 87:80–85.
10. Brauer, H. 1987. Development and efficiency of a new generation of bioreactors. Part 1. *Bioproc. Eng.* 2:149–159.

242   Chapter Ten

11. Brauer, H. 1988. Development and efficiency of a new generation of bioreactors. Part 2, Description of new bioreactors. *Bioproc. Eng.* 3:11–21.
12. Brown, K. A., J. C. Dey, and W. E. McFarland. 1991. Integrated site remediation combining groundwater treatment, soil vapor extraction, and bioremediation, pp. 444–449. *In* R. E. Hinchee, and R. F. Olfenbuttel (eds.), *In Situ* Bioreclamation: Applications and Investigations for Hydrocarbon and Contaminated Site Remediation, Butterworth-Heinemann, Stoneham, Mass.
13. Davis-Hoover, W. J., L. C. Murdoch, S. J. Vesper, H. R. Pahren, O. L. Sprockel, C. L. Chang, A. Hussain, and W. A. Ritschel. 1991. Hydraulic fracturing to improve nutrient and oxygen delivery for *in situ* bioreclamation, pp. 68–82. *In* R. E. Hinchee and R. F. Olfenbuttel (eds.), *In Situ* Bioreclamation: Applications and Investigations for Hydrocarbon and Contaminated Site Remediation, Butterworth-Heinemann, Stoneham, Mass.
14. Doyle, J. D., K. A. Short, G. Stotzky, R. J. King, R. J. Seidler, and R. H. Olsen. 1991. Ecologically significant effects of *Pseudomonas putida* PPO301(pRO103), genetically engineered to degrade 2,4-dichlorophenoxyacetate, on microbial populations and processes in soil. *Can J. Microbiol.* 37:682–691.
15. Doyle, R. C., D. D. Kaufman, and G. W. Burt. 1978. Effect of dairy manure and sewage sludge on ^{14}C-pesticide degradation in soil. *J. Agric. Food Chem.* 26:987–989.
16. Durand, A., and D. Chereau. 1988. A new pilot reactor for solid-state fermentation: Application to the protein enrichment of sugar beet pulp. *Biotechnol. Bioeng.* 31:476–486.
17. Fernando, T., J. A. Bumpus, and S. D. Aust. 1990. Biodegradation of TNT (2,4,6-trinitrotoluene) by *Phanerochaete chrysosporium*. *Appl. Environ. Microbiol.* 56:1666–1671.
18. Hallas, L. E., and M. Alexander. 1983. Microbial transformation of nitroaromatic compounds in sewage effluent. *Appl. Environ. Microbiol.* 45:1234–1241.
19. Harmsen, J. 1991. Possibilities and limitations of landfarming for cleaning contaminated soils, pp. 255–272. *In* R. E. Hinchee and R. F. Olfenbuttel (eds.), On-Site Bioreclamation: Processes for Xenobiotic and Hydrocarbon Treatment, Butterworth-Heinemann, Stoneham, Mass.
20. Hartter, D. R. 1985. The use and importance of nitroaromatic compounds in the chemical industry, pp. 1–14. *In* D. E. Rickert (ed.), Chemical Institute of Toxicology Series: Toxicity of Nitroaromatic Compounds, Hemisphere Publishing, Washington, D.C.
21. Hathaway, J. A. 1985. Subclinical effects of trinitrotoluene: A review of epidemiology studies, pp. 255–274. *In* D. E. Rickert (ed.), Chemical Institute of Toxicology Series: Toxicity of Nitroaromatic Compounds, Hemisphere Publishing, Washington, D.C.
22. Huling, S. G., B. E. Bledsoe, and M. V. White. 1991. The feasibility of using hydrogen peroxide as a source of oxygen in bioremediation, pp. 83–102. *In* R. E. Hinchee and R. F. Olfenbuttel (eds.), *In Situ* Bioreclamation: Applications and Investigations for Hydrocarbon and Contaminated Site Remediation, Butterworth-Heinemann, Stoneham, Mass.
23. Isbister, J. D., G. L. Anspach, J. F. Kitchens, and R. C. Doyle. 1984. Composting for decontamination of soils containing explosives. *Microbiologica* 7:47–73.
24. Jensen, H. L., and Lautrup-Larsen. 1967. Microorganisms that decompose nitroaromatic compounds with special reference to dinitro-*ortho*-cresol. *Acta Agric. Scand.* 17:115–126.
25. Kaake, R. H., D. J. Roberts, T. O. Stevens, R. L. Crawford, and D. L. Crawford. 1992. Bioremediation of soils contaminated with 2-*sec*-butyl-4,6-dinitrophenol (dinoseb). *Appl. Environ. Microbiol.* 58:1683–1689.
26. Kaplan, D. L. 1990. Biotransformation pathways of hazardous energetic organonitro compounds, pp. 155–182. *In* D. Kamely, A. Chakrabarty, and G. S. Omenn (eds.), Advances in Applied Biotechnology, 4: Biotechnology and Biodegradation, Portfolio Publishing, The Woodlands, Tex.
27. Kaplan, D. L., and A. M. Kaplan. 1982. Thermophilic biotransformations of 2,4,6-trinitrotoluene under simulated composting conditions. *Appl. Environ. Microbiol.* 44:757–760.
28. Klausmeier, R. E., J. L. Osmon, and D. R. Walls. 1973. The effect of trinitrotoluene on microorganisms. *Dev. Ind. Microbiol.* 15:309–317.

29. Kleijntjens, R. H., K. C. A. M. Luyben, M. A. Bosse, and L. P. Velthuisen. 1987. Process development for biological soil decontamination in a slurry reactor, pp. 252–255. *In* O. M. Neijssel, R. R. van der Meer, and K. C. A. M. Luyben (eds.), Proc. 4th European Congress on Biotechnology, 1987, vol. 1, Elsevier Science, Amsterdam.
30. Kukor, J. J., and R. H. Olsen. 1989. Diversity of toluene degradation following long term exposure to BTEX *in situ,* pp. 405–421. *In* D. Kamely, A. Chakrabarty, and G. S. Omenn (eds.),Advances in Biotechnology Series, 4: Biotechnology and Biodegradation, Portfolio Publishing, The Woodlands, Tex.
31. Laukevics, J. J., A. F. Apsite, U. E. Viesturs, and R. P. Tengerdy. 1984. Solid substrate fermentation of wheat straw to fungal protein. *Biotechnol. Bioeng.* 26:1465–1474.
32. Lee, M. D., and R. L. S. Raymond. 1991. Case history of the application of hydrogen peroxide as an oxygen source for *in situ* bioreclamation, pp. 429–436. *In* R. E. Hinchee and R. F. Olfenbuttel (eds.), In Situ Bioreclamation: Applications and Investigations for Hydrocarbon and Contaminated Site Remediation, Butterworth-Heinemann, Stoneham, Mass.
33. McCarty, P. L. 1988. Bioengineering issues related to *in situ* remediation of contaminated soils and groundwater, pp. 143–162. *In* G. S. Omenn (ed.), Basic Life Sciences, 45: Environmental Biotechnology: Reducing Risks from Environmental Chemicals through Biotechnology, Plenum Press, new York.
34. McFarland, M. J., J. Q. Xiu, W. A. Aprill, and R. C. Sims. 1989. Biological composting of petroleum waste organics using the white rot fungus *Phanerochaete chrysosporium.* Proceedings of the Institute of Gas Technology's Second International Symposium on Gas, Oil, Coal and Environmental Biotechnology, pp.2.3.1–22.
35. Mikesell, M. D., and S. A. Boyd. 1988. Enhancement of pentachlorophenol degradation in soil through induced anaerobiosis and bioaugmentation with anaerobic sewage sludge. *Environ. Sci. Technol.* 22:1411–1414.
36. van den Munckhof, G. P. M., and M. F. X. Veul. 1991. Production-scale trials on the decontamination of oil-polluted soil in a rotating bioreactor at field capacity, pp. 441–451. *In* R. E. Hinchee and R. F. Olfenbuttel (eds.), On-Site Bioreclamation: Processes for Xenobiotic and Hydrocarbon Treatment, Butterworth-Heinemann, Stoneham, Mass.
37. Naumova, R. P., S. Y. Selivanovskaya, and F. A. Mingatina. 1988. Possibility of deep bacterial destruction of 2,4,6-trinitrotoluene. *Mikrobiologiya* (USSR) 57:218–222.
38. Osmon, J. L., C. C. Andrews, and A. Tatyrek. 1978. The biodegradation of TNT in enhanced soil and compost systems. Report No. ARLCD-TR-77032. U.S. Army Armament Research and Development Command. Large Caliber Weapon Systems Laboratory, Dover, N.J.
39. Porta, A. 1991. A review of European bioreclamation practice, pp. 1–13. *In* R. E. Hinchee and R. F. Olfenbuttel (eds.), In Situ Bioreclamation: Applications and Investigations for Hydrocarbon and Contaminated Site Remediation, Butterworth-Heinemann, Stoneham, Mass.
40. Salkinoja-Salonen, M., P. Middeldorp, M. Briglia, R. Valo, M. Haggblom, and A. McBain. 1989. Cleanup of old industrial sites, pp. 347–368. *In* D. Kamely, A. Chakrabarty, and G. S. Omenn (eds.), Advances in Biotechnology Series, 4: Biotechnology and Biodegradation, Portfolio Publishing, The Woodlands, Tex.
41. Schmitt, E. K., M. T. Lieberman, J. A. Caplan, D. Blaes, P. Keating, and W. Richards. 1991. Bioremediation of soil and groundwater contaminated with stoddard solvent and mop oil using the PetroClean bioremediation system, pp. 581–599. *In* R. E. Hinchee and R. F. Olfenbuttel (eds.), In Situ Bioreclamation: Applications and Investigations for Hydrocarbon and Contaminated Site Remediation, Butterworth-Heinemann, Stoneham, Mass.
42. Semprini, L., G. D. Hopkins, P. V. Roberts, and P. L. McCarty. 1991. *In situ* biotransformation of carbon tetrachloride, freon-113, freon-11, and 1,1,1-TCA under anoxic conditions, pp. 41–58. *In* R. E. Hinchee and R. F. Olfenbuttel (eds.), In Situ Bioreclamation: Applications and Investigations for Hydrocarbon and Contaminated Site Remediation, Butterworth-Heinemann, Stoneham, Mass.
43. Short, K. A., J. D. Doyle, R. J. King, R. J. Seidler, G. Stotzky, and R. H. Olsen. 1991. Effects of 2,4-dichlorophenol, a metabolite of a genetically engineered bacterium, and

2,4-dichlorophenoxyacetate on some microbe-mediated processes in soil. *Appl. Environ. Microbiol.* 57:412–418.

44. Sims, J. L., R. C. Sims, and J. E. Mathews. 1990. Approach to bioremediation of contaminated soil. *Haz. Waste Haz. Mat.* 7:117–149.

45. Skinner, J. H., G. G. Ondich, and T. L. Baugh. 1991. U.S. EPA Bioremediation Research Programs, pp. 1–15. *In* R. E. Hinchee and R. F. Olfenbuttel (eds.), On-Site Bioreclamation: Processes for Xenobiotic and Hydrocarbon Treatment, Butterworth-Heinemann, Stoneham, Mass.

46. Spain, J. C., and D. T. Gibson. 1991. Pathway for biodegradation of p-nitrophenol in a *Moraxella* sp. *Appl. Environ. Microbiol.* 57:812–819.

47. Spain, J. C., O. Wyss, and D. T. Gibson. 1979. Enzymatic oxidation of *p*-nitrophenol. *Biochem. Biophys. Res. Comm.* 88:634–641.

48. Spanggord, R. J., J. C. Spain, S. F. Nishino, and K. E. Mortelmans. 1991. Biodegradation of 2,4-dinitrotoluene by a *Pseudomonas* sp. *Appl. Environ. Microbiol.* 57:3200–3205.

49. Stegmann, R., S. Lotter, and J. Heerenklage. 1991. Biological treatment of oil-contaminated soils in bioreactors, pp. 188–208. *In* R. E. Hinchee and R. F. Olfenbuttel (eds.), *In Situ* Bioreclamation: Applications and Investigations for Hydrocarbon and Contaminated Site Remediation, Butterworth-Heinemann, Stoneham, Mass.

50. Stevens, T. O. 1989. Ph.D. dissertation, University of Idaho, Moscow, Idaho.

51. Stevens, T. O., R. L. Crawford, and D. L. Crawford. 1990. Biodegradation of dinoseb (2-*sec*-butyl-4,6-dinitrophenol) in several Idaho soils varying in dinoseb exposure history. *Appl. Environ. Microbiol.* 56:133–139.

52. Stevens, T. O., R. L. Crawford, and D. L. Crawford. 1991. Selection and isolation of bacteria capable of degrading dinoseb (2-*sec*-butyl-4,6-dinitrophenol). *Biodegradation* 2:1–13.

53. Tewfik, M. S., and W. C. Evans. 1966. The metabolism of 3,5-dinitro-*o*-cresol (DNOC) by soil micro-organisms. *Biochem. J.* 99:31.

54. Visscher, K., J. Brinkman, and E. R. Soczo. 1989. Biotechnology in hazardous waste management in the Netherlands, pp. 389–403. *In* D. Kamely, A. Chakrabarty, and G. S. Omenn (eds.), Advances in Applied Biotechnology Series, 4: Biotechnology and Biodegradation, Portfolio Publishing, The Woodlands, Tex.

55. Wallnoefer, P. R., W. Ziegler, G. Engelhardt, and H. Rothmeier. 1978. Transformation of dinitrophenol herbicides by *Azotobacter* sp. *Chemosphere* 7:967–972.

56. Webb, O. F., T. J. Phelps, P. R. Bienkowski, P. M. Digrazia, G. D. Reed, B. Applegate, D. C. White, and G. S. Sayler. 1991. Development of a differential volume reactor system for soil biodegradation studies. *Appl. Biochem. Biotechnol.* 28/29:5–19.

57. Wesseling, J., and W. R. van Wijk. 1957. Land drainage in relation to soils and crops: I. Soil physical conditions in relation to drain depth, pp. 461–504. *In* L. N. Luthin (ed.), Drainage of Agricultural Lands, American Society of Agronomy, Madison, Wisc.

58. Won, W. D., L. H. DiSalvo, and J. Ng. 1976. Toxicity and mutagenicity of 2,4,6-trinitrotoluene and its microbial metabolites. *Appl. Environ. Microbiol.* 31:576–580.

59. Woodward, R. E. Evaluation of composting implementation: A literature review. Report No. TCN 89363, U.S. Army Toxic and Hazardous Materials Agency, Aberdeen Proving Ground, Maryland.

60. Yare, B. S. 1991. A comparison of soil-phase and slurry-phase bioremediation of PNA-containing soils, pp. 173–187. *In* R. E. Hinchee and R. F. Olfenbuttel (eds.), On-Site Bioreclamation: Processes for Xenobiotic and Hydrocarbon Treatment, Butterworth-Heinemann, Stoneham, Mass.

11

Biological Treatment
of Industrial and
Hazardous Wastewater

Gregory D. Sayles

U.S. Environmental Protection Agency
Risk Reduction Engineering Laboratory
Cincinnati, Ohio

Makram T. Suidan

Department of Civil and Environmental Engineering
University of Cincinnati
Cincinnati, Ohio

Microbial-based technologies for the destruction of naturally occurring organic compounds have historically been the treatment of choice, relative to physical or chemical methods, because of their low cost. These technologies are lower in cost because the destruction reactions, which are mediated by naturally occurring biocatalytic activity, can occur relatively quickly at ambient temperatures. Thus, the energy costs for treatment are relatively very low.

Over the last century, the development of microbial-based technologies for the treatment of liquid domestic waste streams has provided excellent processes for the destruction of waste constituents that are readily biodegradable under aerobic conditions. Therefore, processes similar to those used for conventional domestic wastewater treatment have been applied successfully to the treatment of many industrial and hazardous wastewaters.

In fact, industry relies heavily on treatment of their chemical wastes by publicly owned wastewater facilities. Over 150 million pounds of chemical wastes were sent to domestic treatment plants in 1990 alone (Thayer, 1992).

However, standard classic design of wastewater treatment systems does not promote degradation of many of the chemicals that 20th century humans have invented and that now appear in industrial and domestic waste streams. New or improved approaches employing microbial-based biodegradation processes must be developed.

This chapter provides an overview of the current state of large-scale industrial and hazardous wastewater treatment. First, the capabilities of domestic wastewater treatment facilities to handle typical domestic wastes are summarized. Again, many industrial and hazardous wastewater treatment processes are essentially scaled-down versions of domestic wastewater treatment plants. Next, the shortcomings of conventional treatment processes for destroying unconventional compounds, such as the toxic organics and metals found in some industrial and hazardous wastewater streams, are discussed. Finally, state of the art processes that overcome these weaknesses so that the release of toxic compounds into the environment is minimized and destruction maximized will be examined.

Conventional Biological Treatment of Liquid Wastes

Traditionally, the primary goal of domestic biological wastewater treatment has been to reduce the biological oxygen demand (BOD) of domestic wastewater so that the liquid stream could be released to the environment with minimal impact on the local ecology. In general, this goal is accomplished by (1) removal of the soluble organics by biological utilization whereby microbial cells degrade organic compounds to generate energy and building blocks for more cells, and (2) removal of residual suspended organic matter (suspended solids, biological and nonbiological) by clarification (sedimentation).

These tasks are most typically accomplished for domestic wastewater treatment by the serial processes of primary and secondary treatment (shown schematically in Fig. 11.1). Currently, combined primary and secondary treatment is being utilized in at least 75 percent of the domestic wastewater treatment facilities in the United States (see Table 11.1). Primary treatment consists of screening and sedimentation of the influent wastewater stream to remove the large insoluble materials. Secondary treatment includes a bioreactor that is usually followed by clarification. The goal of the bioreactor process is to mineralize soluble organic matter to carbon dioxide and water. However, no reactor can destroy 100 percent of the influent BOD, and a fraction of the influent BOD is converted into additional cell mass due to cell

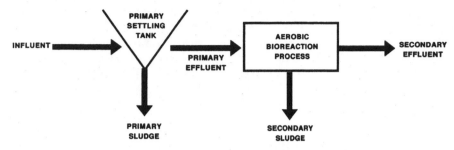

Figure 11.1 Schematic of a typical combined primary and secondary treatment.

TABLE 11.1 Number of Domestic Wastewater Treatment
Facilities in U.S. by Level of Treatment in 1988
(Total = 15,708) (adapted from U.S. EPA, 1989)

Level of treatment	Percent of total facilities
Primary only	11
Primary and secondary	54
Primary, secondary, and more	21
Others	14

growth. In the subsequent clarification (settling tank) step, bioreactor effluent suspended solids (mostly cell material) are separated from the effluent. Typically, the solids-depleted liquid stream exiting the settling tank is disinfected and then discharged to the environment. The solids-enriched stream departing the settling tank can be partially recycled to the bioreactor to enhance the biomass concentration in the reactor. The remaining portion of the solids-enriched stream undergoes further biological treatment to reduce its BOD. Ultimately, the stabilized solids are disposed in a landfill or incinerated. Treatment of industrial wastes may not require primary treatment if the waste stream contains no insoluble material.

The effluent quality goals for U.S. domestic wastewater treatment plants are dictated by law and summarized briefly in Table 11.2.

This chapter will concentrate on the biological reactor (bioreactor) portion of the process. The vast majority of wastewater treatment bioreactors rely on aerobic biological processes, although nonaerobic—

TABLE 11.2 Minimum U.S. Standard for Secondary Treatment Effluent
(Fed. Reg., 1988, 1989)

Effluent characteristic	Average removal (%)	Average 30-day concentration	Average 7-day concentration
BOD_5 (mg/L)	≥85	≤30	≤45
Suspended solids (mg/L)	≥85	≤30	≤45
pH	Between 6.0 and 9.0 at all times		

i.e., anaerobic or anoxic—bioreactors are utilized for special purposes. The general characteristics of aerobic and nonaerobic wastewater treatment bioreactors employed in conventional treatment facilities are discussed separately below. These reactors utilize a mixed population of microorganisms that develop from the microorganisms present in the feed stream and operating conditions in the reactor.

Aerobic bioreactors

Soluble organic sources of biochemical oxygen demand can be removed by any viable microbial process, aerobic, anaerobic, or otherwise. However, aerobic processes are typically used as the principal means of BOD reduction of domestic wastewater because the aerobic microbial reactions are fast, typically 10 times faster than anaerobic microbial reactions. Thus, aerobic reactors can be built relatively small and open to the atmosphere, yielding the most economical means of BOD reduction.

The major disadvantage of aerobic bioprocesses for waste treatment, relative to nonaerobic processes, is the large amount of cell matter produced. A relatively high accumulation of biomass occurs in the aerobic bioreactor because the biomass yield (mass of cell produced per unit mass of biodegradable organic matter) for aerobic organisms is relatively high, roughly 4 times greater than the yield for anaerobic organisms. Residual cell matter and other insoluble materials (sludge) present in the reactor effluent can contain residual BOD that may need to be reduced in an additional process, and must ultimately be disposed of as a solid waste.

The unbalanced stoichiometry of an aerobic microbial process is

$$s_1 C_a O_b H_c N_d S + s_2 O_2 \xrightarrow{\text{cells}} s_3 CO_2 + s_4 H_2O + C_5 H_7 NO_2 +$$

$$s_5 NH_3 + s_6 \text{ (other products)}$$

where $C_a O_b H_c N_d S$ represents the elemental composition of the feed stream where a, b, c, d, and e are determined by the particular waste stream composition, and $C_5 H_7 NO_2$ is an average microbial cell composition (Metcalf and Eddy, Inc., 1991). If the amount of "other products" formed can be neglected, then the stoichiometry (s_1, s_2, etc.) of the reaction is determined by the feed composition (a, b, c, and d) by solving elemental balances for C, H, O, and N. Assuming the reaction goes to completion, the rate of oxygen supply needed and the rate of cell sludge production can then be calculated from the balanced equation given the rate of feed to the reactor.

Several designs for aerobic reactors are commonly utilized and are summarized in Table 11.3. The most common of these designs in sec-

TABLE 11.3 Description of Typical Large-Scale Aerobic Biological Treatment Reactors

Reactor name	Description	Schematic
Suspended-cell processes: Activated sludge	Plug-flow or well-mixed reactor followed by a settling tank; portion of the settled solids (mostly cells) is returned to the reactor; reactor and settling tank open to atmosphere; forced aeration. Used by most medium to large municipalities for secondary treatment. (See Bailey and Ollis, 1986; Metcalf and Eddy, Inc., 1991; Stall and Sherrard, 1978; WPCF, 1987.)	
Ponds, lagoons	Essentially a pond dug in the ground, open to atmosphere. Can operate as fully aerobic or as combined aerobic and anaerobic (facultative) depending on the type of aeration mechanism used and the degree of mixing. Aeration is either passive or by surface aerators. Facultative design is used by small communities and industries for inexpensive secondary treatment; it allows both aerobic BOD reduction and anaerobic sludge stabilization in one basin. Industrial design includes impermeable liner if hazardous waste is present. Aerobic ponds are used for low-strength waste treatment such as polishing of secondary effluent. (See Mancini and Barnhart, 1968; Metcalf and Eddy, Inc., 1991; Middlebrooks et al., 1982; Thirumurthi, 1969.)	

TABLE 11.3 Description of Typical Large-Scale Aerobic Biological Treatment Reactors (Continued)

Reactor name	Description	Schematic
Attached-cell processes: Trickling filters	Towers packed with solid media to support attached cell growth; waste stream distributed over top of media and then trickles down over media where it contacts cells; passive aeration from top and bottom of tower where it is open to atmosphere; media can be rocks ("low-rate filter") or high-surface-area hollow plastic structures ("high-rate filter"). Can be used for secondary treatment. Low-rate filter can also provide denitrification. High-rate filter usually includes partial recycle of effluent. (See Bailey and Ollis, 1986; Aryan and Johnson, 1987; Harrison and Dagger, 1987; Metcalf and Eddy, Inc., 1991.)	

Rotating biological contactor   Solid disks supporting microbial attachment rotate on horizontal axis alternatively contacting liquid waste stream at bottom and the atmosphere at top. Multiple disks used per reactor, often multiple reactors run in series and parallel because size of each reactor is limited by the weight of disk with attached cells. Can be used for secondary treatment. (See Metcalf and Eddy, Inc., 1991; U.S. EPA, 1984a; U.S. EPA, 1984b.)

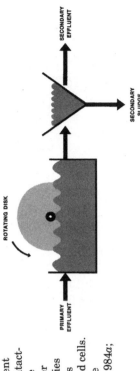

ROTATING DISK

PRIMARY EFFLUENT

SECONDARY EFFLUENT

SECONDARY SLUDGE

TABLE 11.4   Typical Capabilities of Aerobic Biological Treatment Processes (adapted from Metcalf and Eddy, Inc. 1991)

Aerobic process	Typical BOD removal efficiency (%)	Typical BOD removal rate	Liquid residence time
Activated-sludge processes			
Conventional (plug flow)	85–95	15–40*	4–8 h
Completely mixed (CSTR)	85–95	40–110*	3–5 h
Ponds/lagoons			
Aerobic-anaerobic	80–95	0.9–4†	5–30 days
Aerobic (low-rate)	80–95	1.1–2.5†	10–40 days
Trickling filters			
Rock media (low rate)	80–90	4–25†	
Plastic media (super-high-rate)	65–90	20–80†	
Rotating biological contactors	60–95	2–3.5†	0.7–1.5 h

*Measured in $lb/(10^3 ft^3 \cdot day)$.
†Measured in $lb/(10^3 ft^2 \cdot day)$.

ondary treatment is activated sludge (Fig. 11.3). The typical operating parameters and capabilities of the aerobic bioreactors are summarized in Table 11.4.

Nonaerobic bioreactors

Although aerobic reactors typically provide the most economical means of destroying nonhazardous BOD, nonaerobic bioreactors can be employed for special treatment needs in conjunction with conventional treatment. Nonaerobic reactors can play a lead role in the treatment of hazardous wastes, as discussed in a later section of this chapter.

Three nonaerobic microbial processes that are commonly utilized are (1) anaerobic, i.e., methanogenesis; (2) anoxic, i.e., nitrate reduction; and (3) sulfate reduction. Methanogenic processes are characterized by a low cell yield and a tolerance of high organic concentrations, features that should render them a logical choice for waste treatment. However, their low rate of utilization of organic matter, sensitivity to temperature (optimum temperature is 37°C), and high reactor cost limit their use to small-volume waste stream applications such as sludge stabilization (further reduction of the BOD in the waste sludge from secondary treatment), and industrial treatment of highly concentrated organic wastes.

The typical (unbalanced) methanogenic reaction is

$$s_1 C_a H_b O_c N_d S \xrightarrow{cells} s_2 CH_4 + s_3 CO_2 + s_4 H_2O + C_5 H_7 NO_2 + s_5 NH_3 + s_6 H_2S$$

Often methanogenic reactors are heated to increase the reaction rate. Frequently, the heating is accomplished by collecting and burning the methane gas produced.

Anoxic (nitrate-reducing) processes are employed as the final step in ammonia oxidation. High levels of ammonia must be removed from the waste stream because of its high nutritive value. Typically, the aerobic bioreactor portion of the treatment process (e.g., an activated-sludge process) oxidizes the ammonia present in the feed stream, and the ammonia generated in the reactor due to protein decomposition, to nitrate (nitrification). Nitrate retains nutritive value and is a human toxin. Thus, if high levels of nitrate are present in the secondary effluent, an anoxic reactor can be added in series to oxidize nitrate to nitrogen gas (denitrification) according to the unbalanced equation

$$s_1 C_a H_b O_c N_d S + s_2 NO_3^- + s_2 H^+ \overset{cells}{\rightarrow} s_3 N_2 + s_4 CO_2 + s_5 H_2 O + C_5 H_7 NO_2$$

Thus, biodegradable compounds must be present in the anoxic reactor feed, either as a residual from the aerobic bioreactor or added to the waste stream (usually as methanol).

For industrial waste streams that contain high levels of sulfate as well as organics, the microbial sulfate reduction process can be employed: in this case, sulfate is reduced to sulfide by oxidizing the organic feed. The unbalanced reaction is

$$s_1 C_a H_b O_c N_d S + s_2 SO_4^{-2} \overset{cells}{\rightarrow} s_2 S^{-2} + s_3 CO_2 + s_4 H_2 O + C_5 H_7 NO_2$$

Shortcomings of Conventional Biological Processes for Treatment of Hazardous Chemicals

The U.S. Environmental Protection Agency (U.S. EPA) regulates the release of about 120 compounds into the environment from ongoing industrial operations under the Resource Conservation and Recovery Act (RCRA), passed in 1976, and from nonoperational, abandoned, or closed sites under the Comprehensive Environmental Response, Compensation, and Liability Act (CERCLA or Superfund), passed in 1980. These listed compounds have been determined to be hazardous, i.e., they pose risks to the ecosystem or to human health. A representative sample of the listed organic compounds is shown in Table 11.5. Several metals are also listed as hazardous, including arsenic, chromium, mercury, selenium, vanadium, and zinc.

The success of conventional treatment in degrading these hazardous compounds depends on what the principal fate of the compound is in the treatment facility. There are several competing processes contributing to the ultimate fate of a compound:

TABLE 11.5 Some Priority Pollutants and Associated Biological and Physical Properties (from database associated with Govind et al., 1991a)

Compound	Biodegradation rate constant $(k_b \times 10^2$ L/mg $\bullet$ h)	Henry's law constant $(H \times 10^3$ atm $\bullet$ m^3/mole)	Octanol-water partition constant (K_{ow})
Acetone	0.020	2.5×10^{-2}	0.57
Anthracene	0.30	8.6×10^{-2}	2.8×10^4
Benzene	2.9	5.5	140
Chlorobenzene	0.30	3.6	6.9×10^2
Chloroform	$<10^{-3}$	3.4	91
Chlordane	3.0×10^{-4}	9.4×10^{-2}	3.0×10^5
2,4-dichlorophenol	10.	2.8×10^{-3}	790
Endrin	1.0×10^{-2}	4.0×10^{-4}	3.4×10^4
Ethylbenzene	3.0	6.4	1400
Fluorene	1.0	6.4×10^{-2}	1.5×10^4
Hexachlorobenzene	3.0×10^{-4}	0.68	2.6×10^6
Naphthalene	10.	0.46	2000
Nitrobenzene	1.5	1.3×10^{-2}	72
Pyrene	1.0×10^{-2}	5.1×10^{-3}	8.0×10^6
Tetrachloroethane	3.0×10^{-4}	11.0	1100
Toluene	10.	6.7	620
Tricholoromethylene	1.0×10^{-4}	9.1	260
Vinyl chloride	0.30	11.9	140

1. *Biodegradation.*   Compound is at least partially destroyed. Biodegradation may not occur because the compound is not degradable by aerobic organisms.

2. *Volatilization.*   Compounds with relatively high Henry's law constants may exit the facility at the primary settling tank, at the bioreactor, or at the secondary settling tank, by volatilization into the atmosphere at the quiescent air-water interface and by forced aeration or ventilation of the bioreactor.

3. *Sorption to solids and precipitation.*   Organic compounds that preferentially partition from water to solids such as microbial cells (relatively high octanol-water coefficient) in the bioreactor will eventually exit the facility through the sludge disposal activities. Heavy metals are often precipitated and removed in waste sludges from biological treatment processes.

4. *Pass-through.*   Compounds which do not completely biodegrade, volatilize, or partition to solids will exit the bioreactor in the liquid effluent stream, possibly to be discharged into the environment to surface or marine waters.

Fate mechanisms for conventional treatment are illustrated in Fig. 11.2.
   The fate of a particular compound is determined by the relative rates of the above processes. These rates are determined by the biological and

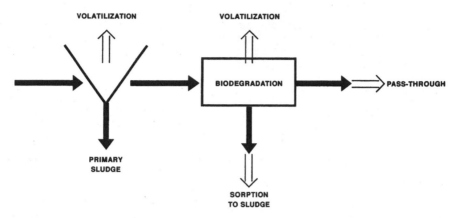

Figure 11.2 Fate of chemicals in conventional wastewater treatment.

physical properties of the compounds: k_b, the activated-sludge biodegradation rate constant; H, the Henry's law constant for water-air partitioning; and K_{ow}, the octanol-water partition constant which has been reported to be a good measure of the water-sludge partition coefficient (Dobbs et al., 1989). These properties are summarized in Table 11.5. The fate of a compound can be estimated by comparing the values of the parameters listed in Table 11.5. For example, by comparing the values of the parameters for 2,4-dicholorophenol, trichloroethylene (TCE), and the pesticide endrin, a reasonable estimate of the principal fates would be biodegradation, volatilization, and sorption to sludge, respectively.

The fate of a compound entering the plant also depends on the operating parameters of the facility, such as aeration rate, liquid residence time, and others. Thus, a more accurate estimation of the fate of hazardous chemicals in conventional treatment requires either a mathematical model of the entire treatment process incorporating the various fate pathways, or experimental measurements of fates during actual treatment or pilot-scale simulation of treatment facilities.

There are several mathematical models available of fate of toxic compounds in conventional treatment processes (e.g., Petrasek et al., 1983; Barton, 1987; Blackburn, 1987; Namkung and Rittman, 1987; Govind et al., 1991a). Models cannot simulate the performance of all treatment facilities equally well because simplifying assumptions built into the models are not always met at the actual facility. However, models allow independent variation of the system parameters and simulation of the resulting performance. For example, analysis of activated-sludge bioreactor optimization with respect to aeration can be accomplished by using a model to calculate biodegradation rate and volatilization rate of organics as a function of aeration rate.

TABLE 11.6   Fate of Selected Compounds Calculated for a Typical Conventional
Wastewater Treatment Facility* Using the FATE Model (Govind et al., 1991a)

Compound	Biodegradation	Volatilization	Sorption	Pass-through
2,4-dichlorophenol	91	4	5	10
Endrin	26	3	34	37
Tetrachloroethylene	2	91	0	7

*Plant size = 1 MGD, includes primary and activated-sludge secondary treatment, diffused, course-bubble aeration, recycle ratio = 0.5, wastage rate = 32 m³/day, flow rate = 3785 m³/day, volume = 646 m³.

When operating parameters were chosen to simulate a typical conventional wastewater treatment plant, one of the more sophisticated models (Govind et al., 1991a) yielded the fates of 2,4-dichlorophenol, TCE, and the pesticide endrin shown in Table 11.6, in good agreement with the estimate above using biological/physical parameters alone.

Hazardous metals were not included in Table 11.5 because little consistent information is available on water-sludge partitioning of heavy metals. As a general rule, however, most heavy metals entering a conventional treatment facility will partition to and be removed with the sludge.

The principal fate of many hazardous chemicals is biodegradation; i.e., the rate of biodegradation is much faster than volatilization or sorption onto solids in an optimized reactor. For example, referring to Table 11.5, the major fate of benzene, nitrobenzene, and toluene would likely be biodegradation. Table 11.6 shows that most 2,4-dichlorophenol is also biodegraded. The introduction of compounds of this type, although hazardous, by an industrial source into the domestic sewer system should pose no problem for treatment by the domestic wastewater treatment facility if concentrations are not inherently toxic to microorganisms. Alternatively, these compounds could be treated by conventional treatment at the industrial source.

However, many of the hazardous chemicals are apparently not readily biodegradable by conventional treatment processes—i.e., volatilization, sorption, and pass-through are the major fates. These fates are undesirable for hazardous compounds because they imply release of these compounds into the environment causing undesirable outcomes, as summarized in Table 11.7. Several experimental studies have been conducted at full scale and pilot scale (e.g., Bell et al., 1988; Bhattacharya et al., 1990; Burns and Roe Industrial Service Corporation, 1982; Hannah et al., 1986; Thompson et al., 1990). Bell et al. (1988) studied volatilization of volatile organic compounds in several Canadian full-scale municipal wastewater treatment facilities. Table 11.8 summarizes a portion of the data collected at the Lakeview treatment plant. The combined re-

TABLE 11.7 Environmental Impact of the Fates of Hazardous Compounds in Conventional Treatment Facilities

Fate of compound	Environmental impact
Biodegradation	None, the desired outcome.
Volatilization	Release of volatile organic compounds into atmosphere; human exposure to carcinogens; contributes to ozone layer degradation.
Sorption	Toxic compounds carried with sludge to landfills; may result in groundwater contamination which can threaten drinking water supplies.
Pass-through	Release of toxic compounds into surface waters (rivers, lakes, oceans); can contaminate drinking water sources, injure local ecosystem.

moval rate by biodegradation and sorption, shown in Table 11.8, was calculated using a mass balance and the measured rate of loss of the compound in the off-gas and in the liquid effluent. Clearly volatilization can be a major fate of hazardous chemicals entering conventional treatment facilities. For example, referring to Table 11.8, over 50 percent of the dichloromethane and over 25 percent of the toluene and TCE entering the plant was released to the atmosphere by volatilization.

Below, biological treatment amendments and alternatives to conventional treatment are discussed to address unacceptable fates of hazardous chemicals in conventional treatment facilities.

TABLE 11.8 Fate of Selected Volatile Organic Compounds (Combined Biodegradation and Sorption to Sludge, Volatilization, and Pass-through) at the Lakeview Treatment Facility* in Ontario, Canada (adapted from Bell et al., 1988)

Compound	Inlet rate (g/day)	Removal rate (g/day) by: Biodegradation and sorption	Volatilizaton	Pass-through	Percent removal by: Biodegradation and sorption	Volatilization	Pass-through
Benzene	31	27	4	ND	87	13	0
Chloroform	435	271	85	79	62	20	18
Dichloromethane	4198	514	2316	1368	12	55	33
Ethylbenzene	215	120	92	3.5	56	43	1
Toluene	1469	1091	361	17	74	25	1
Trichloroethylene	72	36	21	15	50	29	21

*Grit chamber: volume = 683 m^3, air flow rate = 10.4 m^3/min. Activated-sludge aeration basin: volume = 16,070 m^3, air flow rate = 2064 m^3/min, coarse diffuser. Primary and secondary treatment residence time = 6.0 h, overall air rate = 46 m^3 air/m^3 water.

Bioprocess Remedies for Conventional Biological Treatment Shortcomings

The above discussion emphasizes that the limitations of conventional treatment of many hazardous compounds are strongly dependent on the character of the hazardous waste in the waste stream being considered. Therefore, the remedy for these shortcomings is dependent on the compounds in the waste stream.

Existing conventional wastewater treatment facilities—whether operated by municipalities or at industrial plants—that cannot satisfactorily treat the incoming hazardous waste must be modified to treat these compounds to avoid illegal or unethical release into the environment. Process modifications would also be necessary in the design of new facilities to treat such wastes; however, when designing a new facility, the opportunity exists to optimize the design and the integration of the conventional treatment capability with the processes needed for hazardous waste treatment.

Many biologically based approaches have been developed to assist or replace conventional treatment for the removal of hazardous waste. These processes employ mixed microbial consortia taken from conventional treatment facilities and either inoculated directly into the bioreactor or allowed to acclimate to the toxic compounds of concern prior to inoculation. In the future, bioprocesses may be available that utilize pure cultures of microorganisms, either natural or genetically altered. These possibilities will not be discussed here. Table 11.9 lists the classes of biological/physical characteristics of hazardous waste and a

TABLE 11.9 Bioprocess Alternative/Additions to Conventional Treatment for Hazardous Compounds of Various Biological/Physical Characteristics

Characteristic of hazardous compound(s)	Alternatives/additions to conventional treatment
Easily aerobically biodegradable (high k_b)	None necessary
Volatile (high H)	Strip influent with air, treat air with gas-phase biofilter Treat activated-sludge off-gas using gas-phase biofilter PACT* process Anaerobic pretreatment
Partition to sludge (high K_{ow})	Anaerobic sludge stabilization Pretreatment or sole treatment with aerobic or anaerobic attached-film reactor PACT process
Pass-through (low k_b, H, and K_{ow})	Anaerobic pretreatment or sole treatment PACT process

*Powdered Activated Carbon Treatment (Eckenfelder, 1989).

selection of the biotechnologies that have been developed to treat these compounds. Below, each of these technologies is examined.

Anaerobic digestion of sludge

In many municipal and industrial wastewater treatment plants, the sludge effluent from primary and secondary treatment is fed to an anaerobic bioreactor (often termed anaerobic *digestor* or *stabilizer*) to reduce the residual BOD of the sludge (Metcalf and Eddy, 1991). The anaerobic conditions promote methanogenic microbial degradation of the BOD, thus rendering the sludge fit for landfill disposal. If toxic organic compounds are sorbed to the sludge, the methanogenic conditions of the digestor can stimulate degradation of many of these toxic organics (Govind et al., 1991*b*). Thus, if the primary fate of toxic organics entering conventional treatment is sorption to sludge, the ultimate fate of the compounds may be biodegradation in the anaerobic digestor. Typical operating parameters for the anaerobic digestor are 10- to 20-day liquid and solids residence time and a temperature of 35°C. The methane produced from the methanogenic microbial activity is often burned to help heat the bioreactor.

PACT process

The patented *powdered activated-carbon treatment process* (PACT process) involves continuous addition of powdered activated carbon to the activated-sludge bioreactor to adsorb toxic organics, thus avoiding undesirable fates for these compounds (Eckenfelder, 1989). The activated carbon adsorbs organics of all types, i.e., volatile, recalcitrant, or organics that sorb to sludge. Thus, the PACT process allows existing wastewater treatment facilities to operate at a higher influent flow rate and helps stabilize the process to shock loads of BOD or toxic organics in the influent. After the bioreactor, the powdered activated carbon with adsorbed organics settles with the sludge in the clarifier. Municipal treatment facilities that employ the PACT process typically treat the sludge-carbon mixture by incineration, thus destroying the sorbed organics. An alternative disposal technique is wet-air oxidation, which results in either the solubilization or destruction of the biomass and the partial regeneration of the carbon. The sludge-carbon mixture can also be treated by anaerobic digestion, in which many of the organics sorbed on the carbon are destroyed by methanogenic microbial activity. In this case, the carbon would be disposed of with the digested sludge. Typical carbon concentrations utilized by the PACT process are 20 to 200 mg/L (Metcalf and Eddy, 1991).

GAC sorption/anaerobic stabilization

A variation on the PACT process which offers more operational flexibility is the *granular activated-carbon* (GAC) *sorption/anaerobic stabilization* treatment process train. This process is especially applicable to treatment of hazardous wastewater streams such as industrial wastewater or landfill leachates prior to conventional treatment. The process involves two unit processes: (1) a GAC–waste stream contactor where the pre-secondary-treatment influent passes through a bed of GAC to remove soluble organics by sorption, and (2) an anaerobic biological reactor, where the GAC from the contactor is placed to allow anaerobic biological regeneration of the GAC. Regenerated carbon replaces spent carbon in the contactor, and, ideally, handling losses of GAC are limited. The process has been demonstrated at laboratory scale (Kupferle et al., 1992) for nearly a year of continuous operation. Liquid and GAC residence times in the sorption stage were 30 min and 2 days, respectively, and the GAC residence time in the anaerobic bioreactor was 15 days. A complex hazardous feed stream containing nine volatile organics and five semivolatile organics was treated successfully. Removals of chlorobenzene, methylene chloride, TCE, dibutyl phthalate, and phenol were 96, 50, 95, 66, and 85 percent, respectively.

Anaerobic immobilized-cell bioreactors

Anaerobic microbial processes are known to have several important advantages over aerobic microbial processes: (1) lower production rate of sludge, (2) operable at higher influent BOD and toxics levels, (3) no cost associated with delivering oxygen to the reactor, and (4) production of a useful by-product, methane. However, anaerobic processes have higher capital and operating expenses than aerobic processes because the anaerobic systems must be closed and heated. Thus, anaerobic bioprocesses for treatment of hazardous wastewater streams are typically limited to treatment of low-flow-rate streams such as industrial effluent, landfill leachate, and Superfund-type wastewater treatment applications. Boyle and Switzenbaum (1990) have discussed the applicability of anaerobic processes to large-volume treatment.

The intrinsic rate of biological destruction of hazardous organics in anaerobic systems is slower than aerobic processes; however, the overall rate of destruction by an anaerobic bioreactor can be maximized by immobilizing the biomass on support media. Highly porous plastic media are typically used in a packed bed, or GAC in a fluidized column.

The *expanded bed GAC anaerobic reactor* (Fig. 11.3) has been demonstrated to be ideally suited for the treatment of wastes containing mixtures of both readily biodegradable and biologically refractory organic compounds. The GAC medium serves to sequester inhibitory con-

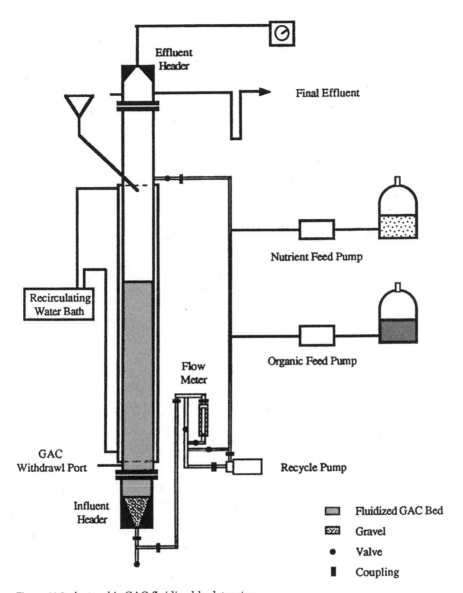

Figure 11.3 Anaerobic GAC fluidized-bed reactor.

stituents from the aqueous phase, thus permitting utilization of readily biodegradable constituents. In instances where the adsorbed inhibitory constituents resist biodegradation, a strategy involving periodic replacement of a portion of the GAC medium may be practiced to replenish GAC adsorptive capacity and render the waste biodegradable. The adsorptive characteristics of GAC permit the retention of slowly biodegradable organics that require acclimation of specially adapted organisms. In these cases, the adsorptive capacity of GAC can be extended indefinitely due to biologically mediated regeneration. GAC has been demonstrated to have superior microbial attachment properties to other media in fluidized beds (Khan et al., 1982; Gardner et al., 1988; Fox et al., 1990). Furthermore, the GAC medium has been shown to provide resistance to shock load and to facilitate startup (Suidan et al., 1981; Khan et al., 1982).

The expanded-bed GAC anaerobic reactor has been demonstrated to be very effective in treating several biologically inhibitory wastes such as paint-stripping baths (Khan et al., 1982), sour water stripper bottoms (Gardner et al., 1988), coal gasification wastewater (Nakhla et al., 1990), and hazardous landfill leachates (Schroeder et al., 1991). The data in Table 11.10 summarize removal efficiencies observed for some compounds of concern in the process.

TABLE 11.10   Performance of the Anaerobic GAC Fluidized Bed (Schroeder et al., 1991)

Compound	Influent concentration (mg/L)	Loading rate (g/kg GAC/day)	Percent removal
Catechol	200–1,000	1.29–6.45	>99
Phenol	3–2,959	0.03–8.0	>97
p-cresol	70–280	0.19–0.76	>99
Indole	50–300	0.14–0.80	>99
Quinoline	50–300	0.14–0.80	>99
o-chlorophenol	2,000	4.0	>98
2,4-dichlorophenol	400	0.8	>99.99
2,4,6-trichlorophenol	150	0.3	>99.99
Di-n-butyl phthalate	0.215–100	0.002–0.8	>83
Naphthalene	30	0.24	>99.9
p-Nitrophenol	100	0.8	>99.9
Nitrobenzene	0.5–100	0.005–0.8	>98
Lindane	10	0.08	>99.9
Acetone	10.0–755	0.10–4.0	>96
Methyl ethyl ketone	12	0.10	>97
Methyl isobutyl ketone	10	0.01	>94
Tetrachloroethylene	20	0.16	>99.99
Trichloroethylene	0.4	0.004	>98
Methylene chloride	1.2–20	0.01–0.16	>96
1,1,1-trichloroethane	20–400	0.04–0.8	>99.97
1,1-dichloroethane	0.1	0.001	>87
Carbon tetrachloride	20	0.16	>99.9
Chloroform	10–20	0.08–0.16	>97
Chlorobenzene	1.1–20	0.01–0.16	>85
1,2,4-Trichlorobenzene	0.2	0.002	>93
Ethylbenzene	0.6	0.006	>86
Toluene	8.2–20	0.08–0.16	>87

Gas phase biofiltration

Gaseous waste streams containing hazardous organic compounds are generated in many processes. In wastewater treatment, volatilization by aeration in secondary treatment is the fate of many volatile organic compounds (VOCs) entering conventional treatment plants. Pump-and-treat remedies for contaminated groundwater can employ air stripping to clean the groundwater of VOCs, thus generating a hazardous air stream. Industrial processes often employ VOCs as reagents or solvents, and their use can produce a contaminated gas stream. In many countries, and in the United States with passage of the 1990 amendments to the Clean Air Act, release of toxic organics are strictly controlled. Biological treatment of VOCs in gas streams can remove and destroy these compounds, thus avoiding release into the environment. The most common alternative, treatment by adsorption to activated carbon, is costly and does not in itself provide destruction of the pollutants.

Historically, gas phase biotreatment has been employed for odor control at conventional treatment facilities where H_2S and other odorous gases produced at the facility are collected and passed through a bed of soil or compost containing microorganisms (Carlson and Leiser, 1966; Metcalf and Eddy, Inc., 1991). Gas residence times in the odor control systems can be as low as 15 to 30 s (ASCE, 1989).

The application of biofiltration to VOC destruction is currently a very active area of development and demonstration in the United States (Leson and Winer, 1991). In Germany and The Netherlands, however, as many as 500 full-scale biofilter operations are on line, treating waste gas streams from such diverse sources as chemical manufacturing, print shops, fish frying, industrial wastewater treatment plants, and landfill gas extraction (Leson and Winer, 1991).

The general design of a biofilter is a packed-bed reactor. However, there is little or no liquid phase (liquid holdup) in the biofilter; most of the volume is the flow-through gas stream or the solid support material with attached microorganisms. The liquid stream provides essential nutrients and moisture to the microbes, although if peat or compost is used for the solid support, nutrients can be leached directly from the support. In Europe, biofilters have followed traditional odor control designs as models, using parallel horizontal beds of soil or compost of about 1 m in height (Leson and Winer, 1991). Mineralization of the media and excessive cell growth often require periodic remixing and eventual replacement of the media. Bohn (Bohn, 1992) reported 90 percent degradation of alcohols and aldehydes in 30 seconds to several minutes of residence time.

Figure 11.4 is a photograph with cutout of a full-scale biofilter in the traditional European design. The biofilter shown is the Bioton system

Figure 11.4 The Bioton Biofilter. (*Courtesy of Ambient Engineering, Inc., Matawan, N.J.*)

built by ClairTech b.v. of The Netherlands and is sold by Ambient Engineering, Inc., in the United States. ClairTech has units in operation with flow rates as high as 75,000 m³/h handling wastewater treatment off-gases for odor and VOC (including methylene chloride and chloroform) control.

Current American biofilter designs have followed the European designs closely; however, other designs utilizing lower-maintenance materials such as GAC in packed beds (Utkigar et al., 1991), or novel support media geometries such as straight passages of extruded celite (Govind et al., 1992) are being investigated.

Conclusion

Although biological treatment of waste organic matter has been studied for about a century, new wastewater treatment challenges, in the form of the biological treatment of hazardous chemicals and of more stringent effluent regulations, continue to motivate the development of new and innovative treatment processes. Many of these new processes have been described in this chapter. As our understanding of the fundamental processes in these systems—microbial metabolism of mixtures of hazardous chemicals, microbial ecology of mixed microbial consortia, transport processes in biofilms, attachment properties of biomass to solid surfaces, fate of toxic chemicals in the reactors—increases, our ability to design cost-effective, high-performance bioprocesses will also increase.

Disclaimer

This chapter has been reviewed in accordance with the U.S. Environmental Protection Agency's peer and administrative review policies and approved for publication. Mention of commercial products does not constitute endorsement or recommendation for use by the U.S. Environmental Protection Agency.

References

American Society of Civil Engineers (ASCE). 1989. *Sulfide in Wastewater Collection and Treatment Systems.* ASCE Manuals and Reports on Engineering Practice No. 69, ASCE,

Aryan, A. F., and S. H. Johnson. 1987. "Discussion Of: A Comparison of Trickling Filter Media." *J. Water Pollut. Contr. Fed.,* vol. 59, no. 10, pp. 180–183.

Bailey, J. E., and D. F. Ollis. 1986. *Biochemcial Engineering Fundamentals,* 2nd edition. Chapter 14. McGraw-Hill, New York.

Barton, D. 1987. "Intermedia Transport of Organic Compounds in Biological Wastewater Treatment Process." *Environ. Progr.,* vol. 6, pp. 246–256.

Bell, J. P., I. Osinga, and J. Melcer. 1988. *Investigation of Stripping of Volatile Organic Contaminants in Municipal Wastewater Treatment Systems, Phase I.* Ontario Ministry of the Environment, Ottawa.

Bhattacharya, S. K., R .V. R. Angara, D. F. Bishop, Jr., R. A. Dobbs, and B. M. Austern. 1990. *Removal and Fate of RCRA and CERCLA Toxic Organic Pollutants in Wastewater Treatment.* U.S. EPA Project Summary, EPA-600/S2-89/026, Washington, D.C.

Blackburn, J. W. 1987. "Prediction of Organic Chemical Fates in Biological Treatment Systems." *Environ. Progr.,* vol. 6, pp. 217–223.

Bohn, H. 1992. "Consider Biofiltration for Decontaminating Gases." *Chem. Eng. Progr.,* April, pp. 34–40.

Boyle, W. C., and M. S. Switzenbaum. 1990. "Anaerobic Treatment of Municipal Wastewaters: Status of the Technology—1990." Report for U.S. EPA Risk Reduction Engineering Laboratory, Cincinnati, under Contract No. 68-03-3429.

Burns and Roe Industrial Services Corporation. 1982. "Fate of Priority Pollutants in Publicly-Owned Treatment Plants." *Final Report,* U.S. EPA Report 1440/1-82/303, Washington, D.C.

Carlson, D. A., and C. P. Leiser. 1966. "Soil Beds for the Control of Sewage Odors." *J. Water Pollut. Contr. Fed.,* vol. 38, p. 829.

Dobbs, R. A., L. Wang, and R. Govind. 1989. "Sorption of Toxic Organic Compounds on Wastewater Solids: Correlation with Fundamental Properties." *Environ. Sci. Technol.,* vol. 23, pp. 1092–1097.

Eckenfelder, W. W., Jr., 1989. *Industrial Water Pollution Control,* 2nd edition. McGraw-Hill, New York.

Federal Register (Fed. Reg.). 1988. "Secondary Treatment Regulation." 40 CFR Part 133, July 1.

Federal Register (Fed. Reg.). 1989. "Amendment to the Secondary Treatment Regulations: Percent Removal Requirements during Dry Weather Periods for Treatment Works Served by Combined Sewers." 40 CFR Part 133, January 27.

Fox, P., M. T. Suidan, and J. T. Bandy. 1990. "A Comparison of Media Types in Acetate Fed Expanded-Bed Anaerobic Reactors." *Water Res.,* vol. 7, no. 7, pp. 827–835.

Gardner, D. A., M. T. Suidan, and H. A. Kobayashi. 1988. "Role of GAC Activity and Particle Size during the Fluidized-Bed Anaerobic Treatment of Refinery Sour Water Stripper Bottoms." *J. Water Pollut. Contr. Fed.,* vol. 60, no. 4, pp. 505–513.

Govind, R., L. Lai, and R. Dobbs. 1991a. "Integrated Model for Predicting the Fate of Organics in Wastewater Treatment Plants," *Environ. Progr.,* vol. 10, no. 1, pp. 13–23.

Govind, R., P. A. Flaherty, and R. A. Dobbs. 1991b. "Fate and Effects of Semivolatile Organic Pollutants during Anaerobic Digestion of Sludge." *Water Res.,* vol. 25, no. 1, pp. 547–556.

Govind, R., V. Utkigar, Y. Shan, W. Zhao, G. D. Sayles, D. F. Bishop, and S. I. Safferman. 1992. "Development of a Novel Biofilter for Aerobic Biodegradation of Volatile Organic Compounds." *18th Annual Risk Reduct. Engin. Lab. Res. Symp.: Abstract Proc.,* EPA/600/R-62/028.

Hannah, S. A., B. M. Austern, A. E. Eralp, and R. H. Wise. 1986. "Comparative Removal of Toxic Pollutants by Six Wastewater Treatment Processes." *J. Water Pollut. Contr. Fed.,* vol. 58, no. 1, pp. 27–34.

Harrison, J. R., and G. T. Dagger. 1987. "A Comparison of Trickling Filter Media." *J. Water Pollut. Contr. Fed.,* vol. 59, no. 7, pp. 276–283.

Khan, K. A., M. T. Suidan, and W. H. Cross. 1982. "Role of Surface Active Media in Anaerobic Filters," *J. Env. Eng., ASCE,* vol. 108, no. EE2, pp. 269–285.

Kupferle, M. J., T. Chen, V. J. Gallardo, D. E. Lindberg, P. L. Bishop, D. F. Bishop, and S. I. Safferman. 1992. "Treatment of Dilute Hazardous Waste Streams by Sorption/Anaerobic Stabilization." *18th Annual Risk Reduct. Engin. Lab. Res. Symp.: Abstract Proc.,* EPA/600/R-62/028.

Leson, G., and A. M. Winer. 1991. "Biofiltration: An Innovative Air Pollution Control Technology for VOC Emission." *J. Air Waste Manage. Assoc.,* vol. 41, no. 8, pp. 1045–1054.

Mancini, J. L., and E. L. Barnhart. 1968. "Industrial Treatment in Aerated Lagoons." In *Advances in Water Quality Improvement.* Edited by E. F. Gloyna and W. W. Eckenfelder, Jr., University of Texas Press, Austin.

Metcalf and Eddy Inc. 1991. *Wastewater Engineering: Treatment, Disposal and Reuse,* 3rd edition. Edited by G. Tchobanoglous and F. L. Burton. McGraw-Hill, New York.

Middlebrooks, E. J., C. H. Middlebrooks, J. H. Reynolds, G. Z. Waters, C. S. Reed, and D. B. George. 1982. *Wastewater Stabilization Lagoon Design, Performance and Upgrading.* Macmillan, New York.

Nakhla, G. F., M. T. Suidan, and J. T. Pfeffer. 1990. "Control of Anaerobic GAC Reactors Treating Inhibitory Wastewaters." *J. Water Pollut. Contr. Fed.,* vol. 62, no. 1, pp. 65–72.

Namkung, E., and B. E. Rittmann. 1987. "Estimating Volatile Organic Compound Emissions from Publicly-Owned Treatment Works." *J. Water Pollut. Contr. Fed.,* vol. 59, pp. 670–678.

Petrasek, A. C., I. J. Kugelman, B. M. Austern, T. A. Pressley, L. A. Winslow, and R. H. Wise. 1983. "Fate of Toxic Organic Compounds in Wastewater Treatment Plants." *J. Water Pollut. Contr. Fed.,* vol. 55, no. 10, pp. 1286–1296.

Schroeder, A. T., M. T. Suidan, R. Nath, E. R. Krishnan, and R. C. Brenner. 1991. "Carbon-Assisted Anaerobic Treatment of Hazardous Leachates." *Proc. of the 17th Annual RREL Haz. Waste Res. Symp.,* pp. 626–648.

Stall, T. R., and J. H. Sherrard. 1978. "Evaluation of Control Parameters for the Activated Sludge Process," *J. Water Pollut. Contr. Fed.,* vol. 56, no. 4, pp. 336–345.

Suidan, M. T., W. H. Cross, M. Fong, and J. W. Calvert, Jr. 1981. "Anaerobic Carbon Filter for Degradation of Phenols." *J. Env. Eng., ASCE,* vol. 107, no. EE3, pp. 563–579.

Thayer, A. M. 1992. "Pollution Reduction." *Chem. Eng. News,* vol. 70, no. 46, pp. 22–52.

Thirumurthi, D. 1969. "Design of Waste Stabilization Ponds," *J. San. Eng. Div., ASCE,* vol. 95, no. SA2, pp. 516–523.

Thompson, D., J. Bell, H. Melcer, and J. Kemp. 1990. *Investigation of Stripping of Volatile Organic Contaminants in Municipal Wastewater Treatment Systems, Phase II.* Ontario Ministry of the Environment, Ottawa.

U.S. Environmental Protection Agency (U.S. EPA). 1984a. *Review of Current RBC Performance and Design Procedures.* EPA-600/2-85-033, Washington, D.C..

U.S. Environmental Protection Agency (U.S. EPA). 1984b. *Summary of Design Information on Rotating Biological Contactors.* EPA-430/9-64-008, Washington, D.C.

U.S. Environmental Protection Agency (U.S. EPA). 1989. *Assessment of Needed Publicly Owned Treatment Facilities in the United States, 1988 Needs Survey Report to Congress.* EPA-430/09-89-0-001, Washington, D.C.

Utkigar, V., Y. Shan, and R. Govind. 1991. "Biodegradation of Volatile Organic Compounds in Aerobic and Anaerobic Biofilters." *Proc. 17th Annual RREL Hazard. Waste Res. Symp.,* EPA/600/9-91/002.

Water Pollution Control Federation (WPCF). 1987. *Activated Sludge.* Manual of Practice OM-9.

12

Effectiveness and Regulatory Issues in Oil Spill Bioremediation: Experiences with the Exxon Valdez Oil Spill in Alaska

P. H. Pritchard

U.S. Environmental Protection Agency
Environmental Research Laboratory
Sabine Island, Gulf Breeze, Florida

Abstract

The use of bioremediation as a supplemental cleanup technology in the *Exxon Valdez* oil spill, in Prince William Sound, Alaska, has proven to be a good example of the problems and successes associated with the practical application of this technology. Field studies conducted by scientists from the U.S. Environmental Protection Agency have demonstrated that oil degradation by indigenous microflora on the beaches of Prince William Sound could be significantly accelerated by adding fertilizer directly to the surfaces of oil-contaminated beaches. Our results from the application of an oleophilic fertilizer are presented as exemplary field and laboratory information. The fertilizer enhanced biodegradation of the oil, as measured by changes in oil composition and bulk oil weight per unit of beach material, by approximately twofold relative to untreated controls.

These studies supported bioremediation as a useful cleanup alternative that was subsequently used by Exxon on a large scale. They have also generated a number of insightful lessons that have significant relevance to future oil bioremediation efforts. This chapter discusses these lessons and examines complications and difficulties in assessing the effectiveness of bioremediation in the field.

Further field studies at a site involving an oil-contaminated beach that was less energetic and higher in nonpetroleum organic matter and using slow-release fertilizer granules applied at different concentrations, contrastingly showed little effect of fertilizer application. Precautions regarding extrapolation either from the laboratory to the field or from field site to field site, are discussed.

As with many types of bioremediation, protocols are needed to generate consistent and relevant data sets for commercial processes that will allow appropriate decisions to be made relative to the use of the process or product in a field cleanup operation. The conceptual basis for these protocols is a complicated matter and its development is significantly influenced by field experiences such as the *Exxon Valdez* oil spill. Discussion of these concepts provides an informative picture of the problems and assumptions faced in making decisions about when and how to apply a bioremediation technology.

The use of bioremediation for the cleanup of soils, sediments, and aquifer materials contaminated with oil and petroleum hydrocarbons has been extensively recognized (Lee and Levy, Bartha and Pramer). Success has been possible because of the relative biodegradability of oil and the knowledge that hydrocarbon degraders can be enriched in many, if not most, types of environments (Levy, Atlas). In addition, bioremediation is gaining acceptance as a viable technology; if used prudently, it can provide efficient, inexpensive, and environmentally safe cleanup of waste chemicals. Thus, the suggestion to use bioremediation as a supplemental cleanup tool in the *Exxon Valdez* oil spill in Prince William Sound, Alaska, was readily accepted as use of a new technology ready for field demonstration (Pritchard and Costa). The implementation of field studies to establish that oil degradation by indigenous microflora on the beaches of Prince William Sound could be significantly accelerated by fertilizer application (Pritchard et al., 1991), and the eventual large-scale application of fertilizer by Exxon as part of their overall cleanup program, provided a number of useful lessons and experiences that, if considered in the proper light, could have considerable influence on future oil bioremediation efforts.

The emphasis of this chapter will be on some of the difficulties and problems associated with the fertilizer application and its effect on oil degradation. I will concentrate primarily on the separate application of an oleophilic fertilizer which occurred at a site called Snug Harbor on Knight Island in Prince William Sound, and on the application of slow-release fertilizer granules which occurred on Disk Island in Prince William Sound. These applications provide contrasting results that are instructive for both their success and failure. (Note that additional fertilizer applications were conducted at these and other sites and summaries of these results are available; Pritchard et al., 1991).

Closely linked to these field applications of fertilizers is the question of which commercial products should be used and how the best ones should be screened out. This includes not only fertilizers but also microorganisms and other oil-biodegradation-stimulating agents and concepts. In Alaska, many of these commercial products could not be considered because of the very short time available for field demonstrations, but also because the data available for each product were so variable and/or insufficient that reasonable selections could not be made in a timely fashion. Subsequent to the *Exxon Valdez* oil spill, however, the Office of Research and Development of the U.S. Environmental Protection Agency (EPA) embarked on the development of effectiveness and environmental safety testing protocols that could be used to establish a consistent and relevant database upon which decisions for the use of particular commercial products might be based in the future. Some of these protocols have been developed and are currently being validated with laboratory studies. The development of a conceptual basis for these protocols has proven to be a useful exercise that depends heavily on the lessons learned in Alaska and other oil spills. Describing, in part, that conceptual development here provides an additional dimension to the regulatory complications that come into play in the application of bioremediation to oil spill cleanup. It is my hope that this information will stimulate others to carefully consider the process by which bioremediation success is measured in the field. It is this success issue, along with environmental safety aspects, that will be the key to good regulatory decision making.

Background

In any bioremediation effort, success will invariably involve a scientifically valid demonstration of process effectiveness and environmental safety. Effectiveness, in the case of oil bioremediation, means establishing that (1) removal or disappearance of the oil is attributable to biodegradation and not other nonbiological processes and (2) enhanced biodegradation rates of oil are sufficiently better than natural rates to justify expenditure of effort to implement the bioremediation process on a large scale. Although environmental safety issues will not be addressed here, they too require considerable effort to verify the absence of any adverse ecological effects associated with the fertilizer application.

In the *Exxon Valdez* oil spill, both aspects were crucial to the eventual acceptance of bioremediation by the public and state and federal regulatory agencies. Quelling skepticism, given the "subtleties" of a biotechnological approach, indeed was and will continue to be a challenge in almost any bioremediation situation, whether it deals with oil or other types of chemicals. Effectiveness and environmental safety issues for bioremediation in Prince William Sound, where treatment was centered on oil-contaminated gravel and cobblestone beaches, will, of course, be considerably different than for oil-contaminated sandy beaches, marshes, and wetlands. However, the lesson learned in terms of demonstrating a viable cleanup technology will have no bounds.

Reflections on the initial assumptions by EPA scientists and their colleagues as planning of the project commenced give rise to important and useful insights. Several assumptions, discussed in the context of what actually occurred in the field, provide instructive lessons that could impact the responses to bioremediation at other spill sites. While confidence provided by almost twenty years of accumulated research data on oil biodegradation laid the groundwork for these assumptions, we would have been naive not to expect some surprises!

Enrichments of oil-degrading microbial communities

Clearly, it was reasonable to expect, even in the cold water temperatures of Alaska, a significant enrichment of oil-degrading microorganisms in the beach material following exposure to the oil. Research by Atlas and his colleagues supported this idea (Atlas, 1981). As it turned out, by early June 1989 (approximately 2 months after the spill), concentrations of oil degraders averaged around 10^6 per gram of oiled beach material, which represented as much as a 10,000-fold increase in the number of degraders relative to beaches that had not been contaminated with the oil. Studies by Lindstrom et al. (1991) have shown similar trends, and an example of their results is shown in Table 12.1. Enrichments of this magnitude suggested that oil was being degraded (previous studies have demonstrated that Prudhoe Bay crude oil is quite biodegradable), that some nitrogen and phosphorus were available to support growth of the hydrocarbon degraders, and that the cold temperatures (10–16°C) were probably not overly restrictive to the indigenous microflora. The information not only implied the possibility of nitrogen-limited biodegradation (i.e., a great excess of degradable organic carbon from the oil in the face of a finite supply of nitrogen in the water), but also opened the possibility of accelerating oil biodegradation by overcoming this limitation alone.

In this case, we believed it was unnecessary to experimentally reverify, through laboratory studies, the stimulatory effect of nitrogen on oil biodegradation, even in oil-contaminated beach samples from Prince William Sound, since previous experiences and numerous published reports in the literature supported this as a sensible approach. Instead,

TABLE 12.1   Median (n = 5 to 9 Samples) Hydrocarbon-Degrader MPN Microbial Counts per Gram (Dry Weight) of Sediment for Treated (T) and Reference (R) Plots*

Beach site	Day sampled	Median MPN cells (10^4) per g of surface sediment		Diff-erent?†	Median MPN cells (10^4) per g of subsurface sediment		Diff-erent?
		T	R		T	R	
KN-135B	0	2.62	4.24	No	1.66	1.63	No
	2	4.79	1.58	No	1.02	0.47	No
	4	15.50	4.20	No	10.30	1.00	Yes
	8	1.56	15.60	No	10.10	2.27	No
	15	15.60	9.75	No	16.20	2.34	No
	52	13.70	23.40	No	75.40	36.00	No
	56	139.00	17.90	Yes	582.00	9.78	Yes
	70	149.00	25.20	Yes	126.00	13.00	Yes
	78	185.00	122.00	No	170.00	117.00	No
KN-211B	0	0.96	4.63	No	4.60	1.63	No
	2	77.00	127.00	No	193.00	2.23	Yes
	4	9.55	48.00	No	81.93	80.35	No
	16	45.40	44.44	No	46.22	97.49	No
	31	23.94	98.78	No	99.80	24.95	No
	45	30.83	53.23	No	33.18	25.32	No
	102	18.10	8.51	No	3.72	0.96	No
	112	3.19	11.70	No	8.51	1.28	Yes
KN-132B	0	24.90	23.00	No			
	2	155.00	21.70	No			
	4	77.70	16.00	No			
	8	160.00	37.10	No			
	16	97.30	15.70	Yes			
	29	28.00	16.00	No			
	43	135.00	0.59	Yes			
	60	84.10	1.78	Yes			
	95	117.00	53.20	Yes			

*All values obtained for each day were subjected to a Mann-Whitney two-sample U test to determine whether the sampled populations were different at the 95% confidence level. KN-135B was initially treated after day 0 and was refertilized after day 52. KN211B was initially treated after day 0 and refertilized on day 42. KN132B was fertilized after day 0 and again on day 40.

†Statistically significant differences are reported for surface and subsurface sediments.

we reasoned that, in light of the magnitude of the problem at the time, it was better to go directly to the field and conduct a practical demonstration of the same principle. Thus, in the very early stages of the oil spill cleanup program in Alaska, the relatively simple measurement of the number of oil degraders, along with the relevant literature information, had provided a reasonable "first-line" indicator of oil bioremediation feasibility.

Mineralization test as an indicator of oil degradation activity

Depending on the environmental situation involved in the oil spill, other first-line indicators of bioremediation potential might be desirable. There is always a tendency, where the local public is involved, to generate "site-specific" information, at the very least to generate a better comfort factor. A discussion of these indicators here, therefore, is worthwhile, even though they were not used initially in Alaska. Circumstances are likely to occur in other spills in which information on the enrichment of oil degraders in the contaminated areas alone is not sufficient for initial decision-making purposes.

Mineralization studies involving measurements of total CO_2 production can provide excellent first-line information. The approach provides, quick, relatively unequivocal time course data suitable for testing different treatment options (e.g., effects of adding nitrogen). If natural oil degradation is occurring in contaminated beach material, then considerable amounts of CO_2 should be produced from oil mineralization relative to a control containing uncontaminated beach material. Several laboratory systems can be used for this type of measurement. Biometer flask systems (Bartha and Pramer, 1965), which are designed to trap CO_2 in side-arms containing an alkaline solution, can be adapted to measure these mineralization rates of oil on contaminated beach material (Mueller et al., 1992). Commercially available respirometric systems, such as the Micro-Oxymax™ (Columbus Instruments, Columbus, Ohio) can also be used. The Micro-Oxymax™ system, which can be adapted to standard laboratory shake flasks, also measures oxygen consumption. The procedure entails placing oil-contaminated beach material and its associated microbial community (mixed sand and gravel in the case of Prince William Sound) directly in the flasks and flushing fresh seawater in and out as a simulation of the tidal exchange (Mueller et al., 1992). An example of the data generated from such a mineralization test is shown in Fig. 12.1. This experiment was performed with oil-contaminated Prince William Sound beach material, taken considerably after our bioremediation field demonstration had begun. The respirometric flask method is indicative of the short-term tests that can be initially conducted as a data-gathering exercise for a particular oil spill. Note the enhancing effect of adding nitrogen fertilizer (Fig. 12.1). By comparing rates of CO_2 production, an estimate of the extent of enhancement of oil biodegradation can be obtained. The system can be easily adapted to test beach material from the other oil spill sites.

Since any other type of organic matter in the beach material can also produce CO_2, care must be taken to assure that one is measuring oil

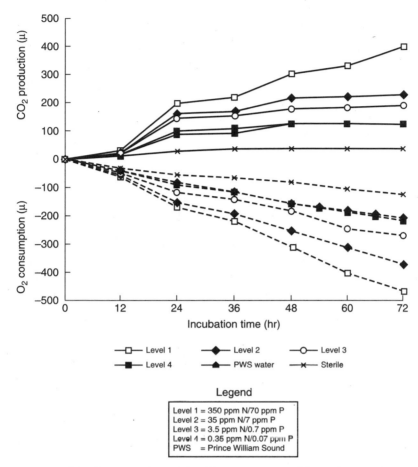

Figure 12.1 Fertilizer specific activity (O_2 consumption, CO_2 production) for six treatments over time.

mineralization. In addition to running control flasks with uncontaminated beach material as indicated above (with and without added nitrogenous nutrients) and ensuring that mineralization in the presence of the oil is considerably above background, one can also add a radiolabeled hydrocarbon. If oil degradation is active, the production of radiolabeled CO_2 should be extensive and immediate. We have found that phenanthrene works well because it was rapidly and completely absorbed to the oil (Mueller et al., 1992; Pritchard et al., 1991). The tidal cycle washed out any phenanthrene remaining in the aqueous phase, and subsequent CO_2 production was then mainly from the bacterial communities associated with oil surfaces.

Oil spill bioremediation as a finishing step

We also made the very important initial assumption that bioremediation would be most effective as a finishing step in the cleanup program. As it turned out, the physical washing procedure employed by Exxon removed the bulk of the oil, but it left the beach material still quite contaminated and aesthetically unpleasing, and the oil ecologically available. As there were very few follow-up alternatives, bioremediation then became intriguing as a finishing step. Without the initial removal of this bulk oil, bioremediation may not have been tenable in the context of Alaska. That is, even if oil biodegradation was quite active, it is largely a surface-oriented process and it would take extended time (there was a relatively small window in the summer months in Alaska when water temperatures are amenable to oil biodegradation). For quicker results, the oil must first be distributed throughout the beach material to increase the surface-to-volume ratio. And since tilling was unreasonable on most of the cobblestone beaches in Alaska, the more bulk oil removed by a physical cleanup process and the more the residual oil dispersed into the beach material, as was accomplished by the physical washing procedure, the more effective bioremediation would likely be. Similar considerations would be required for other types of beach material.

In highly porous beaches, as found in Prince William Sound, oil diffuses to some extent into the gravel, thus increasing the contaminated surface area. The physical washing process may enhance this aspect. On a sand or mud shoreline, however, where porosity is much lower, most of the oil will predictably be concentrated at the surface. Initially, the contaminated beach material may very well be physically removed. But the remaining contaminated beach material can potentially be treated effectively by bioremediation, again as a finishing step. In this case, tilling may be used as a mechanism to further disperse the remaining oil into the beach material, increasing the surface-to-volume ratio, and improving the success of bioremediation. Tilling also aerates beach material and helps disseminate added fertilizers, thus preventing the availability of oxygen or inorganic nutrients from becoming a major limiting factor to bioremediation. In Alaska, because of the highly porous nature of the beaches, the high oxygen concentrations in the cold seawater, and the flushing effect of 5-m tides, oxygen limitation was never a consideration.

Choice of Fertilizer Formulations

We assumed for Alaskan beaches that nitrogen (and phosphorus) had to be applied in a manner which would passively expose the oil-degrading microbial communities to the elevated nutrient concentrations

TABLE 12.2   Description of Fertilizers Tested

	Commercial Name		
	Woodace	Customblen	Inipol EAP 22
Manufacturer	Vigoro Industries	Sierra Chemical Co.	Elf Aquataine
Form	Briquette	Granule	Liquid
Size	5 × 5 × 5 cm	2- to 3-mm diameter	——
N source	Isobutyraldehyde-diurea	Ammonium nitrate	Urea
N:P:K	14:3:3	28:8:0	7.3:2.8:0
Specific gravity	1.8	1.8	0.996 g/mL
Viscosity	——	——	250 cSt
Application rate on 12-m × 35-m plots	986 g/m^2	100 g/m^2	284 g/m^2
Method of application	Net bags (11.8 kg. ea.)	Fertilizer spreader	Backpack sprayer
Test areas	Snug Harbor	Snug Harbor Passage Cove	Snug Harbor Passage Cove

over an extended period. Given the large tidal cycle and significant wave action, fertilizer materials placed on the beach surface would likely wash away in a few days. To overcome this problem, two types of slow-release fertilizers were initially considered; solid pelletized formulations and liquid oleophilic formulations. Characteristics of each fertilizer considered are summarized in Table 12.2.

Summarizing some of the criteria used to make the selections is instructive, as this could help in making fertilizer selections in the future. The three main criteria were (1) ease of application and potential to retain position on the beaches, (2) nutrient release characteristics, and (3) physical durability over time. As it turned out, fertilizer granules seemed to best fit the criteria (Pritchard et al., 1991). They were easy to apply over a large surface area and were found to stick tightly to the oiled beach material and worked their way down under cobble where they were difficult to dislodge. Nutrient release characteristics, which were determined in simple laboratory test systems (Venosa et al., 1990; Glaser et al., 1991), showed that much of the nitrogen (ammonia and nitrate) and phosphorus (phosphate) release from the granules occurred in the first 24 to 72 h (Fig. 12.2). However, sufficient quantities continued to be released steadily for considerable periods thereafter. Thus, as the tides wash in and out of the beach, nutrients should be distributed to the microbial communities associated with the oil for a period of 2 to 3 weeks or longer. Although the physical condi-

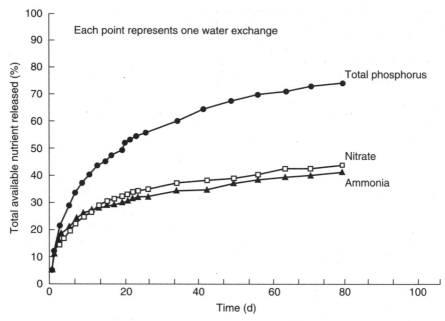

Figure 12.2 Cumulative release of ammonia and nitrate from SIERRA CHEMICAL granules in static flask equipments.

tion of the granules slowly deteriorated on contact with seawater, we observed granules on the beaches with fertilizer inside 2 to 3 weeks after application. The fertilizer granules were ultimately used by Exxon in combination with the liquid oleophilic fertilizer for all of the large-scale applications in Prince William Sound.

Fertilizer briquettes were also found to be satisfactory based on the above criteria (Venosa et al., 1990). These are approximately the size of charcoal briquettes and contain organic sources of nitrogen and phosphorus that slowly hydrolyze to release urea and phosphate over time. Nutrient release characteristics of the briquettes were very similar to those of the granules. Although the briquettes maintained their physical integrity much better, if broadcast over the beach surface, they were easily moved around by the tides and waves, resulting in very heterogeneous distributions. Consequently they had to be packaged in seine net bags and secured to the beach with metal stakes. Although this was effective, it presented significant logistical problems at the time of their consideration for use on a large scale. The briquettes were, however, used as part of our field demonstration in bioremediation, and appeared effective in enhancing oil biodegradation (Pritchard et al., 1991).

Rather than broadcast fertilizer granules and briquettes onto the beach surface, they could also be buried in the beach material, say in trenches running parallel to the water line in the contaminated intertidal zone. Depending on the porosity of the beaches, tides and interstitial water movement could effectively distribute the released nutrients to the bacteria. Because of the physical integrity of the briquettes, they would be most suitable for this type of application. A burial approach was initially considered in Alaska but never really tested, again because of perceived logistical restrictions (later to become unfounded). However, fertilizer granules were applied by Exxon and the State of New York in this manner to a sandy beach on Prall's Island (located in an estuary southwest of New York City) that was contaminated with diesel fuel. Initial reports (Madden, 1991) suggested that the application was successful in enhancing diesel oil biodegradation. Distribution of nutrients will, of course, depend on the hydrodynamics of interstitial water in the beach, and, in many cases, specific information will be lacking. Rather than proceeding based on these limited successes, one could quite easily perform a pilot study, of several days' duration, to actually measure movements of nutrients in interstitial water (Madden, 1991).

Pragmatically, the best criteria for determining how much slow-release fertilizer to place on a given beach were to apply as much as possible without exceeding toxic concentrations of ammonia and/or nitrate. These nutrients are toxic (96-h LC_{50}) to marine invertebrates (good sentinel bioassay for most sensitive species) at concentrations in the water of around 10–15 ppm (Pritchard et al., 1991). Our experience has been that one will face adverse environmental effects thresholds long before the demand for nitrogen by the oil-degrading microbial communities is saturated. Keep in mind that the toxic effect threshold should take into account the initial burst release of nutrients associated with these types of fertilizer formulations. Alternatively, determining fertilizer application rates based on the quantity of oil present in the environment is made difficult because of the tremendous heterogeneity in oil distribution and concentration frequently encountered in the field.

Oleophilic fertilizer

The concept of oleophilic fertilizers is based on the use of organic sources of nitrogen and phosphorus in a liquid carrier that is miscible with oil. In theory, when the liquid carrier is applied, the nutrients essentially dissolve into the oil and thereby keep them in contact, for sustained periods, with the bacteria growing on the oil's surface. Several types of oleophilic fertilizers have been successively tested in both laboratory and small-scale field experiments. Most of these were designed and tested with the idea of treating oil on the surface of water rather than oil on beach material. Pioneering studies by Atlas and Bartha

(Atlas and Bartha, 1973), demonstrated that the addition of paraffinized urea and octylphosphate to Prudhoe Bay crude oil on the water surface significantly enhanced the biodegradation of the oil. Similar success has been reported for a commercial product, Victawet 12, or 2-ethylhexyl-dipolyethylene oxide phosphate (Bergstein and Vestal, 1978), for several natural sources of lipophilic nitrogen and phosphorus, such as soybean lecithin and ethyl allophanate (Olivieri et al., 1978), and for $MgNH_4PO_4$ incorporated into a paraffin support base (Olivieri et al., 1976). Uncertainties related to the specific mechanisms of enhanced oil degradation, the factors affecting physical release rates of the nutrients in situ, and the effect of adding large quantities of organic carbon to the oil, have perhaps limited the application of these products. In addition, several reports comparing enhanced oil degradation by oleophilic and regular fertilizer showed little difference (Lee and Levy, Halmo). Therefore, the advantage of using oleophilic fertilizers has not been clearly established. Nonetheless, the concept of oleophilic fertilizers was intriguing to us given that it was the prevailing method for placing nutrients directly in contact with the microbial communities, and thus further field-testing seemed justified.

We selected the oleophilic fertilizer Inipol EAP 22, produced by Elf Aquataine Company in France, since it was the only commercially available source with large production capability at hand. This unique product is a stable microemulsion consisting of a core of urea (the nitrogen source) surrounded by an oleic acid carrier. Laureth phosphate (a surfactant and the source of phosphorus) acts as an emulsion stabilizer and butoxy ethanol (methyl cellosolve) reduces viscosity. This formulation has shown good promise in tests conducted in the laboratory and in large outdoor tanks using different types of oil-contaminated beach material and environmental locations (Lee and Levy, 1987; Sveum and Ladousse, 1989; Tramier and Sirvins, 1983). Interesting results were obtained from a test conducted on an oil-contaminated beach in Spitsbergen, Norway (Sveum and Ladousse, 1987); Inipol appeared to increase oil biodegradation rates when applied to coarse-grained gravelly beach material but not when applied to fine-grained sandy beach material. Based on these studies, Inipol appeared to have significant potential for bioremediation. When Inipol was applied to oil-contaminated mixed sand and gravel from Prince William Sound in laboratory studies designed to provide intermittent submersion with seawater, approximately 60 percent of the urea (measured as total Kjeldahl nitrogen, TKN) was released within the first few minutes after application (Table 12.3). Measurements of total phosphorus showed a similar percentage. A 6-h wait before the first submersion produced the same results as a 5-min wait, suggesting that the Inipol "sets" quite quickly. However, following this initial burst of TKN and

TABLE 12.3   Release of Ammonia, Total Kjeldahl Nitrogen (TKN), and Total Phosphorus (TP) from Inipol EAP 22 during Intermittent Submersion Experiment

Minutes from start of experiment	5-min contact time*	6-h contact time*
Ammonia released (mg N/L)†		
5	1.1	0.5
15	1.1	0.4
30	1.4	0.5
60	1.3	0.7
120	1.4	0.7
510‡	0.2	
540	0.1	
600	0.0	
TP released (mg P/L)		
5	1.3	1.4
15	1.2	1.2
30	1.0	1.1
60	1.5	1.3
120	1.0	1.0
510‡	1.1	
540	0.9	
600	0.9	
TKN released (mg N/L)†		
5	24.6	29.8
15	26.1	34.8
30	27.2	35.5
60	32.5	34.3
120	29.4	32.3
510‡	4.6	
540	4.6	
600	4.3	

*Time between fertilizer application and initial submersion.

†Initial concentration of nitrogen = 57 mg/L.

‡Water drained; beach material remained unsubmerged for 6 h; seawater replaced.

phosphorus, there was essentially no release of these materials. Presumably, the residual nitrogen and phosphorus, although tightly held to the oiled beach material, was in fact available to the bacteria degrading the oil. Experimentally demonstrating this availability was difficult, as will be discussed below.

Inipol application in Prince William Sound was initially conducted with a backpack sprayer to give a thin coating over the oiled beach material (Pritchard et al., 1991). Appropriate coatings could be controlled because the oil appeared wet and became a deeper black in color following coverage by the Inipol. This visual effect disappeared several hours after application. In the large-scale use of Inipol by Exxon, an application rate of approximately 0.3 L/m^2 was used, based, in large part, on obtaining the surface coverage used in our field demonstration project.

Visual changes. Test beaches at Snug Harbor, where Inipol was applied as part of a field demonstration of its effectiveness, produced some surprising visual results (Pritchard and Costa, 1991). These beaches were moderately contaminated with oil and had not been subjected to the physical washing process at the time of our test. They were selected as representative of those that received the physical washing. Visually, the cobble areas had a thin coating of sticky oil covering the rock surfaces and mixed sand and gravel under the cobble. Oil did not penetrate more than a few centimeters below the gravel surface. In some areas, small patches of thick oil and "mousse" (oil/water/air mixture in colloid form) could be found.

Approximately 10 to 14 days following oleophilic fertilizer application, reductions in the amount of oil on rock surfaces were visually apparent. The change was particularly evident from observations in aircraft where the contrast with oiled areas surrounding the plot was dramatic, etching a "clean" rectangle (12 × 28 m) on the beach surface. The contrast was also impressive at ground level; there was a precise demarcation between fertilizer-treated and untreated areas. At this time, the untreated control plots appeared unaltered visually.

Close examination showed that much of the oil on the surface of the cobble was gone, yet considerable amounts of the oil remained under the cobble and in the mixed sand and gravel below. Remaining oil was not dry and dull as was the oil on the untreated control beach, but appeared softened and wetter. It was also very sticky to the touch, with no tendency to come off the rocks. At the time of these observations, no oil slicks or oily materials were observed leaving the beach during tidal flushing.

We believe that visual disappearance of oil on the cobble surface 2 to 3 weeks following Inipol application was largely due to biodegradation and not a chemical washing phenomena. Chemical data to support this belief are presented below. In addition, we tried to force the chemical washing effects by adding large concentrations of Inipol repeatedly on several miniplots in Snug Harbor, and it did not affect oil removal; a period of at least 2 to 3 weeks was required to see any "cleaning" effect regardless of the amount of Inipol applied. The application of aqueous fertilizer solutions (tested at a different beach), which contained only inorganic chemicals and no organic-surfactant-like materials, also produced the "cleaning" effect in about the same time period, further supporting the role of biodegradation (Pritchard et al., 1991). Finally, experiments performed by Exxon researchers (R. Prince, S. Hinton, and J. Bragg, personal communications) have shown that, in specially designed tests to measure the effectiveness of a variety of commercially available chemical rock washers, Inipol was ineffective. Also, they have observed in microcosm studies that Inipol seemed to become more tightly associated with the beach material; that is, the oil had much more of a tendency to move to the glass walls of the microcosms in the absence of Inipol.

Six to eight weeks after fertilizer application, the contrast between the treated and untreated areas on the cobble beach had lessened. Reoiling of the Inipol-treated beach from oiled subsurface material and/or the concurrent slow removal of oil on the surface of the beach material surrounding the treated areas was probably responsible for this decrease in contrast. Toward the end of the summer season, the area used for the bioremediation studies became steadily cleaner, including the control plots. Several storms and more frequent rainfall, as well as natural biodegradation, undoubtedly contributed to these changes.

Overall, rapid oil disappearance brought on by the application of the oleophilic fertilizer made these beaches more compatible with local wildlife (less tendency for fur and feathers to become oiled). These dramatic changes occurred in a shorter period of time than the limited changes noticed in untreated plots, and possibly helped accelerate biological recovery of the intertidal area.

Measures of Effectiveness

Obtaining definitive information on the role of biodegradation in the removal of oil residues from beach material, or from any complex environmental matrix for that matter, is a difficult task. In general, for oil spill bioremediation, one has to produce *both* qualitative information on changes in oil composition that are indicative of biological processes *and* quantitative information on decay rates of oil, or some of its hydrocarbons, that are also indicative of biological processes. Qualitative information establishes the extent to which biodegradation has occurred; however, with a complex chemical mixture like Prudhoe Bay crude oil, the removal of more than just a few short-chain hydrocarbons (representing only a small percentage of the oil), and removal of more than just the aliphatic fraction of the oil (i.e., leaving behind aromatic, heterocyclic, and branched hydrocarbons, polar chemicals, etc.), is desirable.

Simultaneously the quantitative information establishes that the enhancement of oil biodegradation by the fertilizer treatments was sufficient to merit full-scale operation; generally a two- to threefold enhancement over the untreated controls will probably be acceptable to many decision makers and regulatory groups, but this is not based on a comprehensive database. Both types of information, however, were difficult to obtain in the field because of many different problems encountered.

Any bioremediation testing program that is based on analytical techniques involving the disappearance of oil residues or the disappearance of hydrocarbons resolvable by gas chromatograph is open to scientific criticism because several environmental fate processes (including photosynthesis, physical dissolution, chemical washing, volatility, etc.) can affect or contribute to this disappearance phenomenon. To confront these potential criticisms, we chose an approach that integrated several analytical procedures with several key assumptions. First, we assumed that the disappearance of several target hydrocarbon groups could be used as definitive indicators of biodegradation. We assumed further that strong indicators of biodegradation would be associated with substantial changes in the composition of several fractions in the oil, particularly selected aromatic hydrocarbons. We would thus attribute these compositional changes to biodegradation.

Second, we assumed that if a correlation between changes in hydrocarbon composition and changes in residue weight of the oil could be established, disappearance rates of the residue weights could be used as the primary quantitative measure of fertilizer effect (i.e., significant differences between treated and control plots). The rate of information could then be used to estimate cleanup effectiveness over extended time periods.

However, some discussion as to why these parameters were selected is in order because the criteria for what constitutes biodegradation of oil are complicated and controversial. Many studies have considered measurements of reductions in aliphatic hydrocarbon concentrations as generally indicative of biodegradation, but their value is often questioned because these hydrocarbons are (1) frequently the most readily degradable fraction, (2) the least toxic, and (3) often only a small percentage (by weight) of the oil. Measuring compositional changes in the aromatic fraction adds a further dimension, as these hydrocarbons are less readily degradable and potentially more chronically toxic. However, most of the common procedures for measuring the aromatic hydrocarbons are based on mass spectral analysis which concentrates on only 10 to 20 selected compounds, representing only a very small fraction of the total aromatics. Whether these selected aromatic hydrocarbons are good surrogates for the degradation of the rest of the aromatic compounds has not been established. However, if one concentrates only on the aromatic hydrocarbons and shows that they degrade, one has the advantage of being able to assume that aliphatic hydrocarbons will almost certainly be extensively degraded as well.

Regardless of compositional changes, it seems reasonable to require that bioremediation, as a worthy cleanup tool, should effect the removal of bulk material; changes in composition without much change in oil residue removal seem to present only half the picture. Moreover, oil biodegradation under optimized conditions in the laboratory will result in as much as a 40 to 60 percent reduction in the total weight of oil

(Atlas and Bartha, 1973) and therefore it does not seem unreasonable that some reduction in oil residue can be expected, even under field conditions. The ultimate measure of biodegradation would be to fractionate the oil into aliphatic, aromatic, heterocyclic, polar, and asphaltene fractions and determine weight loss of each of these fractions. Most of these fractions can be analyzed by gas chromatography to determine qualitative changes in composition. This analytical procedure has been used by Westlake and his colleagues in several studies (Jobson et al., 1972).

Changes in the normal-alkane-to-branched-alkane ratios

To provide perspective on measuring effectiveness in bioremediation at the level of a field demonstration, the results from Alaska with the oleophilic fertilizer are instructive. Chemical analysis of the oiled beach material, exposed and unexposed to the oleophilic fertilizer, was accomplished by collecting beach material according to a block design (21 samples taken at each sampling time, 7 each in contiguous blocks along a line in the high-, mid-, and low-tide zones of the beach) and then extracting samples (with methylene chloride) from the cobble surface and from the mixed sand and gravel under the cobble. The weight of oil recovered (measured in milligrams per gram of beach material) was determined gravimetrically and oil composition was determined by injecting extracts into a gas chromatograph following standard analytical procedures (Pritchard et al., 1991).

Changes in composition were determined first by examining the resolvable alkanes. Historically, this has been done by calculating the weight ratio of a hydrocarbon that is known to readily biodegrade (generally the C17 and C18 normal alkanes) to one that is slower to biodegrade (generally the branched alkanes pristane and phytane), which chromatograph very close to the n-C17 and n-C18 alkanes) (Atlas). We generally focused on the n-C18/phytane ratio because pristane is sometimes found naturally in seawater. The ratio concept is based on the idea that most nonbiological fate processes (physical weathering, volatilization, leaching, etc.) will not produce differential losses of aliphatic and branched hydrocarbons that have similar gas chromatographic, and correspondingly, chemical, behavior. Support for this concept can be found in the biogeochemical studies on oil (Kennicutt, 1988). However, since the branched alkanes do in fact biodegrade (Prinik et al., 1977; Mueller et al., 1992), they need only to degrade more slowly than the straight-chain alkane to take advantage of the ratio method. Clearly, in this case, the measure of biodegradation will be conservative.

Focusing on the effects of the oleophilic fertilizer Inipol EAP™ 22, the results in Fig. 12.3a show that, following an initial lag, extensive decay in the n-C18/phytane ratio occurred through time for cobble surface samples. A decay also occurred on the untreated control beach but at about half the rate on the fertilizer-treated beach (Fig. 12.3b). Despite the large variability around the median values, slopes of the decay curves (following the June 17 sampling and not including the September 9 sampling) were statistically different from zero and from each other at the 95% confidence interval. Based on the assumptions described above about the meaning of the ratio changes, biodegradation was occurring on both beaches and was enhanced by the application of the fertilizer. Note that oil had already undergone biodegradation prior to fertilizer application as the ratio for undegraded weathered Prudhoe Bay crude was around 2.0. Also, large decreases in the ratios were invariably linked to considerable reduction, if not complete removal, in the concentrations of the resolvable (by gas chromatography) alkanes, n-C17 to n-C30.

The large variability in the ratios shown in Fig. 12.3 (that is, many samples showed evidence of biodegradation while others showed very little) was a function of the highly heterogeneous distribution of oil on the beach. Possibly the same amount of biodegradation was occurring in each sample, but since biodegradation takes place on the oil's surface, a grab-sampling procedure (which was almost unavoidable in this case) necessarily encompasses sufficient quantities of undegraded oil from below that surface to dilute the measure of biodegradation.

Much less change in the n-C18/phytane ratio, if any, occurred in the mixed sand and gravel under the cobble for the oleophilic-fertilizer-treated beach and the untreated control (Fig. 12.4a and 12.4b). As striking, however, was the unexpected difference in the initial ratio between the cobble surface samples and the mixed sand and gravel samples (t = 0 sampling, June 8, 1989). In both cases, substantial biodegradation of the oil had occurred prior to fertilizer application but the degradation was much more pronounced in the mixed sand and gravel samples. Why the ratio was so much lower in the mixed sand and gravel was not clear. With less total oil concentration overall in these samples initially, biodegradation was possibly more apparent because of less dilution from undegraded oil during sampling.

Biodegradation of phytane

Following the logic set out above, the absence of a change in the n-C18/phytane ratios through time for the mixed sand and gravel samples suggested that oil biodegradation was not occurring despite its degraded state prior to fertilizer application. The initial low ratio may

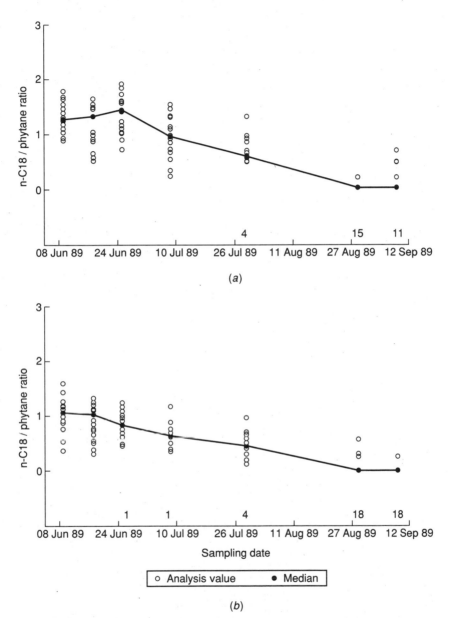

Figure 12.3 Changes in n-C18/phytane relationships over time at treated and control cobble beaches.

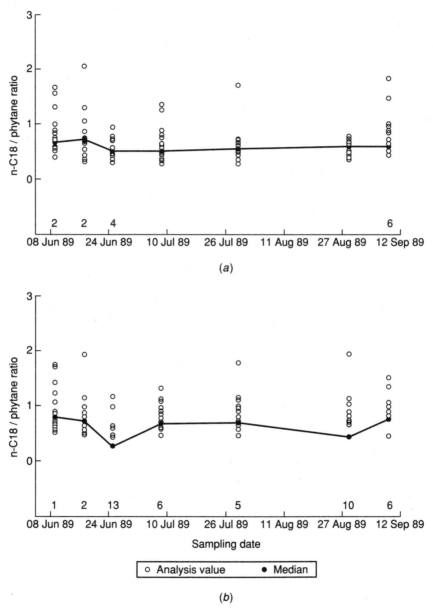

Figure 12.4 Changes in n-C18/phytane relationship over time at treated and control sand and gravel beaches.

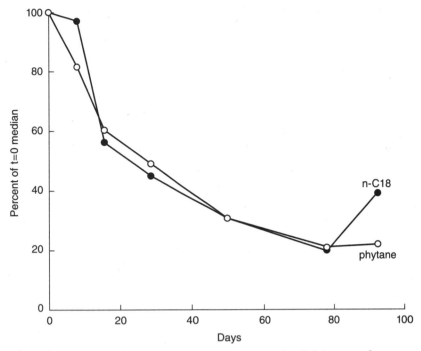

Figure 12.5 Changes in the concentration of phytane and n-C18 (expressed as percent change relative to the t = 0 median concentration) in oil samples from mixed sand and gravel (under the cobble) from Inipol-treated beach plots in Snug Harbor.

have limited subsequent degrees of observable change. However, part of this effect can be explained by another unexpected complication. Examination of phytane itself in the mixed sand and gravel under the cobble on the Inipol-treated beach showed that its decay was as fast as that for n-C18 alkane (Fig. 12.5). Consequently, either biodegradation was not occurring (i.e., some nonbiological process was removing both hydrocarbons simultaneously) or phytane was actually being degraded as fast as the n-C18. Phytane degradation is not common, but it does occur (Pirnik et al., 1977; Mueller et al., 1992), and we have easily isolated phytane-degrading microorganisms from the beach material in Prince William Sound. Thus, microbial communities on Alaskan beaches may have a very pronounced ability to degrade branched alkanes, and the concept of using phytane as an internal biological marker in that case becomes compromised. The cobble surface samples also showed significant decreases in phytane through time, but at a slower rate than the n-C18, thus giving the observed decay in the ratio.

Alternatively, results from the mixed sand and gravel samples under the cobble suggested that possibly Inipol was acting in a chemical manner (surfactant effect) to remove aliphatic and branched hydrocarbons. However, n-C18 and phytane disappeared at essentially the same rate in mixed sand and gravel samples from the untreated control beach (data not shown). Since there was no possibility of a chemical effect on the control plot, one would have to conclude that phytane removal was primarily due to biodegradation.

Compositional changes in aromatic hydrocarbons

At this point we can conclude that biodegradation of the aliphatic fractions of oil was occurring on the samples taken from the cobble surface, and, quite possibly, in the mixed sand and gravel samples as well. As stated above, we believe that biodegradation of the aliphatic fraction was not sufficient in itself to establish that bioremediation was effective. Thus, another approach for examining the overall biodegradation of the oil was to perform selective mass ion spectrometry following gas chromatographic analysis. A variety of aromatic hydrocarbons, which are perhaps more difficult to degrade than the aliphatic hydrocarbons, can be examined with this method (Kennicutt, 1988; Rowland et al., 1986). This is important not only because it tracks another degradable fraction of the oil, but because certain polycyclic aromatic hydrocarbons (PAHs) are known to be procarcinogens under specific conditions. Observing their removal from the oil would therefore imply a reduction in potential adverse ecological effects. Whether this toxicity issue is really relevant (due to improbable exposure scenarios), it was nonetheless a factor to be considered in effectiveness assessments. Furthermore, as the low solubility of the PAHs makes them difficult to degrade, they can be used as the measure of extensiveness of oil degradation. Mass spectral analysis of a variety of aromatic and heterocyclic hydrocarbons in several samples of oil with greatly reduced n-C18/phytane ratios and aliphatic hydrocarbon concentrations is shown in Table 12.4. The selected aromatic hydrocarbons represent a group of methyl-substituted homologs that are found close to the mass number of each parent chemical structure (based on known standards). The values in the table are normalized to hopane (17 alpha, 21 beta), a multiring cyclic alkane (C30). Hopane and its homologs, which are quite resistant to biological attack, have been used for some time as conserved internal biomarkers in oil by the geochemists (Kennicutt, 1988). However, unlike the n-C18/phytane ratio, the relative changes cannot be attributed to biodegradation with as much confidence; differential decay between a hydrocarbon and hopane could be due to nonbiological processes since there may be considerably less chemical similarity between the target hydrocarbon and hopane. Nonetheless, hopane

TABLE 12.4   Relative Concentrations (Mean* and Standard Deviation) of Aromatic, Heterocyclic, and Cyclic Hydrocarbons Normalized to Hopane

	Prudhoe Bay crude	Unfertilized beach	Fertilized beach
n-C18	52.9	1.14 (1.43)	0.96 (0.78)
Phytane	28.3	13.80 (2.30)	6.63 (3.63)
C3/Naphthalenes	31.9	0.15 (0.13)	0.08 (0.04)
C3/Fluorenes	5.30	1.74 (0.38)	1.01 (0.89)
C3/Phenanthrenes	10.0	5.40 (0.25)	3.36 (1.35)
Dibenzothiophene	5.93	0.07 (0.03)	0.04 (0.03)
C3/Dibenzothiophene	9.49	5.34 (0.64)	3.42 (1.43)
Chrysenes	1.22	0.89 (0.04)	0.71 (0.17)
C3/Chrysenes	2.49	1.13 (0.22)	1.03 (0.44)
Norhopane	0.56	0.62 (0.93)	0.59 (0.06)
Stearanes	5.5	5.91 (0.50)	5.15 (0.42)

*N = 8; randomly selected from the 7/29/89 sampling; all samples 50 milligrams of extracted oil residue per milliliter.

provides a consistent standard to normalize the concentrations of aromatic hydrocarbons.

From Table 12.4, the relative difference in the amount of each group of homologs between the beach samples and Prudhoe Bay crude oil indicates the degree of compositional change that occurred in the samples on the beach. Samples from the Inipol-treated beach with very low n-C18/phytane ratios also showed large changes in many of the aromatic hydrocarbons. Norhopane and stearanes are equally as resistant to biodegradation as hopane, and their consistency throughout strongly supports the concept of hopane as a conserved internal biological marker. Again, there was no way to explicitly state that the changes in the aromatic and heterocyclic hydrocarbons were due to biodegradation, but the suggestion is strong. In fact, many of the higher-molecular-weight PAHs are unlikely to be affected by chemical or physical processes, and therefore biodegradation may be the only mechanism that might explain their disappearance. Also, these samples were taken from the beach approximately 78 days after the application of the Inipol fertilizer; any residual chemical effect from the fertilizer that could cause these changes in composition was equally unlikely.

If we assume that nonbiological fate processes (such as leaching or dissolution) remove groups of aromatic hydrocarbons in the order of their solubility, the most soluble being removed first, then any exception to that trend could be attributed to biodegradation since solubility is not the sole characteristic determining susceptibility to microbial attack. Examination of Table 12.4 shows that there are several cases where decreases in aromatic hydrocarbon concentrations in the field samples (relative to Prudhoe Bay crude oil) are greater for hydrocarbons that are more insoluble. For example, there was a greater decrease in C3-chrysenes than there was in chrysene itself, and there was more decrease in the C3-fluorenes than the C3-phenanthrenes. Although these differences could be an artifact of the sampling and/or analysis, it can be argued that biodegradation was the only process where this differential effect is a possibility. Thus, in most samples where substantial degradation of the oil has occurred (as measured by the n-C18/phytane ratios), there was also a concomitant decrease in many high-molecular-weight aromatic hydrocarbons.

Relationship of compositional change to residue weight change

The next step was to establish a relationship between compositional changes and the loss of oil residues. A positive correlation supports the idea that loss of oil residues was due to biodegradation processes. Figure 12.6 shows a plot of the changes in residue weight against changes in the n-C18/phytane ratio. A good positive correlation was ap-

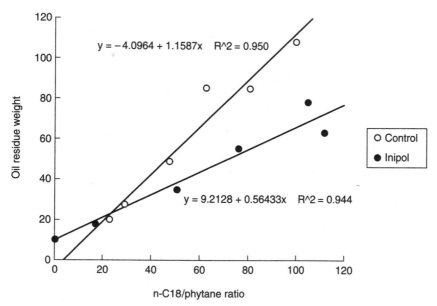

Figure 12.6 Relationship of changes in oil residue weight to changes in the n-C18/phytane ratio (expressed as percent of the median value at t = 0) from oil samples taken from cobble surfaces for both Inipol-treated and untreated beach plots in Snug Harbor. Data points at the lower left corner of the graph represent samples taken late in the sampling period.

TABLE 12.5   Rate Analysis of Natural Log–Transformed Oil Residue Weights (mg/g) in Cobble Surface Samples (July 8, 1989 to July 29, 1989 only) for Test Beaches at Snug Harbor

Beach	Slope (std. dev.)	Significance of slope, greater than zero			Half-life, days	Time to remove 90%, days
		N	T-value	P*		
Inipol treated	−0.016 (0.007)	80	−2.4	0.02	44	146
Untreated control	−0.006 (0.010)	65	−0.56	0.58	124	411

*Only the Inipol rate is significantly different from zero at the 95% confidence level.

parent. The results are interesting because a rapid reduction in the ratio could occur without a significant change in oil residue weight. However, this was not the case, and we can conclude that once degradation of the aliphatic fraction commences, so does biodegradation of many of the other fractions of the oil, at different rates.

Good quantitative information on biodegradation can now be obtained. Since decay rates for the oil residues appeared to be first order, half-lives of the oil can be calculated (Table 12.5). Application of oleophilic fertilizer caused a greater than twofold increase in the disappearance of oil residues on the cobble surfaces as compared to the untreated control. The difference was statistically significant despite the variability in the data. No difference in the oil residue decay rates was detected in the mixed sand and gravel.

Based on the discussion above we would attribute the greater rate of decay on the fertilizer-treated beach to an enhancement of biodegradation from the provision of nitrogen and phosphorus nutrients. Interestingly, the enhancement effect of the fertilizer appeared to be sustained for as long as 90 days. This time period was well beyond that in which nutrients would be released or in which the fertilizer might have a chemical washing effect. Thus, "priming" the biodegradation process with a little bit of nutrient seemed to go a long way. One can generalize and say that over a 120-day period (i.e., the maximum window for Alaska in which water temperatures are >10°C and thereby adequate for oil biodegradation), bioremediation would remove (assuming linearity) approximately 4 times more oil from the cobble surface than would disappear on untreated control beach. Thus with an initial concentration of 1.0 milligram of oil per gram of beach material (cobble surface), biodegradation can potentially remove most of the oil in a single summer season. This was consistent with our visual observations. The absence of any effect on oil residues in the mixed sand and gravel under the cobble suggested that oil may not have been spread in a thin enough layer over the beach material to allow bioremediation to have an effect during this testing period. Or possibly the Inipol was unable to provide nutrients to this area of the beach—i.e., it was primarily acting at the beach surface.

Nutrients and microbial biomass

Following the application of the oleophilic fertilizer, interstitial water samples were taken during several tidal cycles to determine if increased concentrations of nitrogen and phosphorus could be observed. Water samples were taken using a modified root feeder apparatus which sampled water 10–15 cm below the surface of the mixed sand and gravel. Sampling was conducted 2, 10, and 30 days after fertilizer application. Elevated nitrogen concentrations were seen only in the day 2 sampling (Table 12.6), but in areas of the test beach, very high concentrations were observed. However, the variability was quite large with somewhat of a bias toward one side of the treated area. If all of the nitrogen in the fertilizer was released at once into a hypothetical body of water overlying the beach test plot at high tide, one would expect concentrations of approximately 200–300 μm N. Obviously, these concentrations were reached in some areas of the beach. Given that three tidal cycles had occurred prior to this sampling, much of the nitrogen in the Inipol fertilizer was probably released in the first few days. This corresponds with the nutrient release data generated from laboratory studies described above. Thus, the enhancing effect of the oleophilic fertilizer on oil biodegradation may have been the result of an initial pulse of nutrients rather than a sustained concentration of nutrients over extended periods. Other laboratory and field data support this possibility (Pritchard et al., 1991).

Increases in oil biodegradation rates as a result of fertilizer application should also result in increases in the number of oil-degrading bacteria. To determine if this was the case, beach samples (mixed sand and

TABLE 12.6   Ammonia Nitrogen (μM) in Interstitial Water Samples Taken on an Incoming Tide, 2 Days Following Application of Oleophilic Fertilizer on a Cobble Beach in Snug Harbor

	Block*			
	1	3	5	7
High-tide zone	57	300	10	4
Mid-tide zone	410	61	3	6
Low-tide zone	190	3	2	3

*Blocks were 5 m long and 4 m wide running end-to-end parallel to the water line and covering three parallel zones, each 4 m wide running side-to-side up the beach. Blocks 2, 4, and 6 in each zone were not sampled.

gravel) were analyzed using an MPN (most probable number) proce-dure (Pritchard et al., 1991) in which changes in the physical consis-tency of the oil were monitored as an indication of oil biodegradation. There was no significant difference between the control and treated beaches over the 3-month sampling period (data not shown). However, as indicated above, the concentrations of oil degraders were very high to start with, and, with the large variability observed in the data, in-creases of approximately two orders of magnitude were needed to be significant. In addition, increases in biomass could be obscured by sloughing of the cells or predation by protozoa. Field studies the fol-lowing summer (1990) were finally able to demonstrate significant in-creases in hydrocarbon degraders but only in the beach subsurface (Lindstrom et al., 1991).

Disk Island Field Study

A portion of the northwestern shore of Disk Island (located between Ingot and Knight Islands) was chosen as a study site in the summer of 1990, one year after the oil spill. The study was designed to obtain dose-response information for fertilizer application. Fertilizer granules were selected because it was relatively easy to apply different concentrations of the granules in a controlled manner.

The study site was chosen because it was one of the few remaining large areas with moderately to heavily contaminated beach material that could be reasonably used for experimental purposes. The beach area chosen for study, which had not been through the physical wash-ing process used by Exxon, had a shallow slope with little wave activ-ity. Oil contamination was surface and subsurface and was packed into the mixed sand and gravel beach material more densely than observed on other types of beaches in Prince William Sound.

Different amounts of fertilizer granules were applied to plots as shown in Fig. 12.7. The $100 \ g/m^2$ application rate was the concentration of granules applied on a large scale by Exxon. Prior to the fertilizer ap-plication, samples of beach material were homogenized and placed in sampling baskets located in each plot. These sampling baskets were then harvested periodically to determine the effect of the fertilizer on oil biodegradation. The homogenization reduced variability in oil con-centrations and therefore greatly simplified sampling efforts.

Changes in the concentration of ammonia following application of the fertilizer granules are shown in Fig. 12.8. These data were obtained from sampling wells that were driven into the beach material to allow sampling of the interstitial water with incoming and outgoing tides. The highest concentrations of ammonia were seen with the highest

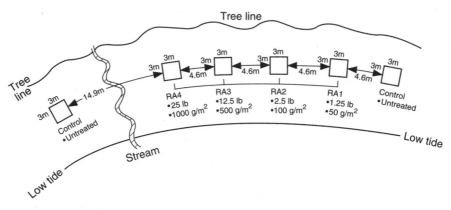

Figure 12.7 Disk Island fertilizer specific activity plot map and rate of CUSTOMBLEN granule application.

concentrations of granules applied (rate 4; 1000 g/m^2), but this concentration was not sustained for more than 1 to 2 days. In fact, ammonia concentrations approached background levels 5 to 10 days after application, regardless of the fertilizer granule concentration applied. Clearly, if a dose response is to be observed, it will result from an initial pulse of nutrients rather than sustained concentrations through time. Results for the release of phosphate and nitrate were similar.

An examination of the decrease in the n-C18/phytane ratio for samples taken from the beaches at different times following initial application of the fertilizer showed that there was essentially no enhancement of biodegradation, as the extent of decrease was not greater than that seen on the control, untreated plots. A comparison of the ratios for a control plot and the plot receiving the highest concentration of fertilizer granules is shown in Fig. 12.9. Biodegradation was obviously occurring (i.e., a steady decrease in the ratios), but it did not seem to be stimulated by the fertilizer. This was a startling result because it reveals that not all beach conditions may be equally amenable to bioremediation. Because of the low-energy features of the beach and the more compact nature of the beach material, mass transport limitations (availability of nutrients and/or oxygen) may have become a significant problem, and this is a key factor to be considered in using bioremediation on other types of oil-contaminated beaches. We are also aware that the beach material contained quantities of humic material; this may have interfered with the oil-degrading microbial communities either as a competing sink for available oxygen or as a degradable carbon source that was preferable to petroleum hydrocarbons.

This experience provides a lesson regarding the use of laboratory tests as an indicator of the potential for bioremediation. If it was a

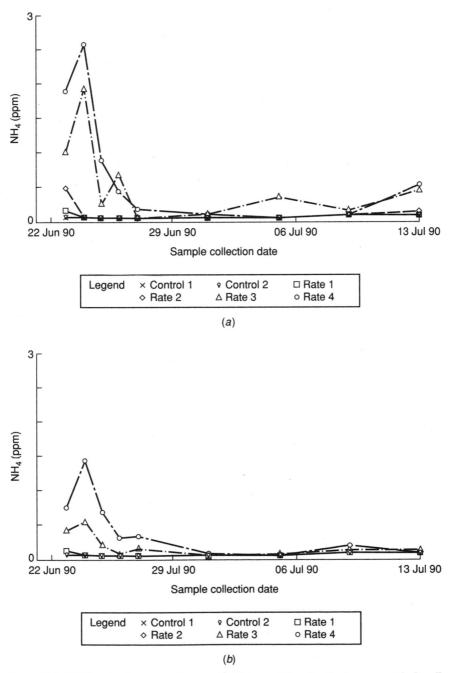

Figure 12.8 (*a*) Changes in ammonia concentration over time for the incoming tide for all plots for the Disk Island Fertilizer Application Rate Study. (*b*) Changes in ammonia concentration over time for the outgoing tide for all plots for the Disk Island Fertilizer Application Rate Study.

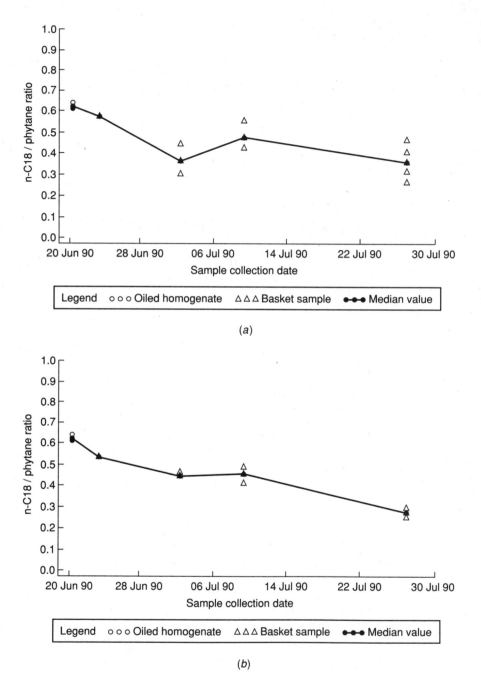

Figure 12.9 (*a*) Changes in the n-C18/phytane ratio over time for the 500 g/m² fertilizer application for the Disk Island Fertilizer Application Rate Study. (*b*) Changes in the n-C18/phytane ratio over time for untreated control plot number 1 for the Disk Island study.

mass transport phenomenon that effected successful bioremediation at Disk Island, removing beach material to the laboratory and conducting tests similar to those described above will likely not reveal the limitations inherent in the field. Most of these laboratory tests involve shake flasks, which by design optimize mass transport, and one would therefore expect that samples may show unrealistically high activities relative to the field. This is illustrated in mineralization studies performed in conjunction with the Disk Island study. Beach material from the sampling baskets, when placed in biometer flasks (see above), showed mineralization activities that reflected an enhancement effect due to the presence of the fertilizer (Fig. 12.10). Some of the total CO_2 production may have been from "nonpetroleum" organic material present in the beach material; however, similar studies using radiolabeled hydrocarbons revealed that stimulation of oil degradation by the fertilizer was probably occurring in these flasks. There also appears to be a dose-response relationship in these results, suggesting that doubling the fertilizer concentrations did not double the oil biodegradation rate as measured by mineralization. Clearly this relationship was not realized in the field.

Thus the stimulatory effect observed in the laboratory was not reflected in the field. The flask studies did, however, indicate that some mass transport limitation was affecting the bioremediation in the field.

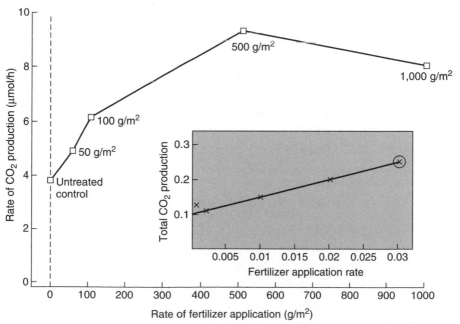

Figure 12.10 Plot of rate of CO_2 production versus rate of fertilizer concentration.

The only way to protect from this extrapolation problem is to perform microcosm studies in which intact samples of beach material can be studied in the laboratory under conditions similar to those in the field. Such systems were in fact developed and tested during the Alaskan oil spill project (Pritchard et al., 1991), but they involve considerably more complexity, time, and expense.

EPA Program in Oil Spill Bioremediation Protocol Development

Closely tied to the field studies just described is the question of which commercial products, whether they are microorganisms, nutrients, surfactants, or others, could or should be used in a bioremediation context. As we mentioned above, in Alaska, many of these commercial products could not be considered because of the very short time frame for field demonstrations, but also because the data available for each product were so variable and/or insufficient that reasonable selections could not be made in a timely fashion. The development of effectiveness and environmental safety testing protocols to be used in establishing a consistent and relevant database, upon which decisions for the use of particular commercial products might be based, is now under way within the Office of Research and Development of the EPA. The conceptual basis for these protocols has proven to be complicated to formulate because of the need to keep the scope of the testing within reason. However, the problems encountered in environmental variability, as illustrated above, make it very difficult to devise protocols that will ultimately provide the "right" kind of information to allow appropriate regulatory decisions to be made. A review of the initial conceptualizing we have carried out to date is in fact quite informative. Obviously, as these protocols are tested and validated, these concepts will have to be modified, so information presented here is not to be considered as final guidelines.

The protocols for determining the effectiveness of oil bioremediation products must ultimately contain the following components:

1. Simplified "expert system" (basically decision trees) for use by regulators that will encourage rational consideration of peripheral factors that are keys to the success of bioremediation on open waters (Tier I—decision trees)

2. A screening test that allows the relative effectiveness of different bioremediation products to be assessed in terms of their ability to promote significant biodegradation of oil under a standard set of laboratory test conditions (Tier II—screening information)

3. A procedure for extrapolating laboratory information to the field on a site-specific basis using definitive kinetic and dose-response information that is integrated with simplified and quantitative predictive frameworks (Tier III—field extrapolation information)

4. A procedure for the use of flow-through microcosm systems to determine the relative effect of different commercial bioremediation products on oil slicks under the environmental conditions that are likely to be experienced during oil spills (Tier III—field extrapolation information)

5. Guidelines for the performance of appropriate controlled field studies in artificial enclosures to clearly establish the fate of the oil during bioremediation (Tier IV—direct field demonstration)

It is prudent to focus on only one portion of the protocol development because of space limitations. I will provide the concept for the testing that would be implemented under the site-specific extrapolation (item 3) and the microcosm testing (item 4) components of the protocol. However, a brief description of the concepts behind a Tier II testing protocol will be given first as a means to develop comparisons.

The purpose of Tier II testing is to determine the ability of a particular bioremediation product to promote significant biodegradation of oil under a standard set of laboratory testing conditions. It is not designed to address effectiveness of a bioremediation product under site-specific conditions.

It is further assumed that the Tier II tests are to be divided into parts: that which measures the activity of biological products and that which examines the effectiveness of nonbiological products. Tests for biological products involve adding the product directly to an oil slick in a proportion recommended by the manufacturer. Other supplements recommended by the manufacturer are also added. It is assumed that the microbial flora of natural water is not required since most of the biological activity is provided by the product. Tests are conducted with and without added nutrients in the water and with sterile (autoclaved) and nonsterile products. Flasks are incubated by shaking at 20°C. If the product is found to be effective, results are compared with a standard data set that is developed from research information specifying the minimum amount of product activity (oxygen uptake profile and changes in oil chemistry) required to make the product effective as a bioremediation agent.

For the testing of nonbiological products, it is assumed that they generally involve some mechanism or procedure for stimulating the oil biodegradation capabilities of natural microbial flora. In most cases this involves the rapid enrichment of the oil-degrading microorganisms within the total bacterial population. Since at the Tier II level of testing it is not appropriate to consider site-specific factors that affect the activity or enrichment of oil-degrading microorganisms, a standard mixture of bacterial pure cultures that degrade all of the major fractions of the oil tested needs to be used. Alternatively, a specific sampling site can be selected as a consistent source of natural populations of bacteria. However, there is no guarantee that the responsiveness of a water sample from this site will remain constant, and thus the use of a mixture of pure cultures has several advantages: (1) It allows one to store the bacteria over long periods and preserve their activity so that it is the same each time it is used; (2) it eliminates the variances that are likely to occur if natural samples are used as inocula, thereby eliminating the requirement to run reference products each time a test is conducted; (3) optimal conditions of enrichment are employed, thus giving an evaluation of the bioremediation product under ideal conditions. If it does not work under these conditions, then it is very unlikely to work in the field, where the conditions will be a lot less ideal.

Effectiveness of the product will be measured against a standard time, as established by practical conditions at a spill site and by the time it will take to enrich the mixed culture to a point where it will affect the oil. The oil may break up on the surface of the water relatively quickly as a result of the product addition. The test system is designed not only to monitor this event but also, as a closed system, to allow the fate of the oil to be followed for several weeks thereafter. If significant degradation is seen during this incubation period, then it will be assumed that oil leaving the water surface as a result of adding the product will in fact be degraded. Research now carried out under the Oil Spill Research Program will provide a verification for this assumption.

Concept and development of Tier III testing protocols for open water

The appropriateness of oil bioremediation products that are proposed for use in open sea bioremediation must be tested in the laboratory under conditions that are reasonably representative of field conditions. Although tests of this nature can become quite complex, microcosm systems that model special features that might be key to the success of oil bioremediation on open waters have been designed. By far the key feature of the microcosms is the incorporation of a dilution capacity in the microcosms that would simulate the field. Flow-through microcosms that contain an oil slick on the water surface are used in this part of the protocol to assess

1. The tendency of a bioremediation product to remain with the oil long enough to be effective

2. The potential of a product to emulsify the oil and cause it to be dispersed

Microcosm systems that allow the application of a bioremediation product and the necessary supplements to a contained oil surface and then allow seawater to flow under the slick at different velocities and turbulences to create the necessary dilution capacity estimated from a particular field situation will be used. This testing will determine the extent to which the product and the supplements will stay associated with the oil. It will not of course indicate that similar dilution will occur in the field, but it will provide a consistent method to screen, on a relative basis, products and the efficacy of their application strategy. If, for example, a product (or the required supplement) is rapidly removed from the oil slick in the microcosm, it will almost certainly be removed much faster in the field. A product that appears to be effective (i.e., appears to stay with the slick and affect the fate of the oil) would be one considered for further testing. In addition, flow and turbulence conditions in the microcosms can be varied to provide a range of conditions under which the effectiveness of the product could be evaluated.

The Tier III testing is based on biodegradation kinetics and involves the use of microcosm studies to determine the effect of certain key environmental parameters on the ability of the bioremediation product to enhance the biodegradation rates of oil. Kinetic information is very important for protocols dealing with the treatment of oil on the open water because it will always be a question of how fast the product will work under a specific set of environmental conditions *relative* to the rate at which the oil slick is being dispersed naturally.

The use of flow-through microcosms is necessitated by a requirement to determine the effectiveness of a product under conditions that cannot be modeled in a flask study or Tier II level testing. These conditions are defined specifically as the following:

1. Presence of an intact oil layer floating on the surface of a water column

2. Ability to impart and control water column turbulence during testing with the intact oil layer

3. Continuous input of water containing significant concentrations of particulates in the water column

4. Flow-through conditions to allow water exchange and dilution under the oil layer

5. Temperature, particularly as it affects the physical nature of the oil

6. Concentrations of inorganic nutrients in the flow-through water

7. Microbial activity in the inflow water

In the consideration of microcosm testing at the Tier III level, several points must be considered carefully.

1. Microcosm testing is expensive for the vendor because it requires rather elaborate testing facilities, extensive analytical chemical analysis, and sophisticated means of interpreting the resulting data. Thus, the testing should be kept to a minimum.

2. Because of this expense and elaborateness, microcosm testing cannot be used to examine the influence of a large number of ecological or oil spill conditions on the effectiveness of the bioremediation product.

3. Microcosms by definition are designed to simulate certain environmental conditions more realistically than simpler tests, such as shake flasks. Therefore, the specific conditions to be modeled in the microcosm must be critically evaluated such that the microcosm testing does not produce data that is more effectively and efficiently obtained in simpler systems.

4. When considering open-water application of a bioremediation product, it must be realized that modeling conditions typical of a spill site will be very complex, to the point that it is questionable how effectively certain conditions can be appropriately modeled in microcosms. For example, the effectiveness of a bioremediation product will likely be a function, in part, of wave and turbulence conditions. Developing these conditions in a microcosm is difficult. In addition, extrapolation of microcosm information to the field is complicated because these conditions will vary on a day-by-day or even hour-by-hour basis. A range of turbulence/wave conditions could be tested in the microcosm, but this must be limited to testing extremes because of the constraints of cost and time when using microcosms.

5. Finally, it must be kept in mind that Tier III testing involves putting information on the "shelf" to be used at the time of the spill (time is too short following the spill to conduct microcosm-type tests). Thus the shelf data must be of such a nature that it can be used effectively in making decisions at the time of the oil spill.

As mentioned above, one of the most critical factors when performing microcosm studies is the ability to interpret the resulting data. The larger the database available on the product performance, particularly in terms of kinetics, the better one will be able to use the microcosm results and extrapolate the information to a site-specific situation. Because each site is different, it is impossible to have information "on the shelf" for a particular product that will deal with every site. Therefore, one has to make a decision on how to generalize the testing approach. This can be accomplished either by testing waters from several designated areas (Atlantic Coast versus Pacific Coast or Gulf Coast, northern waters versus southern water, protected bays versus open bays, wetlands versus marshes, etc.) or by examining the effect of selected environmental parameters that encompass, in a general way, all of the conditions in these different areas and then extrapolating the general results to site-specific conditions that can be determined at the time of the spill. The latter approach seems to be the most reasonable.

Consequently, a protocol must be developed based on an initial decision as to the major factors that will most likely affect the performance of the product in treating oil on open waters wherever the spill takes place. In general, the most important factors will be

1. Temperature

2. Turbulence

3. Salinity

4. Background concentration of inorganic nutrients

5. Suspended particulates

6. Background microbial activities

7. Type and concentration of oil

Then it must be determined if the factors can be measured in the field at the time of the spill (within 1 to 2 days). For those that likely can be measured, quantitative relationships can be established between the factor and the product performance. A good example is turbulence. The effectiveness of the product can be established under three different turbulent conditions: high, medium, and low. The data is graphed (effectiveness versus turbulence expressed in Reynolds number or the equivalent), and a relationship established (linear, exponential, etc.) using statistical techniques. Once the relationship is known, then performance of the product for a particular site condition can be predicted. That is, a general indication of turbulent conditions at the spill site will be obtained and the graph will be used to determine the effectiveness of the product.

Since ultimately one will likely be dealing with several environmental factors at once and several specialized environmental conditions, a simple calculation framework can be used. A protocol should stipulate that, for any product, the key environmental factors that will affect performance for any site will be turbulence, temperature, microbial activity, type of oil, nutrient concentrations in the water column, and particulates in the water column. No other factors need be considered as it is assumed that they will be insignificant in the overall decision to use or not to use the product. Thus every product will have "data on the shelf" that relates these key environmental factors to performance of the product.

Summary and Conclusions

The results from our field demonstration in oil spill bioremediation in Prince William Sound indicate that the oleophilic fertilizer Inipol EAP™ 22 served as an effective nutrient source for oil-degrading microbial communities. It enhanced oil biodegradation, as measured both by changes in oil composition and oil residue weights, by as much as twofold relative to the untreated controls. This was enough of a response to merit incorporation of bioremediation, on a large scale, into the remedial action plan for oil-contaminated beaches in Prince William Sound. Despite this enhancement effect, the importance of its oleophilic nature is still unclear, at least for oil-contaminated beach material from Prince William Sound. However, our studies report the belief that the visual observation of oil removal from the beaches 8 to 10 days following application of the Inipol was due largely to bioremediation and not to a chemical washing effect.

Overall, rapid oil disappearance brought on by the application of the oleophilic fertilizer made these beaches more compatible with local wildlife (less tendency for fur and feathers to become oiled). These changes occurred in a shorter period of time than those limited changes observed in untreated control plots, and possibly helped accelerate biological recovery of the intertidal area.

We have also summarized some of the lessons associated with measuring bioremediation success in the field. Effective measures of biodegradation and interpretation of resulting data, given the highly variable nature of field studies, have been emphasized as a key element in this success. In addition, the complications and difficulties that arose during our use of bioremediation efforts in Alaska were also discussed. Hopefully this will help guide similar applications of bioremediation at future spills.

References

Atlas, R. and R. Bartha. 1973. Stimulated biodegradation of oil slicks using oleophilic fertilizers. *Environ. Sci. Technol.* 7:538–541.

Atlas, R. M. 1981. Microbial degradation of petroleum hydrocarbons: An environmental perspective. *Microbiol. Rev.* 45:180–209.

Bartha, R. and D. Pramer. 1965. Features of a flask and method for measuring the persistence and biological effects of pesticides in soil. *Soil Sci.* 100:68–70.

Chianelli, R. R., T. Aczel, R. E. Bare, G. N. George, M. W. Genowitz, M. J. Grossman, C. E. Haith, F. J. Kaiser, R. R. Lessard, R. Liotta, R. L. Mastracchio, V. Minal-Bernero, R. C. Prince, W. K. Robbins, E. I. Stiefel, J. B. Wilkinson, S. M. Hinton, J. R. Bragg, S.

J. McMillem, and R. M. Atlas. 1991. Bioremediation technology development and application to the Alaskan spill. In *Proceedings 1991 Oil Spill Conf.,* Am. Petroleum Inst., Washington, D.C., pp. 545–555.

Bergstein, P. E. and J. R. Vestal. 1978. Crude oil biodegradation in arctic tundra ponds. *Arctic* 31:159–169.

Glaser, J. A., A. D. Venosa, and E. J. Opatken. 1991. Development and evaluation of application techniques for the delivery of nutrients to contaminated shoreline in Prince William Sound. In *Proceedings 1991 Oil Spill Conf.,* Am. Petroleum Inst., Washington, D.C., pp. 556–562.

Halmo, G. 1985. Enhanced biodegradation of oil. In *1985 Proceedings Oil Spill Conf.,* Am. Petroleum Inst., Washington, D.C., pp. 531–537.

Jobson, A. M., F. D. Cook, and D. W. S. Westlake. 1972. Microbial utilization of crude oil. *Appl. Microbiol.* 23:1082–1089.

Kennicutt, M. C. 1988. The effect of biodegradation on crude oil bulk and molecular composition. *Oil Chem. Pollut.* 4:89–112.

Lee, K. and E. M. Levy. 1987. Enhanced biodegradation of light crude oil in sandy beaches. In *1987 Proceedings, Oil Spill Conf.,* Am. Petroleum Inst., Washington, D.C., pp. 411–479.

Lindstrom, J. E., R. C. Prince, J. C. Clark, M. J. Grossman, T. R. Yeager, J. F. and E. J. Brown. 1991. Microbial populations and hydrocarbons biodegradation potential in fertilized shoreline sediments affected by the T/V *Exxon Valdez* oil spill. *Appl. Environ. Microbiol.* 57:2514–2522.

Madden, P. C. 1991. Final Report Prall's Island Bioremediation Project. Exxon Res. and Engineering, Florham Park, N.J. 82 pp.

Mueller, J. G., S. M. Resnick, M. E. Shelton, and P. H. Pritchard. 1992. Effect of inoculation on the biodegradation of weathered Prudhoe Bay crude oil. *J. Ind. Microbiol.* In press.

Olivieri, R., P. Bacchin, A. Robertiello, N. Oddo, L. Degen, and A. Tonolo. 1976. Microbial degradation of oil spills enhanced by a slow-release fertilizer. *Appl. Environ. Microbiol.* 31:629–634. (57)

Olivieri, R., A. Robertiello, and L. Degen. 1978. Enhancement of microbial degradation of oil pollutants using lipophilic fertilizers. *Marine Pollut. Bull.* 9:217–220. (59)

Pirnik, M. P., R. M. Atlas, and R. Bartha. 1977. Hydrocarbon metabolism by *Brevibacterium erythrogenes:* Normal and branched alkanes. *J. Bacteriol.* 119:868–878.

Pritchard, P. H., C. F. Costa, and L. Suit. 1991. Alaska Oil Spill Bioremediation Project. U.S. EPA, Office of Res. and Dev. Report, EPA/600/9-91/046a, 522 pp, Washington, D.C.

Pritchard, P. H. and C. F. Costa. 1991. EPA's Alaskan oil spill bioremediation project. *Environ. Sci. Technol.* 25:372–379.

Rowland, S. J., R. Alexander, R. I. Kazi, D. M. Jones, and A. G. Douglas. 1986. Microbial degradation of aromatic components of crude oils: A comparison of laboratory and field observations. *Org. Geochem.* 9:153–161.

Sveum, P. and A. Ladousse. 1989. Biodegradation of oil in the Arctic: Enhancement by oil-soluble fertilizer application. In *1989 Proceedings Oil Spill Conf.,* Am. Petroleum Inst., Washington, D.C., pp. 439–446.

Tramier, B. and A. Sirvins. 1983. Enhanced oil biodegradation: A new operational tool to control oil spills. In *Proceedings 1983 Oil Spill Conf.,* Am. Petroleum Inst., Washington, D.C., pp. 155–219.

Venosa, A. D., J. R. Haines, J. A. Glaser, E. J. Opatken, P. H. Pritchard, and C. F. Costa. 1990. Bioremediation treatability trials using nutrient application to enhance cleanup of oil contaminated shoreline. In *Proceedings 83rd Air and Waste Management Association Annual Meeting,* Air and Waste Management Assoc., Pittsburgh, Pa., pp. 90–22.3.

Chapter 13

Future Directions in Bioremediation

Rita Colwell

Maryland Biotechnology Institute
University of Maryland
College Park, Maryland

Morris A. Levin

Maryland Biotechnology Institute
University of Maryland
Baltimore, Maryland

Michael A. Gealt

Department of Bioscience and Biotechnology
Drexel University
Philadelphia, Pennsylvania

Predictions of trends must be based on events and discoveries of the immediate past and the present. The major areas in which current research is moving forward rapidly are the application of molecular genetics to develop new bacterial strains capable either of more complete degradation of compounds or of metabolizing compounds otherwise "refractory" to microbial metabolism, and the development of new engineering technologies to enhance rate and extent of degradation. Similar observations have been made[41,42,47] in which examples of specific applications combining microbial and engineering approaches were discussed. We note at the outset that the regulatory infrastructure must be addressed in parallel with innovations in the science and engineering of biodegradation, since new strains cannot be utilized except within a regulatory framework.

It is well that much progress is being made in advancing the science and technology of biodegradation and bioremediation. In 1991, the market for bioremediation of hazardous wastes was about $60 million and it is estimated that it will grow to between $125 and $300 million by 1995. A growth rate of 65 percent per year has been forecast.[16] Another important indication of the industry's maturation is seen in the efforts of the members of the Applied Biotechnology Association, who are using, to date, only nonengineered microorganisms. The association was formed in 1988 to promote the interests of firms involved in biotreatment.[3] It has prepared a compendium of successes and is currently devising a means of accreditation for waste treatment providers using nonengineered microroganisms to treat a wide variety of wastes.

The significance of the regulatory infrastructure is becoming more evident. For example, specific regulatory factors driving the development of the entire biotreatment industry are (1) federal and state cleanup regulations and (2) real estate transfer issues. Closely related is the need to demonstrate unequivocally that microbial degradation is the major factor in disappearance of pollutants and that the process results in mineralization or production of innocuous products.

Unfortunately, in some areas progress is less rapid. After conducting a literature survey for evidence of successful biotreatment, Madsen[31] was able to list only seven cases in which biodegradation attributed to microbial action could be documented without equivocation. These included three soil, two aquatic, and two marine field applications. Only one of the soil treatments included addition of microorganisms; the others used only addition of nutrients in the form of fertilizer. None of the marine or aquatic trials employed addition of microorganisms. In two cases, indigenous microorganisms were enhanced by addition of fertilizer. Madsen[31] was able to identify a number of other studies in which laboratory work predicted the ability of indigenous microorganisms to degrade particular wastes, but the associated field studies did not generate conclusive data to support the conclusion that biodegradation was a result of microbial action. Clearly, additional data detailing in situ effects of microbial action are required before regulators, the public, and prospective clients accept biotreatment as a viable, safe, and cost-effective alternative.

In a recent article, Alexander[2] observed that areas for high-priority research include (1) increasing bioavailability of pollutants, (2) improving reactor/in situ remediation designs, (3) overcoming scale-up problems, and (4) finding improved bioremediation processes. While, given enough time, many, if not most, compounds are degraded by microorganisms, the thrust for the next decade must be to increase efficiency of indigenous microorganisms and reactors (or in situ practices) so that the rate of degradation will outpace the rate of deposition.

Areas Where Progress Is Observed

Cyanogenic wastes are being treated via degradation by microorganisms. That is, cyanide is produced through a variety of industrial processes, including electroplating and the manufacture of steel and paint. Because of the toxicity of cyanide, degradation is necessary before discharge of cyanide-containing wastes to the environment can be permitted. Since there are approximately 2000 green plant species that produce cyanogens (organic compounds containing cyanide) which are converted to HCN when the plant is wounded or attacked by fungi, it is no surprise to find that fungi, as a defense mechanism, have developed enzymes (e.g., cyanide hydratase) capable of detoxifying HCN by conversion to formamide.[58] Several of these detoxification mechanisms are under investigation for application in the treatment of waste cyanide. Stoichiometric conversion of cyanide to formamide, using an immobilized fungus in a column, has been demonstrated. Imperial Chemical Industries (ICI) has developed an enzyme system that can be used to treat high concentrations of cyanide in continuous flow, ranging from a few to several hundred cubic meters per day. Homestake Chemical Company in South Dakota has developed a biotreatment system using rotating biological contactors (RBCs). The company selected a biological process for treatment of cyanogenic wastes, after careful evaluation of chemical and physical treatments, both of which were rejected because of treatment performance, capital cost, and the high concentration of cyanide in the wastes (up to 20 mg/L). Some success has also been reported using methanogenic bacteria maintained under anaerobic conditions for degradation of cyanide.[12]

It is clear, then, that many examples of the use of living organisms for environmental remediation can be cited. The organisms, in general, include plants as well as the much discussed microorganisms. Plants have been used for the accumulation and removal of toxic wastes from water and soil, especially where they contain heavy metals. This application was first made by Minguzzi and Vergano in 1949, in which plants accumulated nickel from soil (to levels of [sim] 1 percent).[21] Subsequently, the term *hyperaccumulator* was employed for plant species accumulating ≥ 0.1 percent metal, measured in the dried leaves. More than 50 species of alyssum have the ability to hyperaccumulate heavy metals. Homer et al.[21] reported uptake of nickel, copper, and cobalt. Other workers have found that some plants accumulated 0.1 to 5 percent of their total dry weight in metal content. Based on findings such as these, it has been proposed that plants be utilized to accumulate metals and that the plant tissues serve as a source for recovery of these metals. Baker[5] hypothesized that accumulation is linear until a plateau is reached at very high concentrations of metal in the soil.

Research in marine bioremediation using marine microorganisms which possess a wide range of degradative capabilities and abilities to survive adverse conditions is ongoing and has been reviewed.[60] Survival capabilities in the marine environment are being investigated because of the interest in genes coding for enzymes for degradation and the need for genetic capabilities which will remain active under adverse conditions (e.g., high pressure of the deep sea and high temperature of the hydrothermal vent systems).

Development of Strains

Many studies have been done describing bacterial and fungal strains which are able to degrade one or more types of hazardous waste. It is almost axiomatic that new strains can be readily isolated. Recently, for example, *Alcaligenes* sp. O-1 was reported to mineralize orthanilic acid[22] and *Nocardia* sp. HB has been described which degrades 1,2-epoxyalkanes.[57] With additional research, one can expect that new strains will be available for degradation of noxious materials not now readily metabolized by native organisms.[1,45]

Issues of safety and public concern have to be taken into consideration before engineered organisms can be employed in the field.[41,42,43] Initially, laboratory and pilot-scale demonstrations are necessary to show that genetically engineered microorganisms (GEMs) offer a clear advantage over conventional organisms and do not pose hazards. Obviously, compliance with relevant regulations must be assured. In addition, economics, market issues, social obstacles, and the political environment will all play major roles in bioremediation. These issues are covered in depth in Chap. 2.

Several strategies are being developed to employ GEMs to overcome problems encountered with use of indigenous bacteria. For example, it is typical that when many different waste products are present, several pathways must function in concert for degradation to occur. The suggestion has been made by several investigators[1,23,56] that different functions could be engineered into a single species, which could then be released to the environment for the purpose of degrading a complex compound or mixture of compounds. Alternatively, it may be more expedient to control the proportions of different bacterial species within a consortium to achieve the same mix of activities.[34] Unfortunately, knowledge of how species work in consortia is lacking. Clearly, this is an issue that must be resolved, either in principle or on a case-by-case basis, so that the needed information can be obtained.

One of the impediments to introduction of GEMs to the environment is the legal requirement that they be tracked after deliberate release or escape from a reactor. Several methods have been developed for tracking GEMs, with the typical approach being to use unique sequences as targets for either hybridization or the polymerase chain reaction (PCR).[36,51] Selective plating methods have been suggested which employ special characteristics cloned into the GEM. For example, Nakamura et al.[37] proposed a culture plate method for detection of a *Pseudomonas* sp. capable of assimilating monofluoroacetate as sole carbon source.[37]

Monitoring, once the detection method has been developed, is more difficult when the DNA sequence of interest is transferred between bacterial strains or species in the environment. Gene transfer has been demonstrated to occur in aqueous systems[13,14,15,24,32,35,48] and soil.[17,49,59] This transfer necessitates the use of DNA probes and PCR to monitor released GEMs, because these methods are sensitive to the modified DNA and not to the host organism (which may change as a result of gene transfer). Recent work confirms that bacteria can take up naked DNA in the environment.[36]

Perhaps more fruitful than addition of new metabolic functions to a single bacterium—i.e., creation of a "superbug"—is the modification of metabolic activities of indigenous microorganisms and creation of consortia. For example, simple modifications of promoter activity can increase the amount of enzyme produced by an organism. Very small changes in primary amino acid sequence may alter enzyme structure, perhaps enhancing stability, decreasing substrate specificity (increasing the number of possible substrates), or expanding the range of environmental conditions under which the enzyme will function (pH, temperature, etc.). The ability to perform such feats, using genetic engineering, has already been demonstrated, exemplified by the modification of subtilisin (see Ensley and Zylstra, Chap. 3).

Mixtures of toxic materials discharged to waste streams may best be treated with mixtures of microorganisms working together in a consortium. However, there may be times when microorganisms will work more effectively when individual species or strains are applied as needed, rather than the entire mixture at the initiation of remediation. It has been demonstrated[34] that naturally occurring, but selected, microorganisms can be successfully applied on a commercial scale in such situations. Waste Stream Technologies (WST) has reported successful treatment of mixtures of pollutants containing volatile, semivolatile, and aliphatic compounds. The company used sequential introduction of microorganisms, with one organism applied to degrade volatile compounds, followed by a set of microorganisms to continue the process, thereby degrading pyrenes and higher-molecular-weight compounds.

They have selected for microorganisms specifically metabolizing each group of compounds. WST has shown that, using all microbial species simultaneously, degradation is inhibited, an interesting and important observation. The company selected temperature-specific microorganisms (thermophiles, psychrophiles, and/or mesophiles), thereby achieving degradation throughout all seasons of the year. In one case, diesel-fuel-contaminated soil was treated with psychrophilic pseudomonads in the late fall, with the result that diesel fuel was not detectable in the soil in the spring. In another case, 8000 to 9000 yd^3 of soil containing naphthalene at 8000 to 12,000 ppm was inoculated in November. On testing the soil in March, the concentration had dropped to 100 ppm.[33] WST also champions the use of large numbers of microorganisms, with continual replacement in cases where toxicity or nutrient deficiency may be a problem. This approach is based on the assumption that some of the microorganisms will digest a portion of the material and the inoculum itself will generate a nutrient source for a short time.

The effectiveness of degradation depends on pathways and terminal electron acceptors available in the microbial strain or consortium used.[52] The crucial difference will not be whether a reaction is anaerobic or aerobic; rather, success will depend on environmental factors which favor reactions carried out by those microorganisms to which the substrate is accessible. This kind of information, derived from microbial ecology, can be used to model fate and transport of compounds, as well as of the bacteria themselves, and is useful in designing specific treatments. Alexander et al.[2] pointed out that, since adsorption contributes significantly to retention and movement of bacteria through aquifers, manipulation of ionic constituents may facilitate the ability of the microorganisms to function.

In the future, there is no doubt that degradation will be improved by genetic engineering. Degradation rates and range of material which can be treated will be increased. Application of nucleic acid technologies will significantly advance environmental microbiology, as well as biotreatment of waste material.[40] Environmental applications of nonengineered microorganisms, coupled with the use of nucleic acid probes and PCR for identification of microorganisms involved in the bioremediation processes and in environmental monitoring will establish the field of biotreatment as a major component of bioremediation.

Acceptance of biotreatment and the use of engineered microorganisms in bioremediation by the public is essential if the biotreatment industry is to thrive. Acceptance by the industrial community and public is not based solely on applications being successful and without adverse effects, but also requires solid data showing successful treatment at reduced cost (in time and/or money). Laboratory data on ability of a given microorganism to degrade a particular compound under controlled conditions are freely available and are generated practically on a daily basis by many workers. The importance of demonstrating that a process can be scaled up and operated under field conditions cannot be overemphasized. In fact, scale-up feasibility (and actuality) is a major obstacle in the development of biotechnology, and particularly of bioremediation.[39,42]

Specific to use of GEMs in bioremediation is the question of safety, because of possible distribution of either the GEM itself or its genetically engineered DNA sequences into the environment, with the potential for transfer amongst naturally occurring microorganisms. One approach that has gained some support is the incorporation of a "suicide" gene into a bacterium proposed for deliberate release.[9] The killing function can be combined with the engineered degradative functions so that the bacteria die as soon as the hazardous waste is depleted, i.e., as soon as the substrate is metabolized. In one system, a protein (the GEF protein) is inserted into the cell membrane. When degradation ceases, unrestricted flow into and from the cell is facilitated, resulting in cell death. The GEF-based system has been demonstrated to function in several species suggested for deliberate release.

There are data which show that engineered organisms may not thrive, or even persist, under some environmental conditions.[27] They may not survive long enough to accomplish the intended objective. Laboratory tests have been conducted with mutated strains, and such tests are planned for engineered strains.[7] Both Celgene and Envirogen have investigated advantages of using engineered microorganisms and have moved ahead to develop strains for use in the future, anticipating a climate in which GEMs can be released with public approval. In general, distinct advantages have been reported.[10]

An area that has seen immeasurable improvement during the past few years is development of monitoring methods for use in remediation. Calibrated models are being developed to monitor degradation, a good example being hydrocarbons at an actively remediating site.[4] In the future, when use of GEMs in situ is routine, the amount of degradative activity will be measured by constructing the GEM to include the *lux* gene of *Vibrio harveyii*. Such a construct allows light (which can be monitored) to be generated as intracellular ATP increases during degradation.[25] No doubt some waste sites will provide a remarkable view in the night! In any case, the presence of organisms containing specific degrading genes can be monitored by classic hybridization[44] or by PCR[26] methods, as these techniques are improved.

Treatment Methods

New methods of biotreatment being developed include technologies for treating volatile chlorinated and light aromatic hydrocarbons in groundwater and in air streams.[8,55] Canter et al.[8] reviewed a variety of innovative processes for reclamation of contaminated subsurface environments, particularly streams containing chlorinated aliphatic hydrocarbons and the light aromatic constituents of petroleum products. They concluded that 60–90 percent removal was achievable using existing techniques. The extent of removal was dependent on the type of material, type of reactor substrate used, flow rate, temperature, and related environmental parameters. Obviously, further research will yield improved performance.

Venterea and Fogel[55] discuss methods for biofiltration of air streams, pointing out that carbon adsorption of contaminants, such as trichloroethylene (TCE), requires 5 to 10 times as much carbon as the weight of the amount of TCE and that the material is not changed, but simply adsorbed. In contrast, biofiltration results in mineralization of the contaminant, without by-products (e.g., vinyl chloride) which may have adverse properties. The technique has already proved effective for treatment of air streams containing low concentrations of organic compounds.

Bioremediation techniques for highly concentrated process flows are in active development. Biofiltration can also be employed effectively when venting is used as a remediation process. A recently completed study for the U.S. Air Force showed that 50 g/h of jet fuel was removed using a biofilter.[55] Capital and operating costs for biofiltration were estimated to be about the same as for catalytic oxidation systems. Operating costs were about one-third of those for carbon adsorption systems.

Treatment of landfill leachate is an increasingly common biotreatment application. With increased (and mandated) use of lined landfills, collection and treatment of leachates pose difficult problems. Treatment in a semicontinuous mode makes it possible to maintain treatment for prolonged periods of time. Semicontinuous biotreatment of a landfill leachate containing phenoxy herbicide chemical allowed accommodation of an average daily loading of 1.6 kg of phenoxies and 0.5 kg of phenols, with greater than 98 percent degradation of both.[6]

Use of reductive conditions in degradation has allowed new ground to be broken. Currently, many halogenated compounds are dehalogenated using anaerobes, after which ring cleavage occurs rapidly under aerobic conditions. Fluidized-bed systems can be used for anaerobic waste treatment.[50,54] In fact, development of various anaerobic reactors, such as the upflow biomass bed and filter reactor for treating synthetic sugar wastes, is moving ahead.[17] Incidentally, anaerobic conditions also aid degradation of cyanide, the biodegradation of which was discussed above.[12]

Fluidized-bed reactors are also used extensively for aerobically degraded wastes, including degradation of S-triazine-containing industrial wastewater. Up to 18 g/L volatile suspended solids can be removed with the S-triazines as the sole nitrogen source for microorganisms carrying out degradation. A maximum removal efficiency of 80 percent has been reported.[19] Mixed cultures of microorganisms have been shown to degrade 3,4-dichloroaniline in a three-phase draft tube fluidized bioreactor at efficiencies of 95 percent.[29]

While many degradative reactions occur when suspended cultures are employed, it may be advantageous to immobilize cells before starting treatment. Bioprocess engineering has shown that immobilization leads to enhanced reaction stability, perhaps by preventing degradation of the biocatalyzing organisms. An immobilized *Arthrobacter* species (NCIB 11075) has been shown to catalyze the rate-limiting step of trimethyl lead degradation to dialkyl lead.[30] Contaminated groundwater has been treated for over 90 days at a flow rate of 80 gal/day using an immobilized-cell packed-bed reactor.[46] Also, immobilized cells of the fungus *Phanerochaete chrysosporium* have been reported to degrade pentachlorophenol.[28]

While certainly not new, some rather innovative uses of traditional methods of degradation, such as composting and land farming, are proving very useful in achieving degradation of hazardous wastes. Improvement in efficiency and enhancement of efficacy can be achieved by continuing refinement of these methods. Composting has been used to decontaminate sludge containing the explosive trinitrotoluene (TNT), thus cost effectively reducing hazard. Data from a large-scale study in Oregon indicate that composting can be 50–67 percent cheaper than incineration.[20]

Regulatory Processes

Major changes have occurred in the regulatory status of biotreatment procedures; many of these have taken place at the federal level and are described in Chap. 6 by Giamporcaro. The changes have spurred development of new techniques and expansion of field application of existing methods. Acceptance of bioengineering at the federal level has placed greater attention of biodegradation. The pressure of this attention and the costs associated with physical-chemical methods comprise driving forces which create significant interest in biotreatment projects.

At the same time, the importance of the influence of state and local regulations on the current status and future prospects of bioremediation is becoming more evident. The various state and local regulatory structures have, unfortunately, hindered the development of the industry since companies must be prepared to comply with a plethora of environmental statutes. In some instances, federal regulations may aid development simply by requiring cleanup of wastes. In other cases, the effect is mixed or negative. In many states property titles cannot be legally transferred unless the liability for contamination discovered is legally covered,[53] requiring assurance from the treatment firm that treatment has been successful and complete—and, in turn, requiring engineers and biologists to join forces in developing proven, economically feasible scaled-up technologies. Since the alternatives to biotreatment—incineration, burial, or dumping at sea—are no longer in favor with the public and are costly, biotreatment rises to the forefront as the method of choice for remediation of waste sites in the future.

References

1. Abramowicz, D. A. 1989. Biodegradation of PCB contaminated soil using recombinant DNA bacteria. In: Proc. A&WMA/EPA Symp., Cincinnati, OH, Feb. 1989, pp. 301–312.
2. Alexander, M., R. J. Wagenet, P. C. Baveye, J. T. Gannon, U. Mingelgrin, and Y. Tan. 1991. Movement of bacteria through soil and aquifer sand. U.S. Environmental Protection Agency, U.S. EPA 600/s2-91/010, June 1991, Washington, D.C.
3. Applied Biotreatment Association. 1989. Compendium of Biotreatment Applications. ABTA, Washington, D.C.
4. Baehr, A. L., J. M. Fischer, M. A. Lahvis, R. J. Baker, and N. P. Smith. 1991. Method for estimating rates of microbial degradation of hydrocarbons based on gas transport in the unsaturated zone at a gasoline-spill site in Galloway Township, New Jersey. In: The Proceedings of the U.S. Geological Survey Toxic Substance Hydrology Technical Meeting, Monterey, CA, March 1991, pp. 129–141.
5. Baker, A. J. M. 1981. Accumulators and excluders—Strategies in the response of plants to heavy metals. *J. Plant. Nutr.* 3:643–654.
6. Bhamidimarri, S. M. R., D. Catt, and C. Mercer. 1990. Semi-continuous biotreatment of a landfill leachate containing phenoxy herbicide chemicals. In: CHEMECA 90. Processing Pacific Resources. 18th Australasian Chemical Engineering Conference, Auckland, New Zealand, Vol. II, pp. 1039–1044.
7. Bioremediation Report. 1991. Waste Stream Technology Not Slowed by Winter. Dec. 1991, 1:1–2. Cognis, Santa Rosa, CA 95407.
8. Canter, L., L. E. Streebin, M. C. Arquiaga, F. E. Carranza, D. E. Miller, and B. H. Wilson. 1990. Innovative processes for reclamation of contaminated subsurface environments. U.S. Environmental Protection Agency, U.S. EPA 600/S2-90/017, July 1990, Washington, D.C.
9. DePalma, A. 1992. GX Biosystems targets various markets with its suicide gene technology. *Genet. Eng. News,* 12:13, 27.
10. Dwyer, D. F. 1992. Evaluation of GEMs for environmental bioremediation in microcosms and *in situ* (abstract). Biosafety Results of Field Tests with Genetical Modified Organisms, pp. 141–149, Publisher: Biologische Bundesenstalt Fur Land und Forstwirtschast, Braunschweig, Germany.
11. Dwyer, D. F., F. Rojo, and K. N. Timmis. 1988. Fate and behavior in an activated sludge microcosm of a genetically-engineered microorganism designed to degrade unsubstituted aromatic compounds. Presented at First International Conferences on the Release of Genetically-Engineered Microorganisms (REGEM 1). In: The Release of Genetically-Engineered Micro-Organisms. (ed.: M. Sussman, C. H. Collins, A. Skinner, D. E. Stewart-Tull), pp. 77–88, Academic Press, London.
12. Fallon, R. D., D. A. Cooper, R. Speece, and M. Henson. 1991. Anaerobic biodegradation of cyanide under methanogenic conditions. *Appl. Environ. Microbiol.* 57:1656–1662.

13. Fulthorpe, R. R. and R. C. Wyndham. 1991. Transfer and expression of the catabolic plasmid pBRC60 in wild bacterial recipients in a freshwater ecosystem. *Appl. Environ. Microbiol.* 57:1546–1553. Gealt, M. A. 1988. Recombinant DNA plasmid transmission to indigenous organisms during waste treatment. *Water Sci. Technol.* 20:179–184.

14. Gealt, M. A. 1992. Gene transfer in waste treatment. In: Microbial Ecology: Principles, Methods, and Applications (ed.: M. A. Levin, R. J. Seidler, and M. Rogul). McGraw-Hill, New York, pp. 327–343.

15. Gealt, M. A., M. Chai, K. Alpert, and J. Boyer. 1985. Transfer of plasmids pBR322 and pBR325 from laboratory strains of *Escherichia coli* to bacteria indigenous to the waste disposal system. *Appl. Environ. Microbiol.* 49:836–841.

16. Glaser, V. 1992. Strong growth in biotechnology market sectors predicted for 1991–2002. *Genet. Eng. News.* 12:3, 6–7.

17. Guiot, S. R., M. F. Podruzny, and D. D. McLean. 1989. Assessment of macroenergetic parameters for an anaerobic upflow biomass bed and filter reactor. *Biotechnol. Bioeng.* 34:1277–1288.

18. Henschke, R. B., and R. J. Schmidt. 1990. Plasmid mobilization from genetically engineered bacteria to members of the indigenous soil microflora *in situ*. *Curr. Microbiol.* 20:105–110.

19. Hogrefe, W., H. Grossenbacher, A. M. Cook, and R. Hutter. 1990. Biotreatment of s-triazine-containing wastewater in a fluidized bed reactor. *Biotechnol. Bioeng.* 28:1577–1581.

20. Hom, S. S. M. 1992. Composting explosives now competitive. *Bioremediation Rep.* 1(1):1.

21. Homer, F. A., R. S. Morrison, R. R. Brooks, J. Clemens, and R. J. Reeves. Comparative studies of nickel, cobalt and copper uptake by nickel hyperaccumulators of the genus *Alyssum*. In press.

22. Jahnke, M., T. El-Banna, R. Klintworth, and G. Auling. 1990. Mineralization of orthanilic acid is a plasmid-associated trait in Alcaligenes sp. O-1. *J. Gen. Microbiol.* 136:2241–2249.

23. Kellogg, S. T., D. K. Chatterjee, and A. M. Chakrabarty. 1981. Plasmid-assisted molecular breeding: New technique for enhanced biodegradation of persistent toxic chemical. *Science* 214:1133–1135.

24. Khalil, T. and M. A. Gealt. 1987. Effect of exogenous compounds on the mobilization of plasmids in synthetic wastewater. *Can. J. of Microbiol.* 33:733–737.

25. King, J. M. H., P. M. DiGrazia, B. Applegate, R. Burlage, J. Sanseverino, P. Dunbar, F. Larimer, and G. S. Sayler. 1990. Rapid and sensitive bioluminescent reporter technology for naphthalene exposure and biodegradation. *Science* 249:778–781.

26. Knight, I. T., W. E. Holben, J. M. Tiedje, and R. R. Colwell. 1992. Nucleic acid hybridization techniques for detection, identification, and enumeration of microorganisms in the environment. In: Microbial Ecology: Principles, Methods, and Applications (ed.: M. A. Levin, R. J. Seidler, and M. Rogul). McGraw-Hill, New York, pp. 65–91.

27. Lenski, R. E. 1991. Quantifying fitness and gene stability in microorganisms. In: Assessing Ecological Risks of Biotechnology (ed.: L. Ginzburg). Butterworth, Boston, pp. 173–190.

28. Lin, J-E., H. Y. Wang, and R. F. Hickey. 1991. Use of coimmobilized biological systems to degrade toxic organic compounds. *Biotechnol. Bioeng.* 38:273–279.

29. Livingston, A. G. 1991. Biodegradation of 3,4-dichloroaniline in a fluidized bed bioreactor and a steady-state biofilm kinetic model. *Biotechnol. Bioeng.* 38:260–272.

30. Macaskie, L. E. and A. C. R. Dean. 1990. Trimethyl lead degradation by free and immobilized cells of an Arthrobacter species and by the wood decay fungus Phaeolus schweintzii. *Appl. Microbiol. Biotechnol.* 33:81–87.

31. Madsen, E. L. 1991. Determining in situ biodegradation. *Environ. Sci. Technol.* 25:1663–1673.

32. Mancini, P., S. Fertels, D. R. Nave, and M. A. Gealt. 1987. Mobilization of plasmid pHSV106 from *Escherichia coli* in a laboratory waste treatment facility. *Appl. Environ. Microbiol.* 53:665–671.

33. Mayer, W. 1991. Company profile; Waste stream technology not slowed by winter. Bioremediation Report Premiere Issue 1-2, Cognis, Santa Rosa, CA.
34. Mayer, W. 1992. Bioaugmentation: The bugs used do make a difference. *Bioremediation Rep.* 1:3–6. Cognis, Santa Rosa, CA.
35. McPherson, P. and M. A. Gealt. 1986. Isolation of indigenous wastewater bacterial strains capable of mobilizing plasmid pBR325. *Appl. Environ. Microbiol.* 51:904–909.
36. Miller, R. V. 1992. Overview: Methods for the evaluation of genetic transport and stability in the environment. In: Microbial Ecology: Principles, Methods, and Applications (ed.: M. A. Levin, R. J. Seidler, and M. Rogul). McGraw-Hill, New York, pp. 229–246.
37. Nakamura, Y., K. Itoh, N. Mikami, T. Matsuda, R. Kikuchi, M. Matsuo, H. Yamada, and J. Miyamoto. 1991. A selective plating method to enumerate target microorganisms in an environment. *J. Gen. Appl. Microbiol.* 37:85–92.
38. Neidle, E. L., M. K. Shapiro, and L. N. Ornston. 1987. Cloning and expression in Escherichia coli of Acinetobacter calcoaceticus genes for benzoate degradation. *J. Bacteriol.* 169:5496–5503.
39. Office of Technology Assessment. 1988. New developments in Biotechnology 4. U.S. Investment in Biotechnology. Office of Technology Assessment, Washington, D.C.
40. Olson, B. H. 1991. Tracking and using genes in the environment. *Environ. Sci. Technol.* 25(4):604–610.
41. Omenn, G. and A. Hollaender (eds.). 1983. Genetic Control of Environmental Pollutants. Plenum Press, New York.
42. Omenn, G. and A. Hollaender (eds.). 1989. Environmental Biotechnology. Plenum Press, New York.
43. Omenn, G. S. and A. W. Bourquin. 1990. Risk assessment of biodegradation in pollution control and cleanup. *Adv. Appl. Biotechnol. Ser.* 2:443–465, Gulf Pubco, Houston, Texas.
44. Pettigrew, C. and G. S. Sayler. 1986. Application of DNA colony hybridization to the rapid isolation of 4-chlorobiphenyl catabolic phenotypes. *J. Microbiol. Methods* 5:205–213.
45. Pierce, G. E. 1982. Diversity of microbial degradation and its implications in genetic engineering. In: Impact of Applied Genetics in Pollution Control (ed.: C. F. Kulpa, R. L. Irvine, and S. J. Sojka). University of Notre Dame, IN, pp. 20–25.
46. Portier, R. J., J. A. Nelson, J. C. Christianson, J. M. Wilkerson, R. C. Bost, and B. P. Flynn. 1989. Biotreatment of dilute contaminated ground water using an immobilized microorganism packed-bed reactor. *Environ. Progr.* 8:120–125.
47. Roberts, L. 1987. Discovering microbes with a taste for PCBs. *Science* 237:975–977.
48. Saye, D. J. and R. V. Miller. 1989. The aquatic environment: Consideration of horizontal gene transmission in a diversified habitat. In: Gene Transfer in the Environment (ed.: S. B. Levy and R. V. Miller). McGraw-Hill, New York. pp. 223–259.
49. Selvaratnam, S. and M. A. Gealt. 1992. Plasmid gene transfer in amended soil. *Water Res.* 26:39–43.
50. Shieh, W. K., C. T. Li, and S. J. Chen. 1985. Performance evaluation of anaerobic fluidized bed system: III. Process kinetics. *J. Chem. Technol. Biotechnol., Biotechnol.* 35B:229–234.
51. Steffan, R. J. and R. M. Atlas. 1988. DNA amplification to enhance detection of genetically engineered bacteria in natural environments. *Appl. Environ. Microbiol.* 54:2185–2191.
52. Suflita, J. M. and G. W. Sewell. 1991. Anaerobic biotransformation of contaminants in the subsurface. *Env. Res. Brief* U.S. Environmental Protection Agency, U.S. EPA 600/M-90/024, February, Washington, D.C.
53. Thayer, A. M. 1991. Bioremediation: Innovative technology for cleaning up hazardous waste. *Chem. Eng. News,* Apr. 26, 1991, pp. 23–42.
54. Veeramani, H. 1987. Fluidized bed systems for anaerobic biotechnology in waste management. *Chem. Age India* 38:543–546.
55. Venterea, R. T. and S. Fogel. 1991. Cleaning the air with biofiltration. *ABB J.* 4:6–7.

56. Walia, S., A. Khan, and N. Rosenthal. 1990. Construction and applications of DNA probes for the detection of polychlorinated biphenyl-degrading genotypes in toxic organic-contaminated soil environments. *Appl. Environ. Microbiol.* 56:254–259.
57. Weijers, C. A. G. M. and J. A. M. de Bont. 1990. Enantioselective degradation of 1,2-epoxyalkanes by Nocardia HB. *Enzyme Microbiol. Technol.* 13:306–308.
58. Wyatt, J. M. and S. J. Palmer. 1992. Biodegradation of nitriles and cyanide. In: Biodegradation: Natural and Synthetic Materials (ed.: W. T. Betts). Springer Verlag, London, pp. 69–88.
59. Zeph, L. R., M. A. Onaga, and G. Stotzky. 1988. Transduction of *Escherichia coli* by bacteriophage P1 in soil. *Appl. Environ. Microbiol.* 54:1731–1737.
60. Zilinskas, R. A. and C. G. Lundin. 1992. Marine Biotechnology for the Developing Countries. World Bank, Washington, D.C.

Index

ABOUT THE EDITORS

MORRIS LEVIN, Ph.D. is a professor at the Maryland
Biotechnology Institute of the University of Maryland. The
Institute is supported by a consortium of academic,
industrial, and governmental organizations. He is the
coauthor of *Microbial Ecology* and *Risk Assessment in
Genetic Engineering,* also published by McGraw-Hill.
Dr. Levin resides in Baltimore, Maryland.

MICHAEL A. GEALT, Ph.D. is a professor in the Department of
Bioscience and Biotechnology and is Associate Director
of the Environmental Studies Institute at Drexel
University. His recent research has been funded by both
industrial and government organizations, including the U.S.
Environmental Protection Agency. Dr. Gealt resides in
Jenkinstown, Pennsylvania.